Computational Music Science

Series editors

Guerino Mazzola
Moreno Andreatta

More information about this series at http://www.springer.com/series/8349

Gabriel Pareyon · Silvia Pina-Romero
Octavio A. Agustín-Aquino
Emilio Lluis-Puebla
Editors

The Musical-Mathematical Mind

Patterns and Transformations

 Springer

Editors
Gabriel Pareyon
CENIDIM-INBA
Centro Nacional de las Artes (CENART)
Coyoacán, Distrito Federal
Mexico

Silvia Pina-Romero
División de Electrónica y Computación
(CUCEI)
Universidad de Guadalajara
Guadalajara, Jalisco
Mexico

Octavio A. Agustín-Aquino
Universidad de la Cañada
Teotitlan de Flores Magón, Oaxaca
Mexico

Emilio Lluis-Puebla
Departamento de Matemáticas
UNAM Facultad de Ciencias
Coyoacán, Distrito Federal
Mexico

ISSN 1868-0305
Computational Music Science
ISBN 978-3-319-83715-4
DOI 10.1007/978-3-319-47337-6

ISSN 1868-0313   (electronic)

ISBN 978-3-319-47337-6   (eBook)

Printed on acid-free paper

This Springer imprint is published by Springer Nature
The registered company is Springer International Publishing AG
The registered company address is: Gewerbestrasse 11, 6330 Cham, Switzerland

To the memory of Julián Carrillo
(1875–1965) and
Alexander Grothendieck (1928–2014)

Foreword

It is my great honour and pleasure to introduce you to this book which focuses on fundamental challenges and issues in the relatively new field of Mathematical Music Theory, in turn able to be translated into computational practice.

This book, under the title *The Musical-Mathematical Mind: Patterns and Transformations*, collects the efforts of specialists who participated in the four-day International Congress on Music and Mathematics (ICMM, which took place in Puerto Vallarta, Jalisco, Mexico, November 26–29, 2014). Its contents reflect the maturing of a variety of new conceptualisations on music and mathematics. This congress was organised by the Mexican mathematicians, musicians and musicologists Octavio A. Agustín-Aquino, Juan Sebastiàn Lach Lau, Emilio Lluis-Puebla (Congress Head), Roberto Morales-Manzanares, Pablo Padilla-Longoria, and Gabriel Pareyon (Program Chair and Main Editor).

Mexican scholars have been uniquely proactive in the propagation and support of the mathematical aspects of music in theory and practice, in creativity and epistemology. Already in 2000, the First International Seminar on Mathematical Music Theory took place in Saltillo, on the occasion of the annual congress of the Mexican Mathematical Society, and the Fourth International Seminar on Mathematical Music Theory took place in Huatulco, again in Mexico, respectively organised by Lluis-Puebla, and by Agustín-Aquino.

It is remarkable that these Mexican conferences took place in the years when the Society for Mathematics and Computation in Music (SMCM) had no conference: its conferences are biannual and have taken place in the odd years since 2007. It is also remarkable because the Mexican initiative proves that there is an increasing intensity of scholarly and artistic work centred around mathematics and music. It gives us a model of how the future of this mathemusical enterprise could look.

The program of the congress in Puerto Vallarta is not only a testimony of the high level of scientific research achieved in the early years of the 21st century, it also proposed a deep spectrum of musical, mathematical, physical, and philosophical perspectives that have emerged in this field of cultural and scientific integration since its Pythagorean origins. The big difference that we observe when comparing the state of this art to the achievements in the 20th century is the involvement of advanced techniques and concepts of modern mathematics and physics, relating for example to Grothendieck's topos theory and physical string theory. It is not astonishing that the mathematician and philosopher of modern mathematics, Fernando Zalamea, has—among other authors in this book—contributed a beautiful perspective on the philosophy that lies inside the efforts to reunite mathematics with music as approaches to a unified universal knowledge.

Minneapolis, USA   Guerino Mazzola
January 2016   (ICMM 2014, Honorary President)

Preface

Proficiency and enthusiasm are gathered in this volume, as the fruit of a long-awaited conference of international specialists who devote their lives to connect, exchange and mutually involve music with mathematics and mathematics with music. We celebrate this publication at the moment of Julián Carrillo's (1875–1965) one hundred and fortieth anniversary, to whom we also dedicated a special panel (with results to be published separate from this book) during our International Congress on Music and Mathematics (ICMM) held at Puerto Vallarta, Mexico (November, 2014).

Our conference was a unique feast of mind and feelings, sound and meaning, imagination and empiricism, as the continuation and synthesis of a long tradition. The link between music and mathematics is a notorious intersection at a common origin of human civilisation embracing aesthetics, pragmatics and abstract thought. As a matter of fact, aesthetics, pragmatics and abstraction arise as human practice deeply rooted in a primary notion of repetition, rhythm, comparison, measurement, spacialization and transformation, all of them common grounds for music and mathematics.

In every part of the world, "civilisation" is a social complexity that seems to need, from its early sources, the sprout of music and mathematics. Thus, in the context of the original civilisations of Mesoamerica, music and mathematics are also strongly associated. I should mention—at least briefly—some milestones in the long history binding music and mathematics in ancient and modern Mexico: the Olmec and the Maya peoples, so admired today for their architectural, astronomical and mathematical achievements, must also be acknowledged for creating original instruments, orchestras and choirs, as well as for developing their own graphic representation of human sounds and sounds from nature. Thereafter, among the Aztec people, the patron of poetry, symmetry, music and numbers is Xochipilli-Macuilxochitl, a name that relates the number five with the symbolisation of colour, abstraction, geometry, ratio and proportion.

Later, in the Spanish colony, Sister Juana Inés de la Cruz (1651–1695) developed her own research about the connections between harmony, numbers and geometry. Even today Sor Juana's conceptualisations are still valid for the philosophical study of music, such as the study of spirals for harmonic modelling. In 19th century

Mexico, Juan N. Adorno (1807–1880) published his treatise *Harmony of the Universe*, based on principles of physical and mathematical harmony. Later, the Porfirian thinker Juan N. Cordero (1851–1916) in his book *Examen de los acordes de transformación tonal* (*Examination of the Chords of Tonal Transformation*) proposed a principle of musical transformation based on logical axioms. A few decades after, in the 20th century, José Vasconcelos (1882–1959) claimed that "only the musical study of mathematics, and the rhythmic comprehension of numbers, could be useful as effective forms of thought and discovery of the human nature". In the same epoch, another Mexican thinker, Samuel Ramos (1882–1959) wrote that "All kinds of perturbation in the Universe are of a rhythmic nature. The fluency of changes cannot be unarticulated among them; therefore the rhythm of changes is accumulative". Quoting Sor Juana, Adorno, Cordero, Vasconcelos, and Ramos are part of what semiotician Mauricio Beuchot (1950–) —a contemporary of us—acknowledges as "the Mexican devotion of Pythagoreanism and related doctrines".

Indeed, the orientation of Mexican cultures seems to be magnetised by the intuitions of ratio, proportion, analogy, metonymy, and geometrical and algebraic transformation. We may trace this influence in the most famous composers and music theorists of modern Mexico, namely Augusto Novaro, Conlon Nancarrow, Ervin Wilson, Julio Estrada, Manuel Enríquez, Antonio Russek, Roberto Morales-Manzanares, Víctor Rasgado, and Hebert Vázquez, among others. Indeed, they influence nowadays Mexican studies on music and mathematics as a new mixed discipline. This transdisciplinarity also flourished thanks to the effort of mathematician Prof. Emilio Lluis-Puebla, who graduated an internationally active group of specialists.

As I mentioned before, our meeting also devoted a special panel to the discussion of mathematics applied to music, in honour of the great violinist, conductor, composer and maker of new musical instruments, Julián Carrillo, who through a long and very productive life achieved the invention of music that transcended the traditional Western principles of consonance and harmony, as he foresaw a "universe of endless musical scales and chords". Carrillo's project in the domain of physics and mathematics, and its musical output, is an inspiration for current discussion on these subjects, addressed from different viewpoints during our congress.

We may mention some recurring concepts and theoretical approaches that motivated us during our meeting: tessellation in topological-musical spaces, scaling and even distribution, diatonicity, algebraic transformations, networks and geometry, partitions and well-formedness theory, theories of gestures, morphisms, set theory and fuzzy logic, as well as a new discussion on elementary particles and quantum symmetry as interests of systematic musicology. Despite this variety, all our mathematical proposals fell into five general areas: I. Dynamical Systems, II. Logic, Algebra and Algorithmics, III. Gestural Theories, IV. New Methods for Music Analysis, and V. Modern Geometry and Topology. Although we followed this thematic division during our congress, this book is classified by alphabetical order of authors, for the sake of practical consultation and because most of the contributions present developments in more than one subject.

I wish to end this Preface emphasising the fact that the international President of our Congress, Prof. Guerino Mazzola, is one of the leading thinkers in the field of the Mathematical Theory of Music; and our national Head of Congress, Prof. Emilio Lluis-Puebla pioneered systematic musicology in Mexico and Latin America, organising the Seminars on Mathematical Theory of Music in previous years. We completed our group of national and international guests with the best and more original proposals received after almost two years of organisation that reached its climax during the four days of ICMM 2014. We remain grateful to all our contributors.

Guadalajara, Mexico   Gabriel Pareyon
December 2015   (ICMM 2014, Program Chair and Editor)

Acknowledgements

This book would not have been possible without the generous support of many wonderful people and organisations. We would like to thank all authors contributing their knowledge and expertise during our International Congress on Music and Mathematics (ICMM), held in Puerto Vallarta, Mexico, November 26–29, 2014.

We are particularly grateful to Prof. Alonso Castillo-Pérez, Head of the Department of Computer Science and Electronics, University of Guadalajara (UDG, Mexico), for his kind and decisive support of our congress, and for his generous and wise guidance giving rise to the construction of more modern facilities and optimal conditions of research at CUCEI-UDG and CU-Costa-UDG, key to the preparation and development of our conference.

Our ICMM would neither have been possible without the efforts of Dr. Yael Bitran-Goren, head of the National Centre for Music Investigation, Documentation and Information (CENIDIM–INBA, Mexico), who provided institutional support and kept faith with our project.

We also owe our gratitude to our Scientific—Organizing Committee, and to Silvia Pina-Romero, Ph.D. Math., for her generous assistance. We also thank our panel chairs' generous cooperation (R. Brotbeck, D. Clampitt, J.S. Lach-Lau, M. Montiel, S. Pina-Romero).

We acknowledge the support of the Office of International Affairs of the State of Jalisco (Dirección de Relaciones Internacionales, Secretaría de Educación, Gobierno del Estado de Jalisco), particularly the efforts of Prof. Jorge Alberto Quevedo-Flores, that made possible the visit of several of our international guests.

Finally we want to express our most special gratitude to Clarence Barlow, D. Gareth Loy, Guerino Mazzola, and Fernando Zalamea, whose passion and wisdom connecting music and mathematics has been luminous for most of us. Our congress in Puerto Vallarta would not have been possible without your inspiration.

Contents

Contributors

Octavio A. Agustín-Aquino Universidad de la Cañada, Teotitlan de Flores Magón, Oaxaca, Mexico

Emmanuel Amiot Institut de Recherche et Coordination Acoustique/Musique, Paris, France

Juan Sebastián Arias Universidad Nacional de Colombia, Bogotá, Colombia

Clarence Barlow Department of Music, University of California, Santa Barbara, CA, USA

Roberto Cabezas Music Technology Graduate Program, UNAM, Coyoacán, Del Carmen, D.F., Mexico

Teresa Campos-Arcaraz Facultad de Música, UNAM, Coyoacán, Del Carmen, D.F., Mexico

Emilio Erándu Ceja-Cárdenas Departamento de Matemáticas, Centro Universitario de Ciencias Exactas e Ingenierías (CUCEI), Universidad de Guadalajara, Guadalajara, Mexico

Yemile Chávez-Martínez Universidad Nacional Autónoma de México, Coyoacán, D.F., Mexico

Elaine Chew School of Electronic Engineering and Computer Science, Queen Mary University of London, London, UK

David Clampitt The Ohio State University, Columbus, OH, USA

Carlos de Lemos Almada Federal University of Rio de Janeiro, Centro, Rio de Janeiro, Brazil

Daniel Moreira de Sousa Federal University of Rio de Janeiro, Rio de Janeiro, Brazil

Micho Đurđevich Institute of Mathematics, UNAM, Mexico, Mexico

Koichi Fujii NTT DATA Mathematical Systems Inc., Tokyo, Japan

D. Gareth Loy Gareth Inc., San Rafael, CA, USA

Pauxy Gentil-Nunes Federal University of Rio de Janeiro, Rio de Janeiro, RJ, Brazil

Franck Jedrzejewski French Atomic Energy Commission (Commissariat à l'énergie atomique et aux énergies alternatives), Gif-sur-Yvette, France

Juan Sebastián Lach Lau Conservatorio de las Rosas, Morelia, Mexico

Emilio Lluis-Puebla Universidad Nacional Autónoma de México, Coyoacán, D. F., Mexico

Jaime Alonso Lobato-Cardoso SEMIMUTICAS-IIMAS, UNAM, Circuito Escolar 3000, Ciudad Universitaria, Coyoacán, D.F., Mexico

Guerino Mazzola School of Music, University of Minnesota, Minneapolis, MN, USA; Institut Für Informatik, Universität Zürich, Zürich, Switzerland

Mariana Montiel Georgia State University, Atlanta, GA, USA

Roberto Morales-Manzanares Facultad de Música, DAAD, Universidad de Guanajuato, Guanajuato, Mexico

Francisco Mugica Departament de Llenguatges i Sistemes Informátics, Soft Computing Research Group, Technical University of Catalonia, Barcelona, Spain

Àngela Nebot Departament de Llenguatges i Sistemes Informátics, Soft Computing Research Group, Technical University of Catalonia, Barcelona, Spain

Thomas Noll Escola Superior de Música de Catalunya, Departament de Teoria, Composició i Direcció, Barcelona, Spain

Martin Norgaard Georgia State University, Atlanta, GA, USA

Pablo Padilla-Longoria IIMAS, UNAM, Cd. Universitaria, Coyoacán, D.F., Mexico

Gabriel Pareyon CENIDIM-INBA, Torre de Investigacion, CENART, Coyoacán, D.F., Mexico

Iván Paz Departament de Llenguatges i Sistemes Informátics, Soft Computing Research Group, Technical University of Catalonia, Barcelona, Spain

Silvia Pina-Romero División de Electrónica y Computación, CUCEI – Universidad de Guadalajara, Guadalajara, Jalisco, Mexico

Mark Reybrouck University of Leuven, Leuven, Belgium

Julian Rohrhuber Institute for Music and Media, Robert Schumann Hochschule, Duesseldorf, Germany

Enrique Romero Departament de Llenguatges i Sistemes Informátics, Soft Computing Research Group, Technical University of Catalonia, Barcelona, Spain

Edmar Soria Music Technology Graduate Program, UNAM, Coyoacán, Del Carmen, D.F., Mexico

Tsubasa Tanaka IRCAM, Paris, France; Tokyo University of the Arts, Tokyo, Japan

Samuel Vriezen Independent Researcher & Composer, Amsterdam, Netherlands

Michael Winter Independent Researcher, Los Angeles, CA, USA

Jason Yust School of Music, Boston University, Boston, MA, USA

Fernando Zalamea Universidad Nacional de Colombia, Bogotá, Colombia

Acronyms

BWT	Burrows-Wheeler Transform
CC	Combinatorial Constraints
CSV	Comma-Separated Variables
DFT	Discrete Fourier Transform
DV	Developing Variation
FFT	Fast Fourier Transform
GMC	Global Morphological Constraints
Gv/Ga	Gödel-vector/Gödel-address
HSA	Hypothesis of Self-Similarity in Euclidean Axiomatics
ICMM	International Congress on Music and Mathematics (Puerto Vallarta 2014)
LMC	Local Morphological Constraints
MCT	Musical Contour Theory
NDFSA	Non-Deterministic Finite State Automata
OGMC	Optimal Global Morphological Constraints
PA	Partitional Analysis
PYL	Partitional Young Lattice
RTM	Rhythm in arrays notation (from RTM-notation to ENP-score-notation)
SAT	Self-referential Abstract Thought
TIP	Theory of Integer Partitions
TM	Tonal Music
UDP	User Datagram Protocol

Introduction

Emilio Lluis-Puebla

For those who read for the first time or inquire about *music and mathematics*, let me tell you that this field is both a recent area of study and also a very old one. At the beginning of history, there was a connection between numbers and music. Later, Pythagoras made a mathematical effort to say things about music with a certain foundation. The names Descartes, Galileo, Kepler, Leibniz, Euler, d'Alembert, Helmholtz, and some others are relevant here.

In the twentieth century, acoustics and its technology were very successful applying mathematics to music, as well as computer science and some other fields like linguistics. Later, the work of Clough in 1979, Lewin in 1982, and Mazzola in 1985 inspired both music-inclined mathematicians and mathematics-inclined musicians to continue working in mathematics and music.

A big trend in the last three decades in mathematics was to do not only applications but to do new mathematics in a variety of fields of knowledge, and the field of music has been no exception.

So, mathematical music theory is both a recent area of study and also a very old one. From Pythagoras until the 1980s, very little and not very sophisticated mathematics was employed in music. When sufficiently powerful mathematical machinery became available and talented mathematicians used it, modern mathematical music theory was born.

One of the main goals of mathematical music theory (I will state some of Guerino Mazzola's thoughts mainly from [1] and from personal conversations with him) was to develop a scientific framework for musicology. This framework had as its foundation, established scientific fields. It included a formal language for musical and musicological objects and relations. Music is fundamentally rooted within physical, psychological and semiotic realities. But the formal description of musical instances corresponds to mathematical formalism.

Mathematical music theory is based on category theory, algebraic topology, in particular, topos theory, module theory, group theory, homotopy theory, homology theory, algebraic geometry, just to name some areas, that is, on heavy mathematical machinery. Its purpose is to describe musical structures. The philosophy behind it is understanding the aspects of music that are susceptible to reason in the same way as physics does it for natural phenomena.

This theory is based in an appropriate language to manage the relevant concepts of the musical structures, in a group of postulates or theorems with regard to the musical structures subject to the defined conditions, and in the functionality for composition and analysis with or without a computer.

Mazzola also says that music is a central issue in human life, though it affects a different layer of reality than physics. The attempt to understand or to compose a major work of music is as important and difficult as the attempt to unify gravitation, electromagnetism, and weak and strong forces. For sure, the ambitions are comparable and hence the tools should be comparable too.

It is only in the last three decades that there is consistent work in mathematical music theory. Thus I will address this period of time in Mexico's history on this subject, since Gabriel Pareyon [2] summarises the time span before 1980. I will write about this in a personal way.

When I was 21 years old, in 1974, I was listening to the station Radio Universidad (University Radio Station), to a low, magnificent voice that was talking (in Spanish) about the application of finite group theory to the musical analysis of Bach's music, etc. This caught my attention and I went to see the owner of this voice. I located him in the old building of the Escuela Nacional de Música de la UNAM (UNAM Faculty of Music) and this young thin man kindly showed me a bunch of papers he had. I read them for half an hour or so and got the idea of what he was doing. This young man was Julio Estrada, a distinguished Mexican composer and musicologist.

Then I went to Canada to do a Ph.D. on algebraic K-theory. I was in love with pure mathematics like homological algebra, algebraic topology, algebraic geometry, homotopy theory, etc. Nobody could have ever told me that these marvellous pieces of pure mathematics were ever to appear more than thirty years later in the other field of my passion: music.

When I came back to México, in the early 1980s I wanted to do some work in mathematics and music, in particular to guide an undergraduate thesis for a student, but the angry face and terrible gesticulations of a colleague who was in charge of some high position at the department demoralised me. Does this sound familiar to anyone?

Some years later, in the 1990s, a lady from the mathematics undergraduate program at UNAM with a piano background, with great conviction, full of energy, appeared in my office, completely determined to do an undergraduate thesis in mathematics and music, particularly based on the ideas of Julio Estrada which turned into a book that he published in the 1980s [3]. I gave her more papers and books and she started to look for more bibliography. The librarian got some references of Guerino Mazzola. Particularly his book Gruppen und Categorien in der Musik, some articles by him, and others, including Chemiller's papers, plus some from the American School. This lady was Mariana Montiel. Now she is a full professor in the United States.

Mariana decided also to do a master's thesis on mathematical music theory, especially on denotator theory. I invited Guerino Mazzola to México for the first time in 1997 and we began a wonderful friendship.

In 2000, when I was President of the Sociedad Matemática Mexicana (Mathematical Society of Mexico), I dared to organise the First International Seminar on Mathematical Music Theory which took place simultaneously at the Facultad de Ciencias (Faculty of Sciences) and the Escuela Nacional de Música (School of Music) both from UNAM. Thomas Noll and Guerino Mazzola attended, among others.

Some days before the first international seminar, we had a previous special session on mathematical music theory at the annual Congreso Nacional de la Sociedad Matemática Mexicana in Saltillo which had an attendance of about 2000 persons, with great success. As a frame to both meetings we had concerts by Guerino Mazzola in Saltillo, Sala Carlos Chávez and at the Sala Xochipilli in Mexico City which turned into a delightful free jazz recording called Folia: The UNAM Concert with Guerino Mazzola playing Rachmaninoff's Corelli: La Folia theme as motive.

At both meetings, many mathematicians and musicians attended with surprise on their faces. The proceedings of the seminar were published by the Sociedad Matemática Mexicana Electronic Publications and lately were unified with the proceedings of the Second International Seminar which took place in Germany in 2001 and with the third one which took place in Switzerland in 2002 and was published by Epos Music of the University of Osnabruck in 2004 [4]. (I almost did not see this publication because I almost died. I was very ill for six months with an unknown disease which was later believed to be a viral meningitis, for which there was no cure!)

After not dying, six years later, in 2009, a student of mine, a young, impetuous and talented mathematician and musician, Octavio Agustin-Aquino, convinced me to organise the fourth seminar. It took place in Huatulco, Oaxaca, in 2010 as the Fourth International Seminar on Mathematical Music Theory [5]. By the way, Octavio became the first Ph.D. in mathematics graduated in Mexico at UNAM in mathematical music theory in 2011 with a thesis on microtonal counterpoint. He is now a full professor at the Universidad de la Cañada which belongs to the SUNEO in Oaxaca State, Mexico.

Finally, in November 2012 another very talented man (musicologist, also doing systematic musicology) which I admire the most because of his vast culture, ability, organisational capabilities, enormous memory and many other wits, contacted me in order to organise a sequel of the international seminars which turned out to be the International Congress on Music and Mathematics, 2014. This great man is Gabriel Pareyon.

Through the years there were also some more students who did some work with me but they did not continue in this field due to economic or vocational reasons. In 2013 and 2014, two of my students (Yemile Chávez and Santiago Rovira, both with music backgrounds) approached me like Mariana and Octavio before. They presented a lecture at ICMM 2014, and I hope they continue to work in this marvellous field.

Of course there are some other colleagues who have worked in mathematics and music in a rather isolated way, but now we had the opportunity to collect their efforts in this book, and made the connections to have a stronger unified community worldwide.

And well, what relationship does exist between music and mathematics? Or equivalently what connection or correspondence exists? We know, for example, that mathematical concepts were applied several years ago and recently (coming after all from nature or from man's abstract thought, etc.), just to mention four examples I use in my lectures [6]: to the entertainment with a game of dice in Mozart's creations; to aesthetics, as in Birkhoff's theory; to musical composition, for example by Bartók; and to create a precise language for musicology and music by Mazzola, among others. Certainly, there are many other music fields where mathematics contributes to our understanding, like in performance or analysis, etc.

For me, the most important relationship between mathematics and music is that both are "fine arts". They possess similar characteristics. They are related in the sense that mathematics provides a way to understand music, and musicology has a scientific basis in order to be considered a science, not a branch of common poetic literature.

I have worked since the 1970s on homotopy theory, cohomology theory, algebraic topology, homological algebra, among other fields of mathematics. As I wrote before, at the time these were considered pure mathematics. However, thirty years later, these wonderful pieces of mathematics came to be applied mathematics, and guess where? It turned out to be (as I wrote before) in my other passion: music! But not only as an application, you can do new mathematics as well!

Let me tell you an anecdote. In 2001, when I was president of the Sociedad Matemática Mexicana, during a visit to Rio de Janeiro I called a friend of mine, the president of the International Mathematical Union at that time, the Brazilian Jacob Palis. We agreed to meet at the famous Copacabana Palace where I was going to play Rachmaninoff's Second Piano Concerto as a soloist of the Rio de Janeiro Philharmonic Orchestra. He did not know I was a pianist. When he got there, he saw the president of the Sociedad Matemática Mexicana getting out on stage and sitting down to play the concerto. He was thrilled and invited me to dinner. We had a very long talk and having answered all his questions about me as a pianist and about mathematical music theory, he told me almost the same phrase that Guerino Mazzola got from Grothendieck: "the mathematics of the future!". So, in brief words, let me tell you that, for me, mathematics is one of the "fine arts", the purest of them, which has the gift of being the most precise of all sciences.

I was very honoured to meet all of the participants of ICMM in order to stimulate the interchange of visions, thoughts and points of view on this fascinating subject in a very friendly way. I am sure we all have profited from this interaction in such a wonderful place.

As you know, not only in Mexico, the funding for meetings is practically nonexistent. Many persons interested in coming could not join us because they did not have economic support from their universities. We thought we could obtain some funding for it, but once more, as in the Fourth International Seminar, we had to do it with our own personal budgets, energies and personal work and risk. We proudly can say that once more we have done it by ourselves!

Besides the small support (for such a big meeting) of very few institutions (see the acknowledgements in this book) we only had a small contribution from the Sociedad Matemática Mexicana to partially finance two of my own students which we, again, sincerely thank. The rest is exclusively ours and yours.

On Gabriel Pareyon's behalf (I recognise all his tremendous work on the organisation), the other organisers and myself, we thank all the participants of the International Congress on Music and Mathematics. We had a wonderful conference!

References

1. Mazzola, G.: The Topos of Music. Birkhäuser (2002)
2. Pareyon, G.: A Survey on the Mexican Tradition of Music and Mathematics. Opening lecture of ICMM 2014, Puerto Vallarta, Mexico (abridged in the Preface of this book)
3. Estrada, J., Gil, J.: Música y Teora de Grupos Finitos (3 variables booleanas). UNAM (1984)
4. Lluis-Puebla, E., Mazzola, G., Noll, T. (eds.): Perspectives in Mathematical and Computational Music Theory. epOs, Osnabrück (2004)
5. Agustín-Aquino, O., Lluis-Puebla, E. (eds.): Memoirs of the Fourth International Seminar on Mathematical Music Theory. Serie Memorias, vol. 4. Publicaciones Electrónicas de la Sociedad Matemática Mexicana (2011)
6. Lluis-Puebla, E.: Music and Mathematics: Two Fine Arts. Perspectives in Mathematical and Computational Music Theory. epOs, Osnabrück (2004)

Extended Counterpoint Symmetries and Continuous Counterpoint

Octavio A. Agustín-Aquino

Abstract A counterpoint theory for the whole continuum of the octave is obtained from Mazzola's model via extended counterpoint symmetries, and some of its properties are discussed.

1 Introduction

Mazzola's model for first species counterpoint is interesting because it predicts the rules of Fux's theory (in particular, the forbidden parallel fifths) reasonably well. It is also generalizable to microtonal equally tempered scales of even cardinality, and offers alternative understandings of consonance and dissonance distinct from the one explored extensively in Europe. In this paper we take some steps towards an effective *extension* of the whole model from a microtonal equally tempered scale into another, and not just of the mere consonances and dissonances, as it was done by the author in his doctoral dissertation [1].

First, we provide a definition of an *extended* counterpoint symmetry that preserves the characteristics of the counterpoint of one scale in the refined one. Then, we see that the progressive granulation of a specific example suggest an infinite counterpoint with a continuous polarity, different from the one that Mazzola himself proposed; a comparison of both alternatives calls for a deeper examination of the meaning of counterpoint extended to the full continuum of frequencies within the octave.

We must warn the reader that just a minimum exposition of Mazzola's counterpoint model is done, and hence we refer to his treatise *The Topos of Music* [3] (whose notation we use here) and an upcoming comprehensive reference [5] for further details.

O.A. Agustín-Aquino (✉)
Universidad de la Cañada, San Antonio Nanahuatipan Km 1.7 S/n. Paraje
Titlacuatitla, C.P. 68540 Teotitlan de Flores Magón, Oaxaca, Mexico
e-mail: octavioalberto@unca.edu.mx

© Springer International Publishing AG 2017
G. Pareyon et al. (eds.), *The Musical-Mathematical Mind*,
Computational Music Science, DOI 10.1007/978-3-319-47337-6_1

1

2   Some Definitions and Notations

Let R be a finite ring of cardinality $2k$. A subset S of R of such that $|S| = k$ is a *dichotomy*. It is often denoted by $(S/\complement S)$ to make the complement explicit. The group

$$\overrightarrow{GL}(R) = R \ltimes R^\times = \{e^u v : u \in R, v \in R^\times\}$$

is called the *affine group* of R, its members are the *affine symmetries*. It acts on R by

$$e^u v(x) = vx + u;$$

this action is extended to subsets in a pointwise manner. A dichotomy S is called *self-complementary* if there exists an affine symmetry p (its *quasipolarity*) such that $p(S) = \complement S$. A self-complementary dichotomy is *strong* if its quasipolarity p is unique, in which case p is called its *polarity*.

Of particular interest are the strong dichotomies of $\mathbb{Z}_{2k}$, since this ring models very well the equitempered $2k$-tone scales modulo octave and Mazzola discovered that the set of classical consonances is a strong dichotomy. For counterpoint, the self-complementary dichotomies of the dual numbers

$$\mathbb{Z}_{2k}[\epsilon] = \{a + \epsilon.b : a, b \in \mathbb{Z}_{2k}, \epsilon^2 = 0\}$$

are even more interesting, since they are used in Mazzola's counterpoint model as counterpoint intervals. More specifically, given a counterpoint interval $a + \epsilon.b$, a represents the cantus firmus, and b the interval between a and the discantus, and from every strong dichotomy (K/D) with polarity $p = e^u.v$ in $\mathbb{Z}_{2k}$ we can obtain the *induced interval dichotomy*

$$(K[\epsilon]/D[\epsilon]) = \{x + \epsilon.k : x \in \mathbb{Z}_{2k}, k \in K\}$$

in $\mathbb{Z}_{2k}[\epsilon]$. It is easily proved that, for every cantus firmus, there exists a quasipolarity $q_x[\epsilon]$ that leaves its *tangent space* $x + \epsilon.K$ invariant.

A symmetry $g \in \overrightarrow{GL}(\mathbb{Z}_{2k}[\epsilon])$ is a *counterpoint symmetry* of the consonant interval $\xi = x + \epsilon.k \in K[\epsilon]$ if

1. the interval ξ belongs to $g(D[\epsilon])$,
2. it commutes with the quasipolarity $q_x[\epsilon]$,
3. the set $g(K[\epsilon]) \cap K[\epsilon]$ is of maximal cardinality among those obtained with symmetries that satisfy the previous two conditions.

Given a counterpoint symmetry g for a consonant interval ξ, the members of the set $g(K[\epsilon]) \cap K[\epsilon]$ are its *admitted successors*; they represent the rules of counterpoint in Mazzola's model. It must also be noted that it can be proved that the admitted successors only need to be calculated for intervals of the form $0 + \epsilon.k$, and then suitably transposed for the remaining intervals.

3 Extending Counterpoint Symmetries

Let (X_n/Y_n) be a strong dichotomy in $\mathbb{Z}_n$ where

$$g_1 = e^{\epsilon.t_1}(u_1 + \epsilon.u_1 v_1) : \mathbb{Z}_n[\epsilon] \to \mathbb{Z}_n[\epsilon]$$

is a contrapuntal symmetry for the consonant interval $\epsilon.y \in X_n[\epsilon]$, with $p_n = e^{r_1} w_1$ the polarity of (X_n/Y_n). This means that if $s \in X_n$ and $p_n[\epsilon] = e^{\epsilon.r_1} w_1$ is the induced quasipolarity then

$$t_1 = y - u_1 p_n(s) \quad \text{and} \quad p_n[\epsilon](\epsilon.t_1) = g_1(\epsilon.r_1),$$

as it is proved in [3, p. 652]. If $a : X_n \hookrightarrow X_{an} : x \mapsto ax$ is an embedding of dichotomies, then

$$p_{an} \circ a = a \circ p_n$$

(where $p_{an} = e^{r_2} w_2$ is the polarity of (X_{an}/Y_{an})) and, evidently,

$$p_{an}[\epsilon] \circ a = a \circ p_n[\epsilon].$$

In particular, $ar_1 = r_2$.

Suppose there is a symmetry

$$g_2 = e^{\epsilon.t_2}(u_2 + \epsilon.u_2 v_2) : \mathbb{Z}_{an}[\epsilon] \to \mathbb{Z}_{an}[\epsilon]$$

such that $a \circ g_1 = g_2 \circ a$, then

$$t_2 = at_1 \quad \text{and} \quad au_2 = au_1.$$

From this we deduce

$$
\begin{aligned}
t_2 = at_1 &= ay - au_1 p_1(s) \\
&= ay - u_2 a p_n(s) \\
&= ay - u_2 p_{an}(as)
\end{aligned}
$$

where $as \in X_{an}$, and

$$
\begin{aligned}
p_{an}[\epsilon](\epsilon.t_2) &= p_{an}[\epsilon](\epsilon.at_1) \\
&= ap_n[\epsilon](\epsilon.t_1) = ag_1(\epsilon.r_1) \\
&= g_2(\epsilon.ar_1) = g_2(\epsilon.r_2).
\end{aligned}
$$

This means that g_2 is almost a contrapuntal symmetry for $\epsilon.ay$, except for the maximization of the intersection $g_2 X_{an}[\epsilon] \cap X_{an}[\epsilon]$. Now we can define a *extended counterpoint symmetry with respect the embedding a* as a symmetry $g_2 \in \overrightarrow{GL}(\mathbb{Z}_{an}[\epsilon])$ that satisfy

1. $a \circ g_1 = g_2 \circ a$ with g_1 a (extended or not) contrapuntal symmetry for $\epsilon.y$, and
2. $g_2 X_{an}[\epsilon] \cap X_{an}[\epsilon]$ has the maximum cardinality among the symmetries with the above property.

Note that extended counterpoint symmetries preserve the admitted successors of $\epsilon.y \in \mathbb{Z}_n[\epsilon]$, since otherwise the restriction $g_2|_{\mathbb{Z}_n[\epsilon]}$ of a extended counterpoint symmetry would be a symmetry such that the intersection $g_2|_{\mathbb{Z}_n[\epsilon]} X_n[\epsilon] \cap X_n[\epsilon]$ is bigger than the corresponding intersection for any counterpoint symmetry. This is a contradiction.

Remark 1 In particular, extended counterpoint symmetries always exist in the case of the embedding $2 : \mathbb{Z}_n \to \mathbb{Z}_{2n}$, because all the elements of $GL(\mathbb{Z}_n)$ are coprime with 2. Thus, for any $\epsilon.y \in \lim_{k \to \infty} X_{2^k.n}[\epsilon]$, there exist a extended contrapuntal symmetry in the limit $\lim_{k \to \infty} \mathbb{Z}_{2^k.n}[\epsilon]$ which is the limit of extended counterpoint symmetries.

Example 1 Let $X_6 = \{0, 2, 3\} \subseteq \mathbb{Z}_6$. The consonant interval $\epsilon.2 \in \mathbb{Z}_6[\epsilon]$ has $e^{\epsilon.3}$ $(1 + \epsilon.3)$ as its only counterpoint symmetry and 15 admitted successors. The extended counterpoint symmetries of $\epsilon.4 \in X_{12} = \{0, 1, 4, 5, 6, 9\} \subseteq \mathbb{Z}_{12}$ with respect to the embedding 2 are $e^{\epsilon.6}.(1 + \epsilon.6)$ and $e^{\epsilon.6}.(7 + \epsilon.6)$. The number of extended admitted successors is 48.

4   A More Detailed Example

In Example 4.11 of [1], it is shown that there exists a strong dichotomy in $\mathbb{Z}_{24}$ that can be extended progressively (via the embedding Lemma 4.5 of [1]) towards a dense dichotomy in S^1 with polarity $x \mapsto x e^{i\pi}$, which is the antipodal map. Analogously, the dichotomy

$$U_0 = \{0, 1, 3, \ldots, 7, 10\}$$

in $\mathbb{Z}_{16}$ can be completed in each step using the dichotomy

$$V_i = \{0, \ldots, |U_i| - 1\},$$

so we have the inductive definition

$$U_{i+1} = 2U_i \cup (2V_i + 1), \quad i \geq 1,$$

which is a strong dichotomy of $\mathbb{Z}_{2^{4+i}}$, in each case with polarity $e^{2^{3+i}}$. Note that the injective limit of the U_i in S^1 is dense in one hemisphere.

The standard counterpoint symmetries for U_0 and successively extended counterpoint symmetries for $\mathbb{Z}_{512}$ are listed in Table 1. With "successively extended" we mean that they are those who commute with the extended counterpoint symmetries of $\mathbb{Z}_{256}$, which in turn commute with those of $\mathbb{Z}_{128}$, and so on down to $\mathbb{Z}_{16}$. In most cases the linear part is -1, and in fact it is remarkable that all of them have no dual component.

5 A Possible Continuous Counterpoint

The previous calculations suggest the following constructions that enable a continuous and compositionally useful counterpoint. First, we consider the space $S^1 \subseteq \mathbb{C}$ (which represents the continuum of intervals modulo octave), with the action of the group $G = \mathbb{R}/\mathbb{Z} \ltimes \mathbb{Z}_2$ given by

Table 1 A set of consonances in $\mathbb{Z}_{16}$, their respective counterpoint symmetries and number of admitted successors, and their extended counterpoint symmetries when embedded in $\mathbb{Z}_{512}$, with the corresponding number of extended admitted successors

Interval	Symmetries for $\mathbb{Z}_{16}$	$|gX[\epsilon] \cap X[\epsilon]|$	Extended symmetries for $\mathbb{Z}_{512}$	$|gX[\epsilon] \cap X[\epsilon]|$
0	$e^{\epsilon 5}3$	96		
	$e^{\epsilon 6}13$			
	$e^{\epsilon 11}15$		$e^{\epsilon 352}511$	82432
1	$e^{\epsilon 10}15$	112	$e^{\epsilon 320}511$	98816
3	$e^{\epsilon 2}5$	96		
	$e^{\epsilon 9}11$			
	$e^{\epsilon 11}15$		$e^{\epsilon 352}511$	82432
4	7	112	7	75264
			439	
5	$e^{\epsilon 1}3$	96		
	$e^{\epsilon 6}13$			
	$e^{\epsilon 7}15$		$e^{\epsilon 244}511$	124416
6	$e^{\epsilon 3}13$	112	$e^{\epsilon 96}205$	76800
7	$e^{\epsilon 1}5$	112	$e^{\epsilon 16}5$	76800
10	$e^{\epsilon 2}5$	96		
	$e^{\epsilon 5}11$			
	$e^{\epsilon 7}15$		$e^{\epsilon 244}511$	124416

$$e^t v(x) = \begin{cases} x \exp(2\pi i t), & v = 1, \\ \overline{x} \exp(2\pi i t), & v = -1. \end{cases}$$

We define the set of consonances (K/D) as the image of $[0, \frac{1}{2})$ under the map $\phi : [0, 1] \mapsto S^1 : t \mapsto e^{2i\pi t}$, which musically means that we consider as consonant any interval greater or equal than the unison but smaller than the tritone (within an octave). Apart from the identity, no element of G leaves (K/D) invariant, thus it is strong and its polarity is $e^{\frac{1}{2}}$.

Now, for counterpoint, we consider the torus $T = S^1 \times S^1$, with the first component for the cantus firmus and the second for the discantus interval. Let G act on T in the following manner:

$$e^t v(x, y) = (vx, e^t vy);$$

this action is suggested by the fact that all the linear parts of the affine symmetries of counterpoint intervals have no dual component.

Thus the set of consonant intervals is $(K[\epsilon]/D[\epsilon]) = (S^1 \times K/S^1 \times D)$, the self-complementary function for any $\xi \in T$ which fixes its tangent space is $e^{1/2}1$, and it commutes with any element of G'. Also $\xi = (0, k) \in g(D[\epsilon])$ for a $g \in G'$ if and only if

$$g = e^t 1, \quad t \in (k, k + 1/2] \quad \text{or} \quad g = e^t(-1), \quad t \in [k - 1/2, k).$$

And here comes a delicate point. If we wish to preserve the idea of cardinality maximization, it would be reasonable to ask the set of infinite admitted successors to attain certain maximum. A possibility is to gauge these sets in terms of the standard measure in T since, for instance, the affine morphisms

$$g = \begin{cases} e^{k-1/2}(-1), & k \in \phi([0, 1/4]), \\ e^{k-1/2}1, & k \in \phi([1/4, 1/2]), \end{cases}$$

maximize the measure of the intersection $(gX[\epsilon]) \cap X[\epsilon]$. The musical meaning of this alternative is that the admitted successors of consonant intervals below the minor third are all the consonant intervals above it, and vice versa. The minor third is special, because it has any consonant interval as an admitted successor.

But, in terms of the new perspective of homology introduced by Mazzola in [4], we observe first that T is homeomorphic to T itself with respect to the Kuratowski closure operator induced by the quasipolarity $e^{1/2}1$. This is so because, for in each section $x \times S^1$, the self-complementary function is the antipodal morphism, thus each $x \times S^1$ is homeomorphic to the projective line, which in turn is homeomorphic to $x \times S^1$ itself [2, p. 58]. Furthermore, any $g \in G'$ which leaves ξ out of $g(X[\epsilon])$ is such that $(g(X[\epsilon])) \cap X[\epsilon]$ is homotopically equivalent to S^1, except when such intersection is empty.[1] Therefore, $H_1((g(X[\epsilon])) \cap X[\epsilon]) = \mathbb{Z}$ is always the group of maximum rank when it satisfies the rest of the conditions of counterpoint symmetries. This implies that, except for itself, any counterpoint interval can be an admitted counterpoint successor, which is clearly an undesirable outcome.

6 Some Final Remarks

In the version of infinite counterpoint that maximizes measure, we arrive to some peculiar features:

1. Certainly there are no culs-de-sac.
2. The only consonance that has all the other consonances as admitted successors is the minor third.
3. All the intervals smaller than the minor third admit only larger intervals as successors, while all those greater admit only smaller ones.
4. Although it is continuous regarding its induced quasipolarity and the cantus firmus can be chosen to be a continuous function of time, the discantus cannot be continuous in the standard topology.

All of these seem to be very close to the general principles of counterpoint. Unfortunately, this specific instance is not a natural extension of the discrete version; their relation is mainly axiomatic. On the other hand, the restriction of the linear part of the morphisms to $\mathbb{Z}_2$, although not entirely artificial, feels too limited with respect to the original finite model.

In fact, the selected dichotomy for the continuous example is a particularly nice one that permits a simple analysis, but by no means it is the only possible one. Considering that the general linear parts for counterpoint symmetries can be recovered as "windings" of S^1, carefully constructed infinite dichotomies could yield more complicated homology groups that make the algebro-topological approach far more interesting.

References

1. Agustín-Aquino, O.A. (2011). Extensiones microtonales de contrapunto, PhD dissertation, Universidad Nacional Autónoma de México, Mexico City, 2011
2. Crossley, M.D.: Essential Topology. Springer, Berlin (2006)
3. Mazzola, G.: The Topos of Music. Birkhäuser-Verlag, Berlin (2002)
4. Mazzola, G.: Singular homology on hypergestures. J. Math. Music 6(1), 49–60 (2012)
5. Mazzola, G., Agustín-Aquino, O.A., Junod, J.: Computational Counterpoint Worlds. Springer, Berlin (2015)

---

[1] By the way, this happens only with the self-complementary function itself.

Gödel-Vector and Gödel-Address as Tools for Genealogical Determination of Genetically-Produced Musical Variants

Carlos de Lemos Almada

Abstract The present paper integrates a broad research project, based on the principles of developing variation and *Grundgestalt* (both formulated by the Austrian composer Arnold Schoenberg), which aims at a systematical production of musical variants through employment of a group of genetic algorithms. The article examines a specific aspect of the process for production of these variants, namely, the creation of an adequate and efficient method for their genealogical organizing and labeling. This led to the elaboration of a couple of complementary concepts, the *Gödel-vector* and the *Gödel-address*, inspired by a function created by the Austrian mathematician Kurt Gödel. The results obtained by the application of both elements in the process of variant production allowed a decisive improvement of the procedures employed for classiying and retriving the derivative musical forms.

1 Theoretical Foundations of the Research

The research is theoretically grounded on the principles of developing variation and *Grundgestalt*, originally elaborated by the Austrian composer Arnold Schoenberg (1874–1951), being perhaps the most important of his innumerous contributions for the musical theory. Both principles, in turn, were directly derived from the trend of Organicism, which was the most external influence on the musical creation of Romantic Austro-German composers in 19th Century [13, p.190]. According to this conception, a musical piece must be constructed in a similar manner of an organic form, like a tree from a seed (which contains implicitly all the instructions for the formation of the plant), through a continuous growing process, based essentially on variation procedures. A *Grundgestalt* can concisely be defined as a set of basic musical elements, from which —at least, in an idealized case— a composer could extract all of the necessary material to create his or her work. This sort of maximally economical process employed for obtaining the musical material corresponds to the

C. de Lemos Almada (✉)
Federal University of Rio de Janeiro, Rua Do Passeio, 98, Centro, Rio de Janeiro, Brazil
e-mail: carlosalmada@musica.ufrj.br

© Springer International Publishing AG 2017
G. Pareyon et al. (eds.), *The Musical-Mathematical Mind*,
Computational Music Science, DOI 10.1007/978-3-319-47337-6_2

different techniques of developing variation, which consists essentially on variation over variation, resulting in several generations of derived forms.

As though the first formulation by Schoenberg of both concepts was appeared only in 1919 [10], they were certainly present in his mind in more remote epochs, as we can easily verify by analyzing some of his tonal pieces (see, for example, the *First Chamber Symphony* Op.9, composed in 1906).[1] Schoenberg attributed the origins of his organic musical conception from the observation of procedures employed by some of his "great masters" - Bach, Mozart, Beethoven and, especially, Brahms [15, p.173–74]. The great mastery and sophistication of the Brahmsian variation treatment has not only deeply influenced the formation of Schoenberg's style, but also has been inspiring more recent studies related to these seminal principles.[2]

2   The Gr-System and the GeneMus Complex

The starting-point of the present research was the elaboration of an analytical model destined to the exam of organically-constructed music, in other words, pieces based on both principles (like some composed by Beethoven, Brahms, Schoenberg, Berg, among others).[3] In turn, the assumptions, terminology, symbology and graphical resources established during the analytical model's development became the basis for a second line of approach, this time, compositional, by using a reversal engineering strategy.

It was then created the Gr-System for systematical production of musical variants (that were renamed as "theorems") from a basic cell (i.e., the *Grundgestalt* or, in the new terminology, the "axiom" of the system). For this purpose it was elaborated a group of genetic algorithms - the GeneMus Complex (gM) - formed by four computational complementary and sequential modules, which are destined to the systematical production of lineages of variants/theorems.[4] These modules are briefly described as follows:

- ax–gT ("axiom → geno-theorem")

The axiom, a short musical fragment, input as a monophonic MIDI file, becomes a referential form for the production of a first generation of abstract variants (i.e. rhythmic and intervallic separate transformations, labeled as geno-theorems, or gT's) by application of some rules of production, sort of transformational operations or algorithms that someway affect the internal structure of the forms. The gT's of the first generation, in turn, become referential forms for the production of a second generation of theorems, by recursive and/or sequential application of the same set of rules of production. This process, therefore, involves derivation of abstract musical

---

[1] A detailed analysis on this aspect is present in [7].

[2] See, for instance, [8–12].

[3] For some published papers with analytical model studies, see [3, 5, 6].

[4] For some published papers related to this compositional approach, see [1, 2, 4].

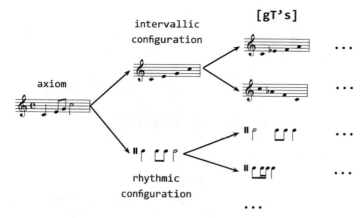

Fig. 1 Basic functioning of ax–gT

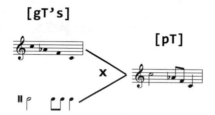

Fig. 2 Basic functioning of gT–pT

structures, being named as "developing variation of first order" (DV1). This sort of procedure can be then indefinitely replicated, resulting in extensive lineages of abstract derived forms. Figure 1 summarizes in musical notation the basic functioning of this module, considering a hypothetical axiom.

- gT–pT ("geno-theorem → pheno-theorem")

In this phase pairs of gT's are crossed over, forming concrete musical unities, named as pheno-theorems (pT's). After the production of pT's (exemplified in Fig. 2), the user may also proceed to a stage of, so to speak, artificial selection (in the Darwinian sense), by applying some fitness filters, which permit to eliminate undesirable results (like redundant forms, among other possibilities).

- pT–axG ("pheno-theorem → axiomatic group")

The third module is responsible for the concatenation of two or more pT's into large and more complex musical structures, labeled as axiomatic groups (axG's), which are created from a series of decisions proposed to the user (about transpositions, metrical displacement and note suppressions). Figure 3 presents an example of a possible axG formed by the combination of two pT's.

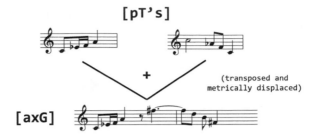

Fig. 3 Basic functioning of pT–axG

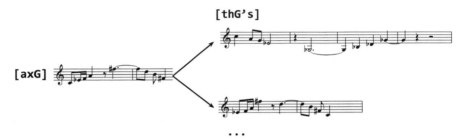

Fig. 4 Basic functioning of axG–thG

- axG–thG ("axiomatic group → group-theorem")

Each previously produced axG can be considered as a potential "patriarch" of a specific lineage of variants (group-theorems, or thG's), which are yield in the fourth module, by sequential and/or recursive application of new rules of production (including some "mutational" ones, i.e., that affect only random selected elements), through at most six generations (this number was arbitrarily chosen and eventually may be expanded in the future). According to the system terminology, this process is labeled "developing variation of second order" (DV2), involving concrete musical structures. Figure 4 shows two possible first-generation group-Theorems derived from the axG of Fig. 3.

3   The Gödel-Vector and the Gödel-Address

An indispensable need that arose in the research was a precise mean for classifying the axG's and thG's produced in the system in such a way that their respective "genealogical" position and derivative order could be adequately preserved and retrieved when desired. For this purpose, it was firstly created the Gödel-vector (Gv). It has seven entries, each one representing one of the possible generations for the groups (the first entry corresponds to generation "zero"). The sequence of integers in the seven Gv's

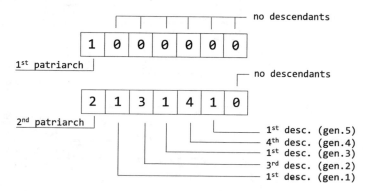

n_0	n_1	n_2	n_3	n_4	n_5	n_6

n_i = group's order number in generation i, $n_i \geq 0$

Fig. 5 Gödel-vector' structure

Fig. 6 Genealogies of two Gödel-vectors: **a** < 1000000 > and **b** < 2131410 >

entries of a given group represents not only its own order of appearance, but also those of all its eventual predecessors (Fig. 5).

Be, for example, the following groups and their respective Gv's: (a) **< 1000000 >** and (b) **< 2131410 >** (Fig. 6). Gv (a) identifies the genealogy the first produced axG (or else, the patriarch of lineage 1), since the zeros indicate that it has no descendants in the subsequent six generations. Gv (b) corresponds to a thG of fifth generation. Its genealogy description is more complex (it must be read, backwards, from right to left): the first descent (of generation 5) of the fourth descent (of generation 4) of the first descent (of generation 3) of the third descent (of generation 2) of the first descent (of generation 1) of the second axG.

A special algorithm was designed to translate a Gödel-vector into a univocal index —the Gödel-address (Ga)— which represents concisely, as a unique integer, the precise genealogical identification of a given group. Their names refer to the mathematician Kurt Gödel (1906–1978). Coincidentally, Gödel, like Schoenberg, was born in Austria and emigrated to US America in the II World War epoch (in his case, in 1940). One of his most elegant creations, the function named "Gödel numbering" inspired the elaboration of the algorithm for the calculation of the Gödel-address. In order to properly explain how a Ga is calculated, it is firstly necessary to describe the application of the above mentioned Gödel's function, taking a simple case as example. Be given a proposition from the Number Theory, like the following:

$$(\exists x)(x = sy), \tag{1}$$

that can be translated in ordinary English as "there exists a number 'x', such that 'x' is the successor of a number 'y'(for example, y = 53 and x = 54)". Looking for

Table 1 Chart of correspondences of typographical signs and Gödel numbers (adapted from [14], p.70)

Constant sign	Gödel-number	Usual meaning
~	1	Not
∨	2	or
⊃	3	if…then
∃	4	there is a …
=	5	equals
0	6	zero
s	7	successor
(	8	punctuation mark
)	9	punctuation mark
…	…	…
x	13	numeric variable 1
…	…	…
y	17	numeric variable 2

a manner to express the meaning of all possible propositions of the theory in their proper terms (or else, without the intermediation of a meta-language – like English or anyone), Gödel had the idea to attribute a number (labeled as "Gödel number") to each of the typographical signs used to write the propositions, according a predefinite chart, shown in Table 1.[5]

The translation of the sequence of signs of the proposition into Gödel numbers produces the following numeric sequence: 8-4-13-9-8-13-5-7-17-9, whose elements become exponents of a product of n first prime numbers (n is equal to the quantity of typographical signs in the proposition, in this case, ten), as follows: $2^8 x 3^4 x 5^{13} x 7^9 x 11^8 x 13^{13} x 17^5 x 19^7 x 23^{17} x 29^9$, resulting in an extremely big number (approximately 1.7×10^{95}), that corresponds to the Gödel number of the exemplified proposition.

This process serves as reference for the calculation of a Gödel-address corresponding to a given Gödel-vector. In short, the seven entries of the respective Gv became exponents of the product of the seven first prime numbers. Be, for instance, the both cases used in Fig. 6. The calculation of their respective Ga's proceeds as follow:

(a) $< 1000000 > \rightarrow Gv_a = 2^1 x 3^0 x 5^0 x 7^0 x 11^0 x 13^0 x 17^0 = 2$
(b) $< 2131410 > \rightarrow Gv_b = 2^2 x 3^1 x 5^3 x 7^1 x 11^4 x 13^1 x 17^0 = 1,998,496,500$

Therefore, we can conclude that there exists a proportional relationship between the genealogical complexity of a given group and the size of its respective Ga. The inverse process (i.e., the retrieval of a Gv from a given Ga) is easily realized with

---

[5]This example as well the corresponding chart are adapted from [14, p.50–88].

a simple factoring algorithm: the exponents of the ordered prime factors obtained become precisely the vector content. Be, for example, the Ga 3,696. Be, for example, the Ga 3,696. Its factoring produces: $2^3x3^1x5^2x7^1x11^1(x13^0x17^0)$, therefore corresponding to the Gv $< 3121100 >$, which displays, therefore, the genealogical description/position of the respective group in the system.

4 Conclusions

The integrated use of the pair of elements described in this article (Gv/Ga) has represented a precise means for organizing the groups produced in the system associated to their genealogical origins. An immediate advantage of their use is to permit non-linear production of variants from distinct axiomatic groups (i.e., is not necessary to proceed according the order of generations), since the algorithms provide the exact data classification and recovery. Recent applications of the both resources in composition of some musical pieces have confirmed their complete efficiency.

References

1. Almada, C.: O Sistema-Gr de composição musical baseada nos princípios de variação progressiva e Grundgestalt [The Gr-System for musical composition based on the principles of developing variation and Grundgestalt]. Música e Linguagem 2(1), 1–16 (2013a)
2. Almada, C.: GENEMUS: ferramenta computacional para composição com Grundgestalt e variação progressiva [GENEMUS: computational tool for composition with Grundgestalt and developing variation]. Anais do XXIII Congresso da ANPPOM. UFRN, Natal (2013b)
3. Almada, C.: Simbologia e hereditariedade na formação de uma Grundgestalt: a primeira das Quatro Canções Op.2 de Berg [Symbology and heredity in the formation of a Grundgestalt: the first of Berg's Four Songs Op.2]. Permusi 27, 75–88 (2013c)
4. Almada, C.: Aplicações composicionais de um modelo analítico para variação progressiva e Grundgestalt [Compositional applications of an analytical model for developing variation and Grundgestalt]. Opus 18(1), 127–152 (2012)
5. Almada, C.: Derivação temática a partir da Grundgestalt da Sonata para Piano op.1, de Alban Berg. [Thematic variation from the Grundgestalçt of Berg's Piano Sonata Op.1]. Anais do II Encontro Internacional de Teoria e Análise Musical. UNESP, São Paulo (2011a)
6. Almada, C.: A variação progressiva aplicada na geração de ideias temáticas [Developing variation applied to the generation of thematic ideas]. Anais do II Simposio Internacional de Musicologia. UFRJ, Rio de Janeiro (2011b)
7. Almada, C. Nas fronteiras da Tonalidade: tradição e inovação na Primeira Sinfonia de Camara op.9 de Arnold Schoenberg [At the frontiers of Tonality: tradition and innovation in Schoenberg's First Chamber Symphony Op.9]. Dissertation (Masters in Music). Federal University of the State of Rio de Janeiro, Rio de Janeiro (2007)
8. Auerbach, B. The analytical Grundgestalt: a new model and methodology based on the music of Johannes Brahms. Thesis (Ph.D. in Music). University of Rochester, Walden (2005)
9. Burts, D. An application of the Grundgestalt concept to the First and Second Sonatas for Clarinet and Piano, Op. 120, no. 1 and no. 2, by Johannes Brahms. Dissertation (Masters in Music). University of South Florida, Tampa (2004)

10. Embry, J.: The role of organicism in the original and revised versions of Brahms' Piano Trio In B Major, Op. 8, Mvt. I: a comparison by means of Grundgestalt analysis. Dissertation (Masters in Music). University of Massachusetts, Amherst (2007)
11. Epstein, D.: Beyond Orpheus: Studies in Music Structure. MIT Press, Cambridge (1980)
12. Frisch, W.: Brahms and the Principle of Developing Variation. University of Chicago Press, Los Angeles (1984)
13. Meyer, L.: Style and Music. University of Chicago Press, Chicago (1989)
14. Nagel, E., Newman, J.: Gödel's Proof. New York University Press, New York (2001)
15. Schoenberg, A.: Style and Idea: Selected Writings of Arnold Schoenberg. Faber and Faber, London (1984)

A Survey of Applications of the Discrete Fourier Transform in Music Theory

Emmanuel Amiot

Abstract Discrete Fourier Transform may well be the most promising track in recent music theory. Though it dates back to David Lewin's first paper (Lewin, J. Music Theory (3), 1959) [33], it was but recently revived by Quinn in his PhD dissertation in 2005 (Quinn, Perspectives of New Music 44(2)–45(1), 2006–2007) [35], with a previous mention in (Vuza, Persp. of New Music, nos. 29(2) pp. 22–49; 30(1), pp. 184–207; 30(2), pp. 102–125; 31(1), pp. 270–305, 1991–1992) [40], and numerous further developments by (Andreatta, Agon, (guest eds), JMM 2009, vol. 3(2). Taylor and Francis, Milton Park) [5], (Amiot, Music Theory Online, 2, 2009) [8], (Amiot, Rahn, (eds.), Perspectives of New Music, special issue 49 (2) on Tiling Rhythmic Canons) [9], (Amiot, Proceedings of SMCM, Montreal. Springer, Berlin, 2013) [10], (Amiot, Sethares, JMM 5, vol. 3. Taylor and Francis, Milton Park (2011) [16], (Callender, J. Music Theory 51(2), 2007) [17], (Hoffman, JMT 52(2), 2008) [29] (Tymoczko, JMT 52(2), 251–272, 2008) [38], (Tymoczko, Proceedings of SMCM, Yale, pp. 258–272. Springer, Berlin, 2009) [39], (Yust, J. Music Theory 59(1) (2015) [42]. I chose to broach this subject because I have had a finger in most, or all, of the pies involved (even using Discrete Fourier Transform without consciously knowing it, in the study of rhythmic tilings).

1 Introduction

Historically Discrete Fourier Transform (hereafter DFT for short) appeared in [33], though its mention in the very end of the paper was as discrete as possible (no pun intended), considering the probable outraged reaction of JMT's readers to the introduction of 'high-level' mathematics in a Music Journal in these benighted times. The paper was devoted to the interesting new notion of Intervallic Relationship between

E. Amiot (✉)
Institut de Recherche et Coordination Acoustique/Musique,
1 Place Igor-Stravinsky, 75004 Paris, France
e-mail: manu.amiot@free.fr

© Springer International Publishing AG 2017
G. Pareyon et al. (eds.), *The Musical-Mathematical Mind*,
Computational Music Science, DOI 10.1007/978-3-319-47337-6_3

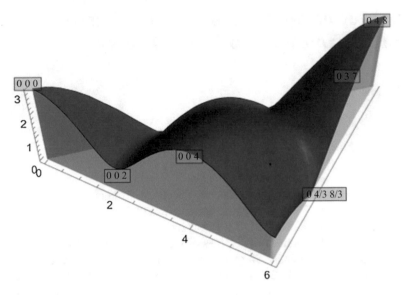

Fig. 1 The continuous landscape of 3-chords

two pc-sets,[1] and its main result was that retrieval of A knowing a fixed set B and IFunc(A, B) was possible provided B did not fall into a hodgepodge of 'special cases' —actually simply those cases when at least one of the Fourier coefficients of B (defined below) is 0.

Lewin himself returned to this notion in some of his latest papers, which may have influenced the brilliant PhD research of Ian Quinn, who encountered DFT and especially large Fourier coefficients as characteristic features of the prominent points of his 'landscape of chords' [35], see Fig. 1 below. Since he had voluntarily left aside for JMT readers the 'stultifying' mathematical work involved in the proof of one of his nicer results, connecting Maximally Even Sets and large Fourier coefficients, I did it in [14], along with a complete discussion of all maximas of Fourier coefficients of all pc-sets.

Interest in DFT having been raised, several researchers commented on it, trying to extend it to continuous pitch-classes [17] or to connect its values to voice-leadings [38, 39]. Another very original development is the study of all Fourier coefficients with a given index of all pc-sets in [29], also oriented towards questions of voice-leadings.

Meanwhile, two completely foreign topics involved a number of researchers in using the very same notion of DFT: homometry (see the state of the art in [2, 34]) and Rhythmic Canons —which are[2] really algebraic decompositions of cyclic groups as direct sums of subsets, and can be used either in the domain of periodic rhythms or

---

[1]I use the modern terms.

[2]In the case of mosaic tilings by translation.

pitches modulo some 'octave' —first extensively studied by [40],[3] then connected to the general theory of tiling by [4, 6] and developed in numerous publications [5, 9, 13] which managed to interest some leading mathematician theorists in the field (Matolcsi, Kolountzakis, Szabo) in musical notions such as Vuza canons.

There were also cross-overs like [16] looking for algebraic decompositions of pc-collections (is a minor scale a sum and difference of major scales?) or an incursion in paleo-musicology, quantifying a quality of temperaments in the search for the tuning favoured by J.S. Bach [8]. The last and quite recent development of Fourier Transform takes up the dimension that Quinn had left aside, the phase (or direction) of Fourier coefficients. The position of pairs of phases (angles) on a torus was only introduced in [10] but has known tremendously interesting developments since, especially for early romantic music analysis [42].

NB: the present survey is per force much abbreviated. Details can be found in an abundant bibliography and will be more lavishly explained in a forthcoming book in Springer's CMS collection [3].

2 Basics

2.1 What is DFT?

The DFT of a pc-set (or multiset) $A \subset \mathbb{Z}_n$ is simply the Fourier transform of its characteristic function, i.e.

$$\mathcal{F}_A = \widehat{1}_A : x \mapsto \sum_{k \in A} e^{-2i\pi kx/n}$$

$\mathcal{F}_A$ is a map on $\mathbb{Z}_n$ whose values $\mathcal{F}_A(0) \ldots \mathcal{F}_A(n-1) \in \mathbb{C}$ are called *Fourier coefficients*. *Inverse Fourier transform* retrieves 1_A from $\mathcal{F}_A$ with a similar formula. For those unfamiliar with Harmonic Analysis (in the mathematical sense!) I suggest reading the illuminating introduction in [17].

Among a number of interesting features that I omit here for lack of space, it should be mentioned that the *magnitude* of $\mathcal{F}_A$ is invariant by transposition, inversion, and even complementation.[4] This is also an immediate consequence of the most important effect of DFT on convolution products, and explains the import of DFT in Sect. 3 among other implications.

---

[3] At the time, probably the only theorist to mention Lewin's use of DFT.
[4] Except for $\mathcal{F}_A(0)$, which is equal to the cardinality of A.

2.2 Convolution and Lewin's Problem

Convolution is familiar to engineers in signal processing and other areas, but many music theorists may not have heard of it. If however I mention Boulezian's "multiplication d'accords"or Cohn's Transpositional Combination [21], it may ring a louder bell: the convolution of chords $(0, 1)$ and $(0, 3, 6, 9)$ is simply the octatonic $(0, 1, 3, 4, 6, 7, 9, 10)$ in $\mathbb{Z}_{12}$. This operation is instrumental in defining rhythmic canons as we will recall *infra*. It also serves in music-theoretic IFunc, IC functions since

$$\text{IFunc}(A, B) = \mathbf{1}_{-A} * \mathbf{1}_B \qquad \text{IC}_A = \mathbf{1}_{-A} * \mathbf{1}_A$$

where the symbol $*$ denotes the convolution product[5] and $\mathbf{1}_A$ is the characteristic function of pc-set A.

Lewin's problem consists in finding A when B and IFunc(A, B) are given. His paper states when this is possible, not how it may be done: for instance if IFunc$(A, B) = (0, 0, 1, 1, 1, 1, 0, 1, 1, 0, 0, 0)$ and $B = \{1, 3, 6\}$ how does one find $A = \{10, 11\}$?

As Lewin had obviously noticed, solving this is much simpler if the DFT is computed, because

Proposition 1 *The Fourier transform of a convolution product is the termwise product of Fourier transforms.*

In other words, IFunc$(A, B) = f \iff \overline{\mathcal{F}_A} \times \mathcal{F}_B = \mathcal{F}_f$. This enables to compute the Fourier coefficients $\mathcal{F}_A(k) = \overline{\mathcal{F}_f(k)/\mathcal{F}_B(k)}$ and thus retrieve A, *except when* $\mathcal{F}_B(k)$ *vanishes*. The pc-sets with at least one nil Fourier coefficient are none other than the 1,502 "Lewin's special cases" which have been so difficult to describe, from [33] to later descriptions by the same author or even the ingenious 'balances' in [35].

Actually, Lewin's problem is easily solved along with many other convolution-related problems by using the matricial formalism that we introduced with Bill Sethares.

2.3 Circulating Matrices

As developed in [16], if one fills the first column of a matrix with the characteristic function of a pc-set, and the other columns are circular permutations of the first one, then the obtained circulating matrix is a very effective representation of pc-sets, since

---

[5]The general definition of $f * g$ is the map $t \mapsto \sum_{k \in \mathbb{Z}_n} f(k)g(t - k)$.

- The eigenvalues of the matrix are the Fourier coefficients of the set, and
- The matrix product corresponds with the convolution product of (the characteristic functions of) pc-sets.

For instance, one computes the Interval Content of a diatonic collection matricially by putting

$$
M_A = \begin{pmatrix}
1&1&0&1&0&1&0&1&1&0&1&0\\
0&1&1&0&1&0&1&0&1&1&0&1\\
1&0&1&1&0&1&0&1&0&1&1&0\\
0&1&0&1&1&0&1&0&1&0&1&1\\
1&0&1&0&1&1&0&1&0&1&0&1\\
1&1&0&1&0&1&1&0&1&0&1&0\\
0&1&1&0&1&0&1&1&0&1&0&1\\
1&0&1&1&0&1&0&1&1&0&1&0\\
0&1&0&1&1&0&1&0&1&1&0&1\\
1&0&1&0&1&1&0&1&0&1&1&0\\
0&1&0&1&0&1&1&0&1&0&1&1\\
1&0&1&0&1&0&1&1&0&1&0&1
\end{pmatrix}
\quad \text{then} \quad
M_{IC(A)} = {}^T A \times A = \begin{pmatrix}
7&2&5&4&3&6&2&6&3&4&5&2\\
2&7&2&5&4&3&6&2&6&3&4&5\\
5&2&7&2&5&4&3&6&2&6&3&4\\
4&5&2&7&2&5&4&3&6&2&6&3\\
3&4&5&2&7&2&5&4&3&6&2&6\\
6&3&4&5&2&7&2&5&4&3&6&2\\
2&6&3&4&5&2&7&2&5&4&3&6\\
6&2&6&3&4&5&2&7&2&5&4&3\\
3&6&2&6&3&4&5&2&7&2&5&4\\
4&3&6&2&6&3&4&5&2&7&2&5\\
5&4&3&6&2&6&3&4&5&2&7&2\\
2&5&4&3&6&2&6&3&4&5&2&7
\end{pmatrix}.
$$

and one reads in the first column the 7 primes, 2 semi-tones, etc...featured in the collection. The solution of Lewin's problem (and also the more general question of Sethares, wishing to decompose a collection in an algebraic combination of translates of another, given one) is then given by solving the simple matricial equation $^T A \times B = M_{IFunc(A,B)}$, thus by-passing the computation of DFT and inverse DFT which is the real reason why this works.

This is also a promising aspect of the study of homometric sets which we will develop in the next section.

3 Homometry and Spectral Units

Homometry is the true name [36] of Z-relation: two pc-sets are homometric whenever they share the same interval content. Since $IC(A) = 1_A * 1_{-A}$ it follows fairly easily that

Proposition 2 *A and B are homometric* $\iff$ $|\mathcal{F}_A| = |\mathcal{F}_B|$ *(the magnitudes of Fourier coefficients are equal).*

This explains and generalizes the invariance of the magnitude of Fourier coefficients under T/I operations (and complementation, i.e. the hexachordal theorem).

Among other developments, this definition by DFT induces the notion of spectral unit: setting $\mathcal{F}_u = \mathcal{F}_A/\mathcal{F}_B$ one gets by inverse Fourier transform $1_A = u * 1_B$ where u has unit length Fourier coefficients, i.e. u is a spectral unit.[6] It is perhaps better seen with the matrices of the last section: the matrix of a spectral unit u is a unitary circulating matrix U i.e. $^T\overline{U}U = I_n$ i.e. the eigenvalues have magnitude one. Hence the group of all spectral units has a simple structure, it is a product of n circles.

---

[6]For instance $j = (0, 1, 0, \ldots 0)$ is the spectral unit that turns any pc-set A into its translate $A + 1$. Its Fourier coefficients are all n^{th} roots of unity.

This presentation enables to solve the question of homometry...in continuous space! Unfortunately it is still unknown how one could restrict the orbits (all continuous distributions homometric to one given pc-set) to pc-sets only, i.e. distributions with values 0 or 1 exclusively. A first difficult step is the classification of all spectral units with rational values and finite order, which I achieved in a constructive way, allowing in principle to apply all such spectral units to all pc-sets and select the pc-sets in the resulting orbits.[7]

Details can be found in [2, 34] and compositional applications in [30].

4 Tilings

A rhythmic canon in the sense of [40] is really a tiling of the integers with translates of one finite tile, and boils down to a direct sum decomposition of some cyclic group:

$$A \oplus B = \mathbb{Z}_n$$

where A is the motif, or inner voice, and B the list of offsets, or outer voice. For instance $\{0, 1, 3, 6\} \oplus \{0, 4\} = \mathbb{Z}_8$. This has been the subject of intense scrutiny from music theorists [1, 5–7, 9, 11–13, 23, 27, 28, 31, 41] which in turn focused the interest of some 'pure maths' specialists of tiling problems, which led eventually to a fruitful collaboration (see [32] for instance).

For the present survey, DFT appears in the definition of tiling that is fashionable today, i.e. A tiles with $B \iff$ for all $k \in \mathbb{Z}_n, k \neq 0$, either $\mathcal{F}_A(k)$ or $\mathcal{F}_B(k)$ is 0 (or equivalently the **zero sets** of $\mathcal{F}_A$, $\mathcal{F}_B$ cover $\mathbb{Z}_n$, 0 excepted).[8] This stems from $1_A * 1_B = 1_{\mathbb{Z}_n}$.

Moreover, the zero set $Z(A)$ of Fourier coefficients of a pc-set A has remarkable structure:

Proposition 3 $Z(A)$ *is stable by the automorphisms of* $\mathbb{Z}_n$, *i.e. if* $k \in Z(A)$ *then all multiples of* k *by any* α *coprime with* n *are also in* $Z(A)$.

In other words, $Z(A)$ is a reunion of orbits of elements sharing the same order in the group $(\mathbb{Z}_n, +)$.[9] Following [22],[10] we set R_A for the collection of the orders of elements in $Z(A)$ and let S_A be the subset of R_A of elements which are prime powers. Then it is possible to give simple sufficient, or necessary, conditions on these two rather abstract but eminently computable sets, for A to tile.

---

[7]There are 6,192 such spectral units for $n = 12$.

[8]With the added technical condition $\mathcal{F}_A(0)\mathcal{F}_B(0) = \#A\#B = n$.

[9]In layman's terms, this means that if motif A tiles, then so does $\alpha \times A \mod n$, for any α coprime with n. This is actually a deep algebraic property, but nonetheless it was rediscovered independently by several music composers.

[10]At the time the authors made use of polynomials, not Fourier coefficients, but this is an isomorphic point of view. We translated their definitions accordingly.

These conditions also reflect on the famous *spectral conjecture* [26, 37] and consideration of the musical notion of *Vuza canons* (originating in wondering what is actually heard while listening to a rhythmic canon) enabled some progress on this still unsolved question [13]. Moreover, new algorithms were developed, based on a classification of possible sets R_A and enhancing the exhaustive search for Vuza canons, see [32]. I skip many fascinating aspects of this beautiful question, which already gave birth to special issues of PNM and JMM [5, 9].

5 Saliency

In this section we look at Fourier coefficients which are large instead of nil.

5.1 Measuring "fifthishness"

In [35], Ian Quinn pursued the quest for a 'landscape of chords' (for some given cardinality k) and realized that most authors agreed on a prevalence of maximally even sets,[11] and that furthermore, these sets could be characterized by a high value of their k^{th} Fourier coefficient:

Theorem 1 *The highest value of* $|\mathcal{F}_A(k)|$ *is reached among k-pcsets for Maximally Even sets and only for them.*

The rigorous mathematical study of this characterization was done in [14]. More generally Quinn links the size of this coefficient, the *saliency* (which is both closeness to an even division in k parts of the chromatic circle, and the quality of being generated by some interval d) to the prevalence of this generating interval.[12] For instance, the magnitude of $\mathcal{F}_A(3)$ can be construed as 'major thirdness' (this coefficient being maximal for augmented triads) and that of $\mathcal{F}_A(5)$ is the 'fifthishness', maximal for pentatonic (or diatonic) collections. In a continuous setting, of course the actual maximums happen for exact divisions of the circle or subsets thereof.

5.2 A Better Approximation of Peaks

Tymoczko [39] improves on remarks by Strauss and others in laying down a connection between voice-leading distances and Fourier saliency: intuitively, since the peaks for saliency culminate for even distributions of the (continuous) circle of pcs,

---

[11] Such as defined in [18–20] and others.

[12] There is a good correlation between this saliency and the saturation of the collection in interval d (Aline Honing, personal communication).

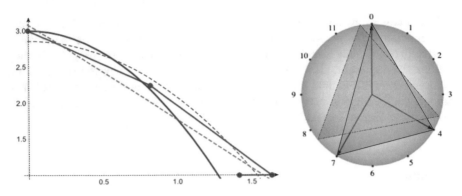

Fig. 2 Linear and quadratic correlation for 3-sets

the closest to one such peak, the largest the Fourier coefficient will be. Acting on this flimsy connection, Tymoczko computed the correlation between this closeness, measured as the standard Euclidean Voice-Leading distance between pc-sets, and was rewarded by extremely good correlation coefficients (between -0.99 and -0.95).

Being dissatisfied both with the heuristicness of the argument and with the result (near a maximum, one expects a curve to be flat, i.e. a 0 slope and not a negative one) I decided to tackle the analytic computation of the saliency of a neighbour of a peak. Not surprisingly the formulas are different,[13] and the true correlation is quadratic, not linear, as expected near a maximum (see Fig. 2 where VL is the Euclidean distance between a 3-set and the closest equilateral triangle). Still this vindicates the use of Euclidean distance for voice-leading instead of taxi-cab metric for instance [39].

6 A Torus of Phases

Another new development of DFT in Music Theory takes up the gauntlet that Ian Quinn had thrown (or rather left aground) in [35], "letting aside the direction component" i.e. focusing on magnitude and leaving aside the phase, or direction, of Fourier coefficients. [29] was probably the first to tackle the whole complex value of a given Fourier coefficient for different pc-sets (with a given cardinality), providing intriguing pictures with almost complete symmetries, see Fig. 3. His paper shows a clear understanding of the meaning of the missing phase component, stating that

> The direction of a vector indicates which of the transpositions of the even chord associated with a space predominates within the set under analysis.

---

[13] For 3-sets, $|\mathcal{F}_A(3)| = 3 - \dfrac{\pi^2}{8} V L^2 + o(V L^4)$, best near the maximum, whereas the linear regression yields $|\mathcal{F}_A(3)| \approx 3.39 - 1.57 \times V L$. The formula is different from the one in [39] because of a different convention in the definition of DFT.

Fig. 3 a_5 coefficient for all 3-sets in $\mathbb{Z}_{12}$

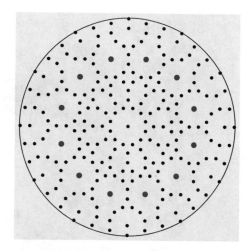

It is perhaps even clearer to measure the phase of a coefficient by how much it changes under basic operations:

Lemma 1 *Transposition of a pc-set by t semitones rotates its kth Fourier coefficient a_k by a $-2kt\pi/n$ angle, i.e. $\theta_k \mapsto \theta_k - 2kt\pi/n$.*

Any inversion of a pc-set similarly rotates the conjugates *of the Fourier coefficients.*

For instance, moving a diatonic collection by a fifth changes the direction of its fifth coefficient by $\pi/6$. Hoffman's pictures are particularly useful in considering close neighbours and parsimonious voice-leadings. But since they do not allow, for instance, to distinguish between all 24 major/minor triads, the following space deserves a closer look.

In [10] I introduced a 2D-space, torus shaped, defined by the pair of phases of two Fourier coefficients.[14] This space enables to project (almost) all pc-sets and is not limited to a given cardinality, this major drawback of most existing models. As it was since developed by J. Yust, it is most advantageous to feature simultaneously on the same simple 2D-model triads, dyads, single notes, diatonic collections, and whatever chords are necessary for the analysis of a given piece of music of even musical style (see [42] for a convincing utilisation of the Torus of Phases in early romantic music). Another striking advantage appears when one focuses on triads, which are disposed in this space with the same topology as the classical Tonnetz, see Fig. 4.[15]

---

[14]The $3^r d$ and 5th were chosen for stringent reasons. It was also the choice independently made by [42]).

[15]Please remember that this picture is a torus, i.e. opposite sides should be construed as glued together.

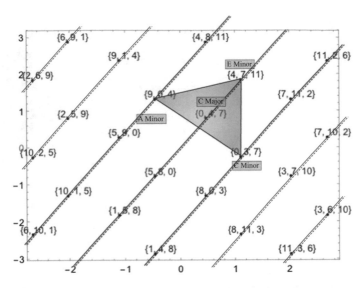

Fig. 4 The neighbours of a triad are its images by L, P and R

A particularly seductive feature of this model discovered by Yust is that central symmetries around a single pc or around dyads appears just like that, as a central symmetry on the planar representation of the torus: the T/I group and its induced action on pc-sets embeds itself in the Euclidean (quotient) group on the torus. For instance the dyad $(0, 4)$ would appear as the middle point of triads $(0, 4, 9$ and $(0, 4, 7)$ on Fig. 4. More specifically,

Proposition 4 *If A and B are symmetrical around a center c (resp. a dyad (a, b)), then their torus projections are symmetrical around the torus image of c (resp. the image of the dyad).*

This makes for especially concise and convincing representations of movements between chords, see again [42] for examples. Among other things, it enabled to explain the strange closeness of the lines connecting chromatically major and minor triads respectively (part of it in red and blue on Fig. 4) that I had presented as a baffling mystery in [10] barely a year before.

Acknowledgements My heartiest thanks to the organizers of this beautiful event for the opportunity of exposing this rich subject to a learned audience.

References

1. Agon, C., Amiot, E., Andreatta, M.: Tiling the line with polynomials, In: Proceedings ICMC (2005)
2. Agon, C., Amiot, E., Andreatta, M., Ghisi, D., Mandereau, J.: Z-relation and homometry in musical distributions. In: JMM 2011, vol. 5. Taylor and Francis, Milton Park
3. Amiot E.: Music through Fourier Space. Springer (2016)
4. Andreatta, M., *On group-theoretical methods applied to music: some compositional and implementational aspects.* In: Lluis-Puebla, E., Mazzola G., Noll T. et al. (eds.) Perspectives of Mathematical and Computer-Aided Music Theory, EpOs, Universität Osnabrück, pp. 122–162 (2004)
5. Andreatta, M., Agon, C.: (guest eds), Special Issue Tiling Problems in Music. JMM 2009, vol. 3(2). Taylor and Francis, Milton Park
6. Andreatta, M.: De la conjecture de Minkowski aux canons rythmiques mosaïques, L'Ouvert, n°114, March 2007, pp. 51–61
7. Amiot, E.: Why rhythmic canons are interesting. In: Lluis-Puebla, E., Mazzola G., Noll, T. et al. (eds.) Perspectives of Mathematical and Computer-Aided Music Theory, EpOs, 190–209, Universität Osnabrück (2004)
8. Amiot, E.: Discrete Fourier Transform and Bach's Good Temperament. Music Theory Online, **2** (2009)
9. Amiot, E., Rahn, J. (eds.), Perspectives of New Music, special issue 49 (**2**) on Tiling Rhythmic Canons
10. Amiot, E.: The Torii of phases. In: Proceedings of SMCM, Montreal. Springer, Berlin (2013)
11. Amiot, E.: Rhythmic canons and Galois theory. Grazer Math. Ber. **347**, 1–25 (2005)
12. Amiot, E.: À propos des canons rythmiques. Gazette des Mathématiciens, SMF Ed. **106**, 43–67 (2005)
13. Amiot, E.: New perspectives on rhythmic canons and the spectral conjecture. In: Special Issue "Tiling Problems in Music", JMM 3, vol. 2. Taylor and Francis, Milton Park (2009)
14. Amiot, E.: David Lewin and maximally even sets. In: JMM 1, vol. 3, pp. 157–172, Taylor and Francis, Milton Park (2007)
15. Amiot, E.: Structures, algorithms, and algebraic tools for rhythmic canons. Perspectives of New Music 49(2), 93–143 (2011)
16. Amiot, E., Sethares, W.: An algebra for periodic rhythms and scales. In: JMM 5, vol. 3. Taylor and Francis, Milton Park (2011)
17. Callender, C.: Continuous harmonic spaces. J. Music Theory **51**(2) (2007)
18. Clough, J., Douthett, J.: Maximally even sets. J. Music Theory **35**, 93–173 (1991)
19. Clough, J., Myerson, G.: Variety and multiplicity in diatonic systems. J. Music Theory **29**, 249–270 (1985)
20. Clough, J., Myerson, G.: Musical scales and the generalized circle of fifths. AMM **93**(9), 695–701 (1986)
21. Cohn, R.: Properties and generability of transpositionally invariant sets. J. Music Theory **35**(1), 1–32 (1991)
22. Coven, E., Meyerowitz, A.: Tiling the integers with one finite set. J. Alg. **212**, 161–174 (1999)
23. Fidanza, G.: Canoni ritmici, tesa di Laurea, U. Pisa (2008)
24. Fripertinger, H.: Remarks on Rhythmical Canons. Grazer Math. Ber. **347**, 55–68 (2005)
25. Fripertinger, H.: Tiling problems in music theory. In: Lluis-Puebla, E., Mazzola, G., Noll, T. (eds.) Perspectives of Mathematical and Computer-Aided Music Theory, pp. 149–164. Universität Osnabrück, EpOs (2004)
26. Fuglede, H.: Commuting self-adjoint partial differential operators and a group theoretic problem. J. Func. Anal. **16**, 101–121 (1974)
27. Gilbert, E.: Polynômes cyclotomiques, canons mosaïques et rythmes k-asymétriques, mémoire de Master ATIAM, Ircam, May 2007
28. Hall, R., Klinsberg, P.: Asymmetric rhythms and tiling canons. Am. Math. Mon. **113**(10), 887–896 (2006)

29. Hoffman, J.: On pitch-class set cartography relations between voice-leading spaces and fourier spaces. JMT **52**(2) (2008)
30. Jedrzejewski, F.: The structure of Z-related sets. In: Proceedings of MCM 204th International Conference in Montreal, pp. 128–137. Springer, Berlin (2013)
31. Johnson, T.: Tiling the line. In: Proceedings of J.I.M., Royan (2001)
32. Kolountzakis, M. Matolcsi, M.: Algorithms for translational tiling, in special issue *tiling problems in music*. J. Math. Music **3**(2) (2009). Taylor and Francis
33. Lewin, D.: Intervalic relations between two collections of notes. J. Music Theory (3) (1959)
34. Mandereau, J., Ghisi, D., Amiot, E., Andreatta, M., Agon, C.: Discrete phase retrieval in musical distributions. In: JMM 2011, (5). Taylor and Francis, Milton Park
35. Quinn, I.: General equal-tempered harmony. Perspectives of New Music **44**(2)–45(1) (2006–2007)
36. Rosenblatt, J., Seymour P.D.: The structure of homometric sets. SIAM J. Algebraic Discret. Methods **3**(3) (1982)
37. Tao, T.: Fuglede's Conjecture is False in 5 and Higher Dimensions. http://arxiv.org/abs/math. CO/0306134
38. Tymoczko, D.: Set-class similarity. Voice leading, and the Fourier Transform. JMT **52**(2), 251–272 (2008)
39. Tymoczko, D.: Three conceptions of musical distance. In: Proceedings of SMCM, Yale, pp. 258–272. Springer, Berlin (2009)
40. Vuza, D.T.: Supplementary sets and regular complementary unending canons, in four parts in: Canons. Persp. of New Music, nos. **29**(2) pp. 22–49; **30**(1), pp. 184–207; **30**(2), pp. 102–125; **31**(1), pp. 270–305 (1991–1992)
41. Wild, J.: Tessellating the chromatic. Perspectives of New Music (2002)
42. Yust, J.: Schubert's Harmonic language and fourier phase space. J. Music Theory **59**(1) (2015)

Gestures on Locales and Localic Topoi

Juan Sebastián Arias

Abstract We present a motivation and a proposal for the definition of gestures and hypergestures on locales and localic topoi. Further possible generalizations, including gestures on Grothendieck topoi are discussed.

1 Introduction

The theory of gestures has meant a revolution for the mathematical music theory established by Guerino Mazzola in his famous book *The Topos of Music* [7] in 2002. In several publications, he has presented a solid framework for the definition of mathematical gestures from three points of view: music, philosophy and mathematics. This definition is formulated originally for topological spaces and topological categories [8, 9]. The iteration of gestures leads to the construction of hypergestures, using tools from classical homotopy theory.

In this article we expose a generalization of mathematical gestures on topological spaces introduced by Mazzola in [8], to locales and categories of sheaves on locales. In first place, we consider a recapitulation of Mazzola's construction in terms of exponentials and limits in the category of topological spaces. Second, we show how these constructions are possible in the category of locales and the category of localic topoi (categories of sheaves on locales). The constructions of exponentials for locales are based on an article of Hyland [2] and Johnstone's *Stone Spaces* [3, VII 4.11]. The constructions of limits for locales can be found in the book *Categorical foundations* [11, II.3] and in [3, II.2.12]. The respective constructions in the category of localic topoi are a consequence of the equivalence of this category and the category of locales. See *Sketches of an Elephant* [5, C1.4] for the details of the construction of this equivalence. It is remarkable that our general construction of gestures takes account of that of hypergestures.

J.S. Arias (✉)
Universidad Nacional de Colombia, Cra 45, Bogotá, Colombia
e-mail: jsariasv1@gmail.com

© Springer International Publishing AG 2017
G. Pareyon et al. (eds.), *The Musical-Mathematical Mind*,
Computational Music Science, DOI 10.1007/978-3-319-47337-6_4

Finally, we comment some possible generalizations to sites, Grothendieck topoi and elementary topoi, which are our ongoing work. In this way, we will discuss subsequent implications for Mazzola's architecture of mathematical music theory based on the topos structure [7] and the diamond conjecture stated on [8]. The emergence of Grothendieck topoi and elementary topoi in gesture theory could help to recast the diamond conjecture in an abstract setting, perhaps easier to handle.

2 Gestures on Topological Spaces

In this section we restate the construction of gestures on topological spaces presented in [8], in a more categorical setting.

The construction runs as follows. Let Δ be a digraph, X a topological space, and I the *interval* $[0, 1]$ in $\mathbb{R}$.

The first step is to construct the set $I@X$ consisting of *paths* on X. The space I is an exponentiable object in **Top** since it is a Hausdorff locally compact space, and the exponential X^I coincides with $I@X$, so $I@X$ is a topological space endowed with the compact-open topology. The subbasic opens of this topology are those of the form $W(K, U) := \{c \in I@X \mid c(K) \subseteq U\}$, where K is compact in I and U is open in X. For a detailed exposition about topologies for function spaces in **Top**, see [1].

Second, we consider the *spatial digraph* $\overrightarrow{X}$ of the topological space X. It is the tuple $(I@X, X, e_0, e_1)$, where e_0 and e_1 correspond to evaluating a path at 0 and 1 respectively. Actually, both e_0 and e_1 are continuous functions; in fact, $e_0^{-1}(U) = \{c \in I@X \mid c(0) \in U\} = W(\{0\}, U)$, and similarly $e_1^{-1}(U) = W(\{1\}, U)$, which are subbasics of the topology of $I@X$.

Third, recall from [8] that a gesture with skeleton $\Delta = (A, V, t, h)$ and body in X is a morphism of digraphs from Δ to $\overrightarrow{X}$. The space of all *gestures with skeleton Δ and body in X*, denoted by $\Delta@X$, can be obtained as the limit of the following diagram $\mathcal{B}$ in **Top**: assign to each arrow a in A the space $I@X$, to each vertex x in V the space X, take a copy of the morphism $e_0 : I@X \to X$ whenever $t(a) = x$, and a copy of the morphism $e_1 : I@X \to X$ whenever $h(a) = x$.

To see that the limit coincides with the set of gestures $\Delta@\overrightarrow{X}$, first check that Δ is the colimit of the diagram $\mathcal{D}$ of digraphs obtained by taking an arrow digraph $\overset{a}{\to}$ for each arrow a in A, a vertex digraph $\bullet x$ for each vertex x in V, and an inclusion morphism $\bullet x \hookrightarrow \overset{a}{\to}$ whenever $x = t(a)$ or $x = h(a)$. The contravariant Hom functor $\_@\overrightarrow{X}$ from the category of digraphs to **Set** carries colimits of digraphs to limits in **Set**, so $\Delta@\overrightarrow{X}$ is the limit in **Set** of the image diagram $\mathcal{D}@\overrightarrow{X}$ of $\mathcal{D}$ under $\_@\overrightarrow{X}$. Finally, note that we can identify naturally $\mathcal{D}@\overrightarrow{X}$ with the diagram of topological spaces $\mathcal{B}$.

Now, we reinterpret the preceding construction of topological gestures in terms of correspondences between sets of opens. It will help us to extend the preceding construction to the category of locales. In the sequel, we shall denote the set of opens of the topological space X by $\mathcal{O}(X)$.

Each path c in $I @ X$ induces a correspondence $c^{-1}(\_)$ from $\mathcal{O}(X)$ to $\mathcal{O}(I)$.

From the evaluation maps e_0, e_1 we obtain correspondences $e_0', e_1' : \mathcal{O}(X) \to \mathcal{O}(I @ X)$ defined by $e_0'(U) = W(\{0\}, U) = \{c \in I @ X \mid 0 \in c^{-1}(U)\}$ and $e_1'(U) = W(\{1\}, U) = \{c \in I @ X \mid 1 \in c^{-1}(U)\}$. Note also that e_0, e_1 are the composites of continuous maps

$$I @ X \xrightarrow{\cong} I @ X \times \{0\} \xrightarrow{id \times i_0} I @ X \times I \xrightarrow{e} X$$

and

$$I @ X \xrightarrow{\cong} I @ X \times \{1\} \xrightarrow{id \times i_1} I @ X \times I \xrightarrow{e} X$$

respectively, where i_0, i_1 are the inclusions of the endpoints of I and e is the evaluation map. Thus, e_0', e_1' are also the composites, in the reverse order, of the correspondences induced by the maps in the above diagrams.

In general, the map from $\mathbf{Top}(X, Y)$ to the set $\mathbf{Frm}(\mathcal{O}(Y), \mathcal{O}(X))$ of all the functions from $\mathcal{O}(Y)$ to $\mathcal{O}(X)$ that preserve finite meets and arbitrary joins, assigning to each continuous function from X to Y the associate inverse image function from $\mathcal{O}(Y)$ to $\mathcal{O}(X)$, is not bijective. But if the space Y is sober, this correspondence is a bijection; see [11, II.1.3]. Next we introduce a short overview of sober spaces taken from [7, p. 1110].

2.1 Sober Spaces

A topological space X is *irreducible* iff every non-empty open subset is dense, or, equivalently, if any two non-empty open sets have a non-empty intersection. A subset of a topological space is called irreducible if it is so with its relative topology. A point x of an irreducible space X is said to be *generic* iff its (always irreducible) closure $\overline{\{x\}}$ is equal to X. We say that *a point x dominates a point y, in signs $x > y$*, iff $\overline{\{y\}} \subseteq \overline{\{x\}}$. This is a partial order relation on X. An irreducible component of a space X is a maximal irreducible subset.

Proposition 1 *We have the following properties concerning irreducibility:*

(i) *A subset of a topological space is irreducible iff its closure is.*

(ii) *Irreducible components are closed.*

(iii) *Every irreducible subset is contained in an irreducible component, in particular, a topological space is the union of its irreducible components.*

(iv) *The image $f(E)$ of an irreducible subset $E \subseteq X$ under a continuous map $f : X \to Y$ is irreducible.*

Definition 1 A topological space X is *sober* iff each closed irreducible subset has a unique generic point. Call **Sob** the full subcategory of the category **Top** of topological spaces consisting of sober spaces. Equivalently X is sober if it is a T_0 space and every irreducible closed set C ($C = C_1 \cup C_2$, for C_1, C_2 closed, implies $C = C_1$ or

$C = C_2$) is the closure of a point. We shall discuss later in Sect. 3.4 the concept of sobriety in the language of adjoints.

Example 1 If A is a commutative ring, and if $E \subseteq Spec(A)$, then we denote $J(E) = \bigcap_{p \in E} p$, and $\overline{E} = V(J(E))$. This ideal is prime iff E is irreducible (exercise). In this case, $\overline{E} = \overline{\{J(E)\}}$. In fact, for two points p, q in $Spec(A)$, $p > q$ iff $p \subseteq q$. In particular, $Spec(A)$ *is a sober space.* Its irreducible components correspond to the minimal prime ideals.

Proposition 2 ([3, II.1.7], [7, p. 1111]) *The canonical injection*

$$j : \textbf{\textit{Sob}} \longrightarrow \textbf{\textit{Top}}$$

have a left adjoint $pt\mathcal{O}$. $pt\mathcal{O}(X)$ *is called the soberification of* X.

3   Gestures on Locales

3.1   Locales and Frames

The category of frames denoted by **Frm** is obtained by taking the essential algebraic properties of both the sets of opens $\mathcal{O}(X)$ in topological spaces X and the correspondences $f^{-1}(\_) : \mathcal{O}(Y) \to \mathcal{O}(X)$ induced by continuous functions $f : X \to Y$. Formally, the objects of **Frm** are the complete Heyting algebras, i.e. complete lattices L satisfying the infinite distributive law

$$a \wedge \bigvee_{s \in S} s = \bigvee_{s \in S} a \wedge s$$

for all $a \in L$ and $S \subseteq L$. The morphisms of frames are the functions that preserve finite meets including **1** and arbitrary joins including **0**. In particular these functions preserve the order.

The category **Loc** of locales is the opposite of **Frm**. If $f : L \to M$ is a morphism of locales, we denote the corresponding morphism of frames by $f^* : M \to L$. **Loc** has products and equalizers, and therefore it has all small limits; see [11, II.3].

Informally, if L and M are locales, the product $L \times_l M$ in **Loc** is the frame generated by the elements of the (Cartesian) product of meet-semilattices $L \times M$ subject to the relations

$$\left(\bigvee_{a \in S} a, b\right) = \bigvee_{a \in S} (a, b),$$

for all $S \subseteq L, b \in M$; and

$$\left(a, \bigvee_{b \in T} b\right) = \bigvee_{b \in T} (a, b),$$

for all $T \subseteq M, a \in L$. In particular $(0, b) = (\bigvee \varnothing, b) = \bigvee \varnothing = (0, 0)$, and similarly $(a, 0) = (0, 0)$ for all $a \in L, b \in M$. The projections $\pi_1 : L \times_l M \to L$, $\pi_2 : L \times_l M \to M$ correspond to the inclusions $\pi_1^* : L \to L \times_l M, \pi_2^* : M \to L \times_l M$ in **Frm** given by $\pi_1^*(a) = (a, 1)$ and $\pi_2^*(b) = (1, b)$. Note that the elements of the form $(a, 1)$, $(1, b)$ can be regarded as the subbasic opens generating the topology of the product of the 'spaces' with sets of opens L and M. See [3, II.2.12] for a formal construction of products based on C-ideals, or [11, II.3] for an one based on equivalence relations.

The equalizer of a pair $f, g : L \rightrightarrows M$ in **Loc** corresponds to the coequalizer of $f^*, g^* : M \rightrightarrows L$ in **Frm**; see [11, II.3] for details.

3.2 Motivation

As we have already seen, the construction of the space of gestures with skeleton Δ and body in X is done in three steps:

1. Construction of the space $I @ X$ of *paths* in X.
2. Construction of the *spatial digraph* $\overrightarrow{X}$ of X.
3. *Gluing* of spatial digraphs and copies of X according to the skeleton Δ.

Thus, if we want to extend the construction to the category of locales, we should try to follow these steps. Note that the step 3 only depends on the existence of both objects in the two preceding steps and limits in the category, so we shall focus on the first two steps.

From now on we identify the locale $\mathcal{O}(I)$ with I.

For paths, in the first instance, one is tempted to define the space of paths in a locale L as the set $\mathbf{Loc}(I, L) = \mathbf{Frm}(L, I)$. On the other hand, note that I is exponentiable in **Loc** since I is locally compact and therefore a continuous lattice. See [2] or [3, VII 4.11]. But, in general, the exponential L^I not necessarily coincides with $\mathbf{Loc}(I, L)$, even in the case that L is a spatial locale [4, p. 100]. Spatial locales will be discussed in Sect. 3.4. In this way, from now on, we shall denote the exponential L^I by $I @ L$ and call it the locale (of opens) of *paths* in L.

Analogously to the case of topological spaces, we should be able to define the localic digraph $\overrightarrow{L}$ of L by means of the evaluation maps $e_0, e_1 : I @ L \to L$ corresponding to morphisms of frames e_0^*, e_1^* sending an element $a \in L$ to 'all the paths whose inverse images of a contain the respective endpoint':

$$e_0^* : L \longrightarrow \qquad I @ L \qquad e_1^* : L \longrightarrow \qquad I @ L$$
$$a \longmapsto \bigvee_{0 \in U \in \mathcal{O}(I)} [W(U, a)] , \qquad a \longmapsto \bigvee_{1 \in U \in \mathcal{O}(I)} [W(U, a)] .$$

The objects $[W(U, a)]$ ($[U \ll f^*(a)]$ in Hyland's terminology) are the equivalence classes of the symbols $W(U, a)$ ($U \ll f^*(a)$) in the construction of the exponential L^I; see [2, p. 270] and [3, VII 4.11]. In fact, e_0, e_1 so defined are morphisms of frames, because they are the composites (analogous to these of Sect. 2)

$$I @ L \overset{\cong}{\to} I @ L \times 2 \xrightarrow{id \times i_0} I @ L \times I \overset{e}{\to} L$$

and

$$I @ L \overset{\cong}{\to} I @ L \times 2 \xrightarrow{id \times i_1} I @ L \times I \overset{e}{\to} L$$

respectively, where $2 = \{\varnothing, \{*\}\}$ is the locale of opens of the point space, i_0, i_1 are the morphisms induced by the inclusions of the endpoints of I, and e is the evaluation map defined by $e^*(a) = \bigvee_{U \in \mathcal{O}(I)}([U \ll f^*(a)], U)$. See [2, p. 275].

3.3   Construction

Following the steps done in Sect. 3.2, we show next the construction of gestures on locales. The key point that enables us to formulate the concept is the possibility of constructing both the locale of paths in a locale and arbitrary limits in **Loc**.

Let $\Delta = (A, V, t, h)$ be a digraph and L a locale.

As we have already noticed, the locale I is a continuous lattice since I is locally compact, and therefore I is exponentiable in **Loc**. Thus, we have the locale $I @ L$ of *continuous paths* in L.

The *localic digraph* $\overrightarrow{L}$ of L is the tuple $(I @ L, L, e_0, e_1)$ where e_0, e_1 are 'the evaluation at the endpoints morphisms'.

In the first instance, it is difficult to define a punctual gesture since L not necessarily has points. So we define the locale of all the *localic gestures* $\Delta @ L$ as the limit of the diagram (analogous to that in Sect. 2) $\mathcal{B}$ defined as follows: assign to each arrow a in A the locale $I @ L$, to each vertex x in V the locale L, take a copy of the morphism $e_0 : I @ L \to L$ whenever $x = t(a)$, and a copy of $e_1 : I @ L \to L$ whenever $x = h(a)$.

Remark 1   (Fundamental Example)

The construction of gestures on locales carries the same construction for complete Heyting algebras. On the other hand, for a sheaf F on a site, the lattice $Sub(F)$ of all subsheaves of F is a complete Heyting algebra ([6, p. 146]), and moreover, every complete Heyting algebra is the lattice $Sub(F)$ for a sheaf on a certain site ([6, p. 149]). So the gestures on locales are precisely the gestures on the complete Heyting algebras of subobjects of sheaves. This remarkable fact is a first step towards a possible definition of gestures on (in) a Grothendieck topos.

3.4   Points and Gestures

The construction of gestures that we have done is a little more abstract of those presented in [8, 9] in the sense that we are not defining punctual gestures by patching curves with matching endpoints in L according to the digraph Δ; in fact, note that, in general, L is not composed of points and the exponential $I @ L$ is not the set **Loc**(I, L); see the discussion in Sect. 3.2.

However, we show next, by means of characterizing the points of the locale $\Delta @ L$, how it is possible to regard topological gestures on reasonable spaces (sober spaces) in terms of localic gestures, i.e. we are showing a possible way to rebuild gestural movements inside an algebraic context. See the question posed in [9, p. 33 (4)].

From [3, pp. 41–44], we have functors:

$$\textbf{Top} \xrightarrow{\;\mathcal{O}\;} \textbf{Loc} \qquad \textbf{Frm} \qquad \textbf{Loc} \xrightarrow{\;pt\;} \textbf{Top}$$

$$
\begin{array}{ccc}
X \longmapsto \mathcal{O}(X) & \mathcal{O}(X) & L \longmapsto pt(L) = \textbf{Loc}(2, L) \\
f \downarrow \quad \downarrow (f^*)^{op} & \uparrow f^* = f^{-1}(\_) & f \downarrow \quad \downarrow f\circ\_ \\
Y \longmapsto \mathcal{O}(Y) & \mathcal{O}(Y) & M \longmapsto pt(M) = \textbf{Loc}(2, M)
\end{array}
$$

where pt is the right adjoint to $\mathcal{O}$. So we have a natural correspondence

$$\textbf{Loc}(\mathcal{O}(X), L) \cong \textbf{Top}(X, pt(L))$$

for all X in $\textbf{Top}$ and L in $\textbf{Loc}$. Moreover, the adjunction restricts to an equivalence between the full subcategories $\textbf{Sob}$ of sober spaces and $\textbf{Sloc}$ of spatial locales; specifically:

$\textbf{Sob}$: 'Spaces isomorphic to $pt(L)$ for some locale L'='Spaces X such that $pt(\mathcal{O}(X)) \cong X$'='fixed points of $pt\mathcal{O}$'.

$\textbf{SLoc}$: 'Locales isomorphic to $\mathcal{O}(X)$ for some space X'='Locales L such that $\mathcal{O}(pt(L)) \cong L$'='fixed points of $\mathcal{O}pt$'.

Note that pt preserves limits since it is a right adjoint, also, we have the following proposition.

Proposition 3 *If L is a locale, then $pt(I @ L)$ is homeomorphic to the exponential $I @ pt(L)$ in* $\textbf{Top}$.

Proof Since I is exponentiable in $\textbf{Top}$ and $\textbf{Loc}$, we have the following diagrams of adjoint functors:

$$\textbf{Top} \overset{pt}{\underset{\mathcal{O}}{\leftrightarrows}} \textbf{Loc} \overset{(\_)^{\mathcal{O}(I)}}{\underset{\_\times\mathcal{O}(I)}{\leftrightarrows}} \textbf{Loc}, \quad \textbf{Top} \overset{(\_)^I}{\underset{\_\times I}{\leftrightarrows}} \textbf{Top} \overset{pt}{\underset{\mathcal{O}}{\leftrightarrows}} \textbf{Loc}.$$

But $\mathcal{O}(X) \times \mathcal{O}(I) \cong \mathcal{O}(X \times I)$ for all spaces, since I is locally compact; see [3, II.2.13]. Also, it can be checked that this isomorphism is natural in X. Thus, by the uniqueness of adjoints, $pt(L^I) \cong pt(L)^I$ for any locale L, i.e. $pt(I@L) \cong I @ pt(L)$. ◆

The isomorphism $pt(L^I) \cong pt(L)^I$ is obtained as follows. Let $\mathbf{2} \to L^I$ be a point of L^I, so we have a corresponding morphism $\mathcal{O}(I) \cong \mathbf{2} \times \mathcal{O}(I) \to L$ given by the universal property of exponentials. Then, the later arrow induces a path $I \to pt(L)$ by the adjunction between pt and $\mathcal{O}$.

Also, it can be checked that pt carries the evaluation morphisms $e_0, e_1 : I@L \to L$ to the evaluation morphisms $e_0, e_1 : I@pt(L) \to L$ by means of the isomorphism $pt(L^I) \cong pt(L)^I$. So pt carries the diagram $\mathcal{B}(L)$ of Sect. 3.3 to a diagram naturally isomorphic to the corresponding diagram $\mathcal{B}(pt(L))$ for $pt(L)$ in **Top**. Thus, since pt preserves limits, we have the following proposition.

Proposition 4 $pt(\Delta@L) \cong \Delta@pt(L)$ *in* **Top**.

By taking $L = \mathcal{O}(X)$, for X a sober space, we obtain the following corollary.

Corollary 1 *If X is a sober space, then*

$$\Delta@X \cong pt(\Delta@\mathcal{O}(X)).$$

So every gesture with skeleton Δ and body in a sober space X is a point of a locale, namely the locale of gestures with skeleton Δ and body in $\mathcal{O}(X)$. This fact could help us to see gestures on sufficiently well behaved spaces (for example Hausdorff spaces) as 'nets' of morphisms of locales, the later being defined in 'purely' algebraic terms. Let us explain it better. Let $\mathcal{B}(\mathcal{O}(X))$ be the diagram of Sect. 3.3 corresponding to $\mathcal{O}(X)$ with limit $\Delta@\mathcal{O}(X)$ in **Loc**. Every point of $\Delta@\mathcal{O}(X)$ is a morphism $\mathbf{2} \to \Delta@\mathcal{O}(X)$ which induces a cone on the diagram $\mathcal{B}(\mathcal{O}(X))$ with vertex $\mathbf{2}$, and conversely, every cone on $\mathcal{B}(\mathcal{O}(X))$ with vertex $\mathbf{2}$ induces a point $\mathbf{2} \to \Delta@\mathcal{O}(X)$. Thus, we have a correspondence between gestures on X, points of $\Delta@\mathcal{O}(X)$, and cones on $\mathcal{B}(\mathcal{O}(X))$ with vertex $\mathbf{2}$, i.e. 'nets' on $pt(\mathcal{B}(\mathcal{O}(X)))$. This characterization of topological gestures in the category of locales is a contribution of the present article to the theory of gestures.

The inverse problem, i.e. that of rebuilding algebraic structures in gestural terms, has been addressed successfully in [8] for finitely generated abelian groups.

From the preceding discussion one may ask for which locales L the locale of localic gestures $\Delta@L$ can be regarded as $\mathcal{O}(\Delta@X)$ (or $\Delta@\mathcal{O}(X)$) for some space X. It seems to be a difficult problem since, for example, $I@L$ may differ from $\mathcal{O}(I@pt(L))$ even in the case when L is spatial; see [4, p. 100]. Moreover, the functor $\mathcal{O}$ not necessarily preserves limits.

4   Gestures on Localic Topoi

This article pretends to be a step towards a possible definition of (hyper)gestures on topoi (both Grothendieck and elementary).[1] In this way, a natural intermediate stage

---

[1] The statements made in this section must be interpreted in a suitable 2-categorical sense.

is the exploration of the concept of gesture in the category of localic topoi. As we shall see later, a wider geometric intuition can be displayed by replacing a site $\mathcal{C}$ (of which a locale is an example) by the Grothendieck topos of sheaves on $\mathcal{C}$.

The following construction is a simple consequence of the equivalence between the category of locales and the category of localic topoi $\mathfrak{LTop}$ (topoi equivalent to $Sh(L)$ for some locale L). This equivalence is given by the functor $Sh(\_)$ from **Loc** to $\mathfrak{LTop}$, which sends a morphism f of locales to the induced geometric morphism (f_*, f^*). Actually, it is an equivalence of 2-categories, but we shall not give the details here; see [5, C1.4]. Thus, note that the constructions of limits and exponentials in $\mathfrak{LTop}$ needed to define gestures are available, as in the case of **Loc**.

Through this section, only the functorial properties of $Sh(\_)$ will be used.

Let $\mathcal{E}$ in $\mathfrak{LTop}$. Note that $I = Sh(\mathcal{O}(I))$ is exponentiable in $\mathfrak{LTop}$ since I is exponentiable in **Loc**, so we have the localic topos $\mathcal{E}^I = I @ \mathcal{E}$ of paths in $\mathcal{E}$.

Now, first suppose that $\mathcal{E} = Sh(L)$ for some locale L. We can assume that $I @ \mathcal{E} = Sh(I @ L)$ since $Sh(\_)$ preserves exponentials, so the localic digraph $(I @ L, L, e_0, e_1)$ induces a digraph $\overrightarrow{\mathcal{E}} = (I @ Sh(L), Sh(L), e_0, e_1)$ in $\mathfrak{LTop}$, where e_0, e_1 are geometric morphisms, by applying the functor $Sh(\_)$. Thus, we can define the topos of gestures with skeleton Δ and body in $\mathcal{E}$ as the limit of the diagram $Sh(\mathcal{B})$ where $\mathcal{B}$ is the diagram defined in Sect. 3.3, which we shall denote by $\Delta @ \mathcal{E}$. In fact, $\Delta @ \mathcal{E} = Sh(\Delta @ L)$, since $Sh(\_)$ preserves limits. In the case when $\mathcal{E}$ is equivalent to $Sh(L)$, the later equivalence induces a corresponding digraph $\overrightarrow{\mathcal{E}} = (I @ \mathcal{E}, \mathcal{E}, e_0, e_1)$ from the above digraph $\overrightarrow{Sh(L)} = (I @ Sh(L), Sh(L), e_0, e_1)$; so we define $\Delta @ \mathcal{E}$ as the limit in $\mathfrak{LTop}$ of a diagram $\mathcal{B}$ defined as in Sect. 3.3, obtained from Δ and $\overrightarrow{\mathcal{E}}$. In fact, the topos $\Delta @ \mathcal{E}$ is equivalent to $\Delta @ Sh(L)$.

5 Comments About Gestures on Sites and Topoi, and Conclusions

Of the three kinds of structures where we are attempting to define (hyper)gestures (sites, Grothendieck topoi, elementary topoi), Grothendieck topoi seem to be the most appropriate to deal with. Indeed, according to [5, B.4.1.1], the 2-category $\mathfrak{BTop}/\mathbf{Set}$ of Grothendieck topoi has finite limits. On the other hand, the interval $I = Sh(\mathcal{O}(I))$ is exponentiable in $\mathfrak{BTop}/\mathbf{Set}$; see [10] and [5, C.4.4.12]. It is possible to define 'paths' and the 'evaluation at endpoints' in a similar way as was done in Sects. 3.3 and 3.4. The agreement between the notion of path (and evaluation points) given in [10] and our notion, supports the pertinence of our generalization.

The full construction of (hyper)gestures in the category of sites and the category of topoi is more difficult, at least from the point of view of the construction proposed in this article as a generalization of topological gestures. The knowledge about the construction of limits in these categories seems to be small, and likely his development will need a giant machinery.

Likely, the category of sites is not well endowed with limits and exponentials, but this situation improves when we deal with the category of Grothendieck topoi, as we have commented; thus, it seems to be suitable to define gestures on sites by means of gestures on the associate category of sheaves (if such a definition is possible). Another reason to think that, is the narrow relation between localic gestures and gestures on localic topoi discussed on Sect. 4. Indeed, the continuous functions between spaces are generalized to morphisms of locales, which are generalized to morphisms of sites, which induce geometric morphisms by applying the functor $Sh(\_)$; so geometric morphisms are the analogues of continuous functions for Grothendieck topoi.

Despite these obstacles to defining gestures on elementary topoi, the possibility of defining gestures on Grothendieck topoi (at least for digraphs with finite vertices) is general enough to embrace the construction on categories of sheaves on sites and, in particular, on categories of presheaves (sheaves on the trivial topology), which are the paradigmatic structure in *The Topos of Music* [7]; specifically the category of presheaves on the category of modules.

On another hand, with respect to the category of elementary topoi, we ignore whether other characterization for gestures on a topos, that does not deal with limits, is possible. Other difficult to deal with, when we try to define topological objects (for example paths) inside an elementary topos, is the fact that the Heyting algebras of subobjects are not complete in general, so that the topological intuition of thinking subobjects as opens vanishes.

In this article, we have presented a generalization of the construction of gestures on topological spaces to locales and localic topoi. More generally, suppose that Δ is a digraph and C is a category having the following properties:

1. C has an interval (or simplex; see [9]) object denoted by I.
2. The interval object is exponentiable in C.
3. The limit of the diagram B analogous to that in Sect. 3.3 exists in C.

Then we define the object (in C) of gestures with skeleton Δ and body in C as the limit of B denoted by $\Delta @ C$. The advantage of this generalization is the possibility of building hypergestures from $\Delta @ C$, which is an object in C. Indeed, the generalizations in this article were done thinking of that. So, if Γ is another digraph we can construct the object $\Gamma @ \Delta @ C$, and so on, depending on the existence of suitable limits in C. For example, though it is not a necessary condition, the constructions of hypergestures are always possible if C is complete.

However, the preceding constructions seem to be difficult, though possible in certain cases, specially in the category of Grothendieck topoi. Perhaps it will be necessary to explore other ways of defining homotopies inside abstract categories that do not deal with paths or interval objects.

Finally, the construction of gestures on locales and localic topoi has shown a way for 'reconstructing gestural instances from abstract categories', namely the category of locales, and therefore the equivalent category of localic topoi. See [9, p. 33] where the problem is stated. Thus, the long adventure towards a definition of hypergestures in the topos structure started in this article promises to contribute to the understanding of the diamond conjecture [8, p. 43]. Also, we have constructed gestures on the lattice of subsheaves of a sheaf, inside any category of sheaves on a site; see Sect. 3.3. Thus, this could open another way to define gestures on Grothendieck Topoi.

Acknowledgements I sincerely thank Professors Fernando Zalamea and Guerino Mazzola for allowing me to work in MaMuTh under their guidance, offering me their confidence and marvellous insights in mathematics, art, and philosophy.

References

1. Escardó, M.H., Heckmann, R.: Topologies on spaces of continuous functions. In: Topology Proceedings, vol. 26, edn. 2, pp. 545–564 (2001–2002)
2. Hyland, J.M.E.: Function Spaces in the Category of Locales, Lecture Notes in Mathematics, vol. 871, pp 264–281. Springer, Berlin (1981)
3. Johnstone, P.T.: Stone Spaces, Cambridge Studies in Advanced Mathematics, vol. 3. Cambridge University Press, Cambridge (1982)
4. Johnstone, P.T.: The Art of Pointless Thinking: A Student's Guide to the Category of Locale. In: Herrlich, H., Porst, H.-E. (eds.) Category Theory at Work, pp. 85–107. Heldermann, Berlin (1991)
5. Johnstone, P.T.: Sketches of an Elephant: A Topos Theory Compendium. vol. 2. Oxford University Press, Oxford (2002)
6. Mac Lane, S., Moerdijk, I.: Sheaves in Geometry and Logic. Springer, Berlin (1992)
7. Mazzola, G., et al.: The Topos of Music: Geometric Logic of Concepts, Theory, and Performance. Birkhäuser, Basel et al (2002)
8. Mazzola, G., Andreatta, M.: Diagrams, gestures and formulae in music. J. Math. Music **1**(1), 23–46 (2007)
9. Mazzola, G.: Categorical gestures, the diamond conjecture, lewin's question, and the hammerklavier sonata. J. Math. Music **3**(1), 31–58 (2009)
10. Moerdijk, I., Wraith, G.: Connected locally connected toposes are path connected. Trans. Am. Math. Soc. **295**, 849–859 (1986)
11. Pedicchio, M.C., Tholen, W. (eds.): Categorical Foundations: Special Topics in Order, Topology, Algebra, and Sheaf Theory, Encyclopedia of Mathematics and its Applications. Cambridge University Press, Cambridge (2004)

On the Structural and the Abstract in My Compositional Work

Clarence Barlow

Abstract From 1959 to 1969 I composed music as most others do and have done—by direct transference from the imagination to a musical instrument (in my case the piano) and from there to a written score. During this period, I found myself relying increasingly on traditionally structured techniques such as canon, fugue, dodecaphony, serialism and electronics. In 1970 I was struck for the first time by a mathematical rule-based idea for an ensemble piece, which necessitated my learning to program a computer. Since then I have composed over fifty works (half my total output) with computer help—works for piano, organ, chamber ensemble, orchestra and electronics. Of these fifty-odd pieces, about half are partially and sometimes wholly based on abstract mathematical principles. This paper describes eight of these pieces or relevant sections of them in varying detail.

For reasons of space, I have omitted a system of quantified harmony and meter I developed in 1978, used in several pieces (*Çoğluotobüsişletmesi, Variazioni, documissa '87, Orchideæ Ordinariæ, Otodeblu, Talkmaster's Choice, Amaludus, Estudio Siete and Für Simon Jonassohn-Stein*) and have written about (see references). Examples of the algebraic formulae used in these pieces are illustrated here below in Fig. 1 without explanation.

$$\xi(N) = 2\sum_{r=1}^{\infty}\left(\frac{n_r(p_r-1)^2}{p_r}\right)$$

$$H(P,Q) = \frac{sgn\left(\xi(Q)-\xi(P)\right)}{\xi(P)+\xi(Q)}$$

If $p=2$, then $\Psi_p(n) = p-n$;

otherwise if $n=p-1$, then $\Psi_p(n) = \lfloor p/4 \rfloor$

or else $\Psi_p(n) = \left\lfloor q+2\sqrt{\frac{q+1}{p}} \right\rfloor$

$$\Psi_z(n) = \sum_{r=0}^{z-1}\left(\prod_{i=0}^{z-r-1}p_i\Psi_{p_{z-r}}\left(1+\left(\left\lfloor 1+\frac{(n-2)\bmod\prod_{j=1}^{z}p_j}{\prod_{k=0}^{r}p_{z+1-k}}\right\rfloor\bmod p_{z-r}\right)\right)\right)$$

Fig. 1 Formulae for quantified harmony (*left*) and meter (*right*)

C. Barlow (✉)
Department of Music, University of California, Santa Barbara, CA, USA
e-mail: b@rlow.org

© Springer International Publishing AG 2017
G. Pareyon et al. (eds.), *The Musical-Mathematical Mind*,
Computational Music Science, DOI 10.1007/978-3-319-47337-6_5

1 *Cheltrovype* (1968–71) for Cello, Trombone, Vibraphone and Percussion

I was given the task of writing a piece for this instrumentation, from which the title derives, by my teacher Bernd Alois Zimmermann (1918–1970). Completed only after his untimely death, it consists of six parts, each of which is characterized by a totally different compositional technique. In Part V, the music for the three melody instruments follow a probabilistic pitch distribution system based on an exponential curve, a sine curve and a transformed parabola.

For Part V, I imagined the cello starting on a repeated low open-string C_2, occasionally interspersing the D-flat above it, then later adding the D, the E-flat, the E and so on chromatically upwards through a range of three and a half octaves to the note F_5 on the top line of the treble staff. At the same time, while keeping the lowest note at C_2, the pitch centroid of the melody gradually rose, reaching the highest note F_5 at the end. Sometime after the start of the cello, the trombone would enter with a repeated C_3, following the same procedure and reaching the same high F together with the cello. Somewhat later than the trombone, the vibraphone would follow suit, starting on a repeated C_4 (Middle C) and ending with the cello and the trombone on the same high F.

Right from the start, I realized that this music could not be composed spontaneously but would have to be subject to a set of rules, finally formalized in the shape of the formulae and corresponding curves shown in Fig. 2 at left. The lowest pitch is seen to be fixed at the MIDI value 36 ($=C_2$) and for the highest value a half-period of a sine curve was chosen. The pitch centroid was determined by an exponential curve for the most frequent pitch, simultaneously marking the peak of a transformed parabola curve in the y-z plane, not shown here. In this plane, this "parabola" has the value zero a half step below the lowest and a half step above the highest pitch, reaching its maximum at the most frequent pitch.

For the cello part, I decided to generate 500 notes, for each of which the probability of every pitch in the 42-half-step range from C_2 to F_5 (totalling $500 \times 42 = 21,000$ values) was to be determined by the transformed parabola. I first tried to do this with logarithmic tables (the year was 1970 and there were no electronic calculators), but soon gave it up due to the time-consuming nature of the process. A second attempt with a 50-pound electric office calculator proved to be also very time-consuming. This is what led me to learn to program in Fortran at the computer center of Cologne University; the cello part was complete within a week of my starting the Fortran course in the form of a 500-page table of probabilities, one page for the choice of each note.

The actual resulting notes were picked by the use of random numbers –see the dots in Fig. 2 at left– and written as a score, seen in Fig. 2 at right. The process was repeated for the trombone and the vibraphone with 222 and 115 notes, respectively, the range being 30 and 18 half-steps. According to my sources, Part V of *Cheltrovype* seems to have been the earliest computer music score composed in Germany.

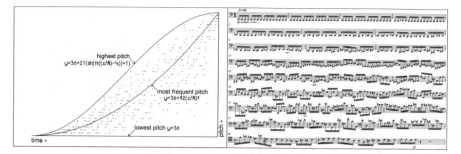

Fig. 2 Algebraic curves and formulae (*left*), score of cello part for Cheltrovype Part V (*right*)

2 Sinophony II (1969–72) for Eight-Channel Electronics

After having composed a four-channel analog electronic piece called *Sinophony* in 1970, consisting in the main of sine-tones (hence the title) but also containing noise bands, impulse-generated sounds and ring modulation, I decided in 1971 to compose a sequel consisting exclusively of sinusoids, not only as sound waves but also as form-shaping parameters, according to which theoretically infinitely many sine tones would move along predetermined sinusoidal paths in the domains of pitch, amplitude and duration. After fruitless attempts with the newly acquired ARP synthesizer in the electronic music studio of Cologne Music University (*Hochschule für Musik*), where I was a student, I drove two days to Stockholm, to the EMS Studio, where I worked at the PDP computer completely alone for two weeks during the Christmas period of 1972.

Figure 3 shows the function of the sine curve in shaping not only the sound wave, but its pitch, loudness and even its duration: in this last case the length of an event periodically increases and decreases within a time period fixed differently for various pitch groups in various tracks, the remainder of the period being occupied by silence.

Figure 4 shows a map of eight different tracks with time on the x axis, pitch on the y axis and the vertical width of each line reflecting the loudness. In some tracks the pitch remains constant (e.g. 1, 3 and 5), in which case individual pitches

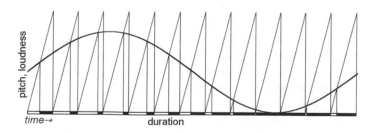

Fig. 3 A sine curve shaping the main parameters

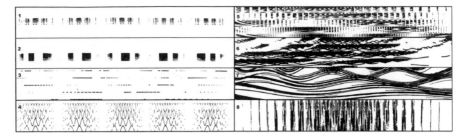

Fig. 4 A pitch-time-loudness map of the eight tracks of *Sinophony II*

are grouped in the form of an overtone series compressed to about 7/12ths of the intervals of a normal harmonic series, i.e. the frequency of the nth partial relative to the fundamental f_1 in each group is given by $f_n = f_1 \cdot n^{\log(1.5)/\log(2)}$. Here the second partial is a perfect fifth above the first instead of an octave, all other intervals being compressed by the same factor. In track 7, each pitch group retains this relationship while undulating sinusoidally as a whole. In tracks 1 and 5 it is the loudness that undulates. Tracks 1 and 2 have only one pitch group each, tracks 3 and 4 only one sine tone per group and the remaining tracks have multiple groups with fundamentals spaced five perfect fifths apart and multiple sine tones in each. In tracks 1, 5 and 8, the duration undulates. Conceptually, *Sinophony II* has an infinite duration, frequency range and loudness ranging downwards from a fixed maximum value; for practical reasons, I factually generated about 800 sine tones within the range of 17 Hz to 17 KHz, 0 to −60 dB and with a total duration of 24'38".

3 *Stochroma* (1972) for Solo Piano

In 1972 my composition teacher Karlheinz Stockhausen gave each of us in the class the task of writing a piano piece starting on the lowest note A, gradually increasing chromatically in range. I did not write the piece, but preferred to improvise on Stockhausen's piano. He was not very impressed. Later I planned a conceptual piano piece in which pitch, loudness and duration are probabilistically determined, allowing duration and dynamic values to randomly and exponentially diverge (as powers of 0.5 and 2) from a central value to rare but great extremes (durations for instance range in seconds from the yocto to the yotta range and beyond in both directions). Figure 5 shows at left a short excerpt of the computer printout as sound number (there are 5000 sounds in total), pitches (German notation; '----' denotes silence), current range in half-steps above the lowest A, duration as multiples and divisions of powers of 2, and dynamics as degrees downwards from *fff* (=0). At right one sees a matching score on one of four systems notated in 2001 for an exhibition of conceptual art curated by composer Tom Johnson in the Queen Sofia Museum in Madrid.

Note the durations in the 2nd (4×2^{18} s) and 4th (17×2^{83} s) bars. Understandably, this piece has never been performed.

Fig. 5 *Stochroma*: part of page 267 of the 1972 computer printout (*left*) and the corresponding score (*right*)

4 *Bachanal* for Jim Tenney and Tom Johnson (1990) for Solo Piano

Once, discussing musically numerical issues with Tom Johnson in a Paris café, I told him of a discovery of mine: if the odd-numbered bits in a series of binary representations of the natural numbers were made negative, the resulting values would start not with 0, 1, 2, 3 ... but with 0, −1, +2, +1 .... This sequence expressed in half-steps corresponds to the notes of the B-A-C-H theme used ever since the composer J. S. Bach himself used it ("B" in German means B-flat in English and "H" means B-natural). But more than that, the next four values, instead of 5, 6, 7, 8, would now be −4, −5, −2, −3, again B-A-C-H transposed down a major 3rd. The next four (+8, +7, +10, +9) are again B-A-C-H transposed up 8 half-steps or two major 3rds. And the next four (+4, +3, +6, +5) are again B-A-C-H transposed up a major 3rd. These four sets of major 3rd transpositions (0, −1, +2, +1) are again in the form B-A-C-H. And so on. I used this phenomenon to generate *Bachanal* ("B-A-C-H analysis") while exponentially accelerating the process in such a way that the tempo of the higher-level transpositions (major 3rds, 10ths, 40ths etc.), bear a relationship to that of the first four notes. Figure 6 shows at left the transformation of the first 40 natural numbers into this "odd-bit negative" form, and also a pitch-time map of the accelerated notes.

Figure 7 shows the piece as a score.

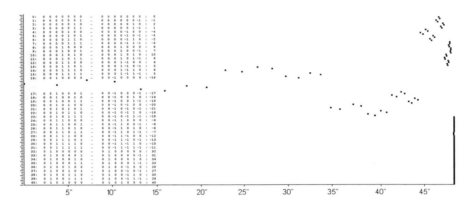

Fig. 6 Exemplified conversion of regular binary numbers into "odd-bit negatives", with a pitch-time map

Fig. 7 Score of *Bachanal for Jim Tenney and Tom Johnson*

5 *Piano Concerto #2* (1961–1998) for Piano and Orchestra

37 years lay between when I began and ended work on this piece. Its style in 1961, when I was 15, resembled European classical music of around 1800, whereas in 1963 its style had advanced to 1900, resembling Rachmaninov. By 1965, about 8 min from the beginning and 3 min before the end were complete. However my style in other pieces developed further, through Prokofiev and Hindemith to Schoenberg and Webern: at a loss as to how to continue the work, I laid it temporarily aside.

In 1975, finally notating that which had been done, I noticed an accelerando I had been unaware of between 3'48" (at MM 60) and 7'49" (at MM 119) where the music broke off, and another of the same rate (0.289%/s) but a much higher speed for about a minute after the recommencement of the music (MM 244–MM288). In 1998 I completed the piece, continuing the accelerando across the unfinished portion, which according to formulae I developed in 1975 for an accelerando (see Fig. 8), would be from bars 145 to 322, lasting 4'4".

$$s = SQ^{\frac{t}{T}}$$

$$n = \frac{TS(Q^{\frac{t}{T}}-1)}{\ln(Q)}$$

$$T = \frac{N\ln(Q)}{(S'-S)}$$

$$t = \frac{T\ln(s/S)}{\ln(Q)}$$

$$= \frac{T\ln((n\ln(Q)/TS)+1)}{\ln(Q)}$$

where S = initial tempo

S' = final tempo

$Q = S'/S$

T = total time

t = current time
[i.e. time at tempo S
or at beat n - see below]

s = tempo at time t

n = beat at time t

N = total number of beats

$\ln(x)$ = natural logarithm of x

Fig. 8 Formulae for acceleration/deceleration

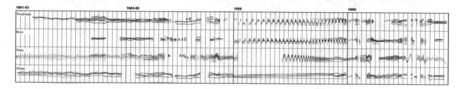

Fig. 9 Map of pitch (y) against time (x) for *Piano Concerto #2*

In this new section, different groups of instruments perform simultaneous but independent rising *accelerandi* and falling *decelerandi*, the pitch and the rhythm deriving from the shape of an inverted cosine. Figure 9 shows a pitch-time map of the piece. Notice the distinctly different shapes in the "1998" section.

6   *Les Ciseaux de Tom Johnson* (1998) for Chamber Ensemble

Written to celebrate Tom Johnson's 60th birthday, this piece is based on the successive positions of six sets of three points derived from the name of the dedicatee, each set moving along a differently sized circle (see Fig. 10).

The letters T O M J O HN S O N were first plotted from left to right on an alphabetically upwards-reaching uniform grid. Six arbitrarily chosen three-letter sets TOM, SOJ, JNS, SON, MJH and OOO were then each transected by a circle, one of them a horizontal straight line through the OOO set. Next, each set was made to rotate in an anticlockwise manner along its circle by a distance equal to the segment of the circumference of the smallest circle SON subtending an angle of 4°. All sets move concurrently 90 times before SON returns to its original position. Each shifted

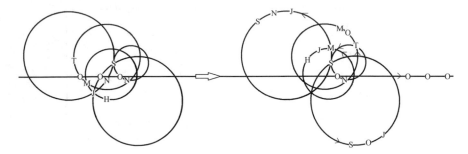

Fig. 10 Rotation states 1 (*left*) and 91 (*right*) of six three-letter sets along *six circles*

Fig. 11 The opening bars of *Les Ciseaux* de Tom Johnson

state is then scanned by a vertical line from left to right representing time, the letters it transected rendered by their height as pitch, yielding a total of 91 "mini-scores". Figure 10 shows states 1 (left) and 91 (right). These scores are then overlapped such that the horizontal (time) distance between one OOO set and the next equals the distance between two successive 'O's in the set. The result is shown in Fig. 11 as a score excerpt comprising states 1, 2, and part of 3 – the repeated Middle Cs derived from the OOO set. The title, literally "the scissors of Tom Johnson" is a reference to his then age ("six-O") and to the six O-shaped circles.

7 "*...or a Cherish'd Bard...*" (1999) for Solo Piano

This piece was written to celebrate the 50th birthday of the pianist Deborah Richards (who by the way premiered my *Piano Concerto #2* with the Icelandic Symphony Orchestra in Reykjavik in 2002). First, the letters DEB and AH were interpreted as German-named pitches and as hexadecimal numbers for the rhythms, yielding an infinitely long chain each, as shown in Fig. 12.

Fig. 12 DEB and AH pitch and rhythm chains, the basic material of *"...or a cherish'd bard..."*

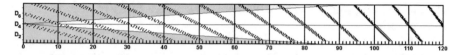

Fig. 13 Pitch-time map showing every tenth DEB chain in *"...or a cherish'd bard..."*

Fig. 14 Bar 93 of *"...or a cherish'd bard..."*

These chains were then repeated at a constant mutual time distance at their midpoints D4 and C♯4 respectively, but with a successively increasing gradient – see Fig. 13, in which every tenth DEB chain is shown for the full 120 bars over the full piano range. Additionally, a wedge-like filter encompasses an increasing number of pitches, as seen in the non-grey area of Fig. 13.

Finally, the probability that a note of the DEB chain is chosen for the piece was made to decrease continuously from 100 to 0% over the duration of the piece, while the complementary AH-chain probability increased from 0 to 100%, i.e. each note is taken from the DEB or AH chain. Since each chain derives from a different whole-tone scale, the music is whole-tone at the beginning and end, but chromatic in the middle. The title is an anagram of the dedicatee's name. Figure 14 shows bar 93 of the score with diamond-shaped note heads an octave higher (treble) or lower (bass) than written.

8   *Approximating Pi* (2007) for up to 16 Channels of Electronics

The Madhava–Leibniz converging series for the constant π begins thus:

$$\pi = 4 \left(1 - \tfrac{1}{3} + \tfrac{1}{5} - \tfrac{1}{7} + \tfrac{1}{9} - \dots \right).$$

Figure 15 shows convergences #1 to #10 (left) and #29,991 to #30,000 (right) to 10 decimal places, with which the 4 billionth convergence finally reaches the correct ten places of π.

For this piece, each convergence is allocated a time window of 5040 samples (twice the lowest common multiple of the numbers 1–10), in which ten square wave partials of frequencies $8\tfrac{3}{4}n$ Hz and basic amplitude $2^{\wedge}d_n$ are set up, '$8\tfrac{3}{4}$' deriving from the 5040 samples, 'n' being the partial number and 'd_n' the nth digit in the convergence's decimal representation; e.g. for '3.141592654', the ten partials' basic amplitudes are 2^3, 2^1, 2^4, 2^1, 2^5, 2^9 etc., thereafter rescaled by the arbitrary sawtooth-spectral factor $2\pi/n$, where n is still the partial number. The convergences stabilize the digits from left to right to a value approaching π, the resultant timbre moving from

1	4.0000000000	29991	3.1411104826
2	2.6666666667	29992	3.1418308139
3	3.4666666667	29993	3.1411102231
4	2.8952380952	29994	3.1418310737
5	3.3396825397	29995	3.1411099631
6	2.9760461761	29996	3.1418313338
7	3.2837384837	29997	3.1411097029
8	3.0170718171	29998	3.1418315942
9	3.2523659347	29999	3.1411094422
10	3.0418396189	30000	3.1418318551

Fig. 15 Some π convergences (Madhava–Leibniz)

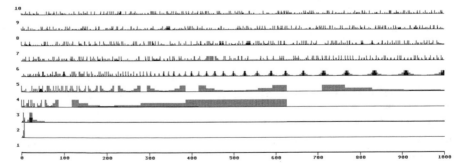

Fig. 16 The first 1000 convergences of the Madhava–Leibniz series as powers-of-2 spectral amplitudes

turbulence to constancy over a period of $4 \times 10^9 \times 5040 = 20.16 \times 10^{12}$ samples or about $14\frac{1}{2}$ years. The installation can be pitch-shifted (by sample dropping) and/or time-truncated. The fundamentals of the sixteen sound channels are transposed from $8\frac{3}{4}$ Hz to frequencies ranging from 9 to 402 times higher. Different versions with 2, 5, 8 and 16 channels have been realized, with durations ranging from about 8 to 74 min. Figure 16 shows the first 1000 convergences as spectral amplitudes in *Approximating Pi*.

References

1. Barlow, C.: SINOPHONIE II, Reprint 1–16, Feedback Papers, Feedback Studio Cologne, pp. 138–141 (out of print) (1979)
2. Barlow, C.: A short essay on musical time: four forms as manifest in my Piano Concerto no. 2. Time in Electroacoustic Music, Mnémosyne, Bourges (2001). ISBN 2-9511363-3-1
3. Barlow, C.: On Musiquantics, Report No. 51, Musikinformatik & Medientechnik, Musikwissenschaftliches Institut der Johannes Gutenberg-Universität Mainz (2012). ISSN 0941-0309
4. Johnson, T.: Minimalism in music: in search of a definition, Minimalismos, un signo de los tiempos, Museo Nacional Centro de Arte Reina Sofia, pp. 164–172 (2001)

A Proposal for a Music Writing for the Visually Impaired

Teresa Campos-Arcaraz

Abstract Braille Musicography is the most used system by blind people for reading and writing music in the world. It is a transcription from the conventional music notation to Braille system, which symbols are generated by a matrix of raised dots of 2 columns and 3 rows. It shows two main difficulties that make it a hard tool for the blind musicians: (1) The number of music symbols exceeds by far the number of possible Braille dots combinations and (2) it is a linear system representing a bidimensional system. These two problems result in the need of using combinations of up to 4 Braille boxes to represent one musical symbol, and the repetition of Braille symbols that change meaning depending of the context. In order to give more clarity or simplicity in Braille scores, abbreviations and contractions are used, thus a fragment of music can sometimes be written in several ways. Because of all this, automatic transcription to Braille is complicated, sometimes not possible at all, and as a consequence blind people do not have full access to Braille scores. Besides the many efforts of people around the world, music scores transcription to Braille musicography is still a problem. In this work some of the Braille musicography problems are identified, and the need of a more efficient musicography is established. In order to create a new set of symbols for the blind and a useful system, which is an objective of a later stage of this work, it is important to notice that our fingertips have a delimited zone in which the density of receptors is high and allows a clear reading of a symbol. Outside this zone, the produced mental image is unclear and makes the reading tiresome and difficult. This and some other physiological and cognitive considerations have to be taken into account. Experience and ideas from the blind must be always regarded.

T. Campos-Arcaraz (✉)
Facultad de Música, UNAM, Calle Xicotencatl 126, 04100 Coyoacán,
Del Carmen, D.F., Mexico
e-mail: tkmpos@comunidad.unam.mx

© Springer International Publishing AG 2017
G. Pareyon et al. (eds.), *The Musical-Mathematical Mind*,
Computational Music Science, DOI 10.1007/978-3-319-47337-6_6

53

1 Introduction

According to the World Health Organization, blindness is a condition where a person's visual acuity test is less than 20/400 in the better eye with the best possible correction, when a normal value is given by 20/20. Visual impairment occurs when the visual acuity is less than 20/60, including people with low vision and blindness [7, p. 97]. In Mexico, according to the census done by the National Institute for Statistics and Geography (INEGI) in 2000, there were more than 467,000 visually impaired persons, 32.2% of them living in rural areas [7, p. 98]. There have been several ways in which the blind people have tried to access information, for example coin values. The first written evidence of these events was made in the 14th century, in the University of Moustansiryeh in Iraq, when a professor called Zain-Din Al Amidi invented a method to identify books in the library. He rolled paper and bent it over the Arabic characters [14, p. 16]. Some other methods to teach the blind to read were letters carved in wood, cut letters, knots tied in strings, dots enclosed in squares, movable raised letters in lead, letters made of tin or metal, pins stocked in cushions, letters cut out of paper, etc. [14, p. 16]. It was until 1821 when Charles Barbier de la Serre, a former Captain of Artillery, visited the Institution Royale des Jeunes Aveugles in Paris to give a conference about a system he invented, called Night Writing. In his system he used a matrix of 12 raised points, and different combinations of raised dots meant different sounds, forming words and sentences. As the code was to be read with the fingers, soldiers needed no light to read it, so they could share messages without putting themselves in danger in the night. Louis Braille was a blind student and a teacher in the Institute. He learnt Barbier's system and modified it until he found a way he considered easy to read and understand. He used a matrix of only 6 points, 2 columns of 3 dots each. Each letter of the alphabet was represented by a specific combination of dots, so words could be read letter by letter. The students in the Institute found this code very useful and much easier to learn and use that any of the other methods [14, p. 9]. Louis Braille established the same code to write mathematics and music, using the same symbols in different contexts, so blind students could have access to any kind of information they needed. Of all the methods mentioned above, the blind people found raised dots symbols were easier to read. Every other method was too difficult or even impossible to write, but raised dots allowed blind people to read and write [14, p. 9].

2 Braille Code

2.1 *Literary Braille*

In the Braille Code, the points are numbered: the first column up-down as 1, 2 and 3 and the second column up-down as 4, 5 and 6. Letter "a" is represented by the raised dot 1, letter "b" by points 1 and 2, "c" by points 1 and 4, etc. The Braille alphabet is

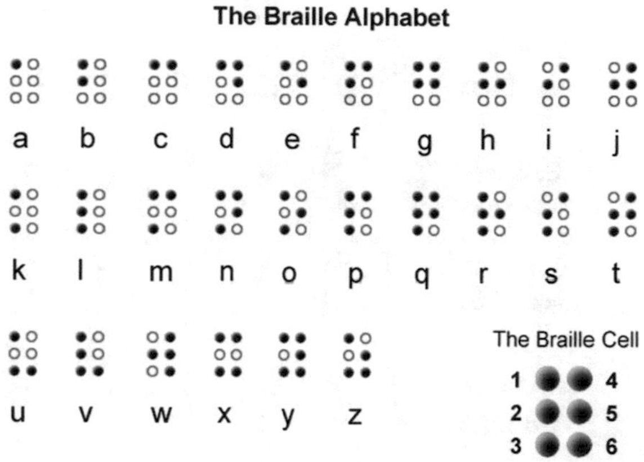

Fig. 1 Braille alphabet.
Taken from http://www.
newyorkgeek.com/2012/01/
braille-morse-code/,
November 12, 2014

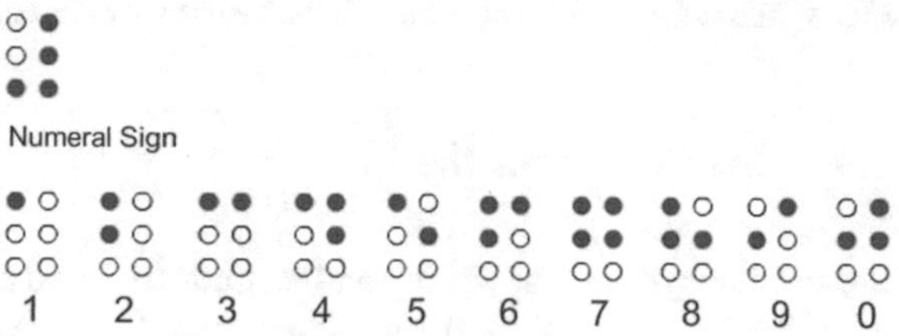

The Braille Alphabet

Fig. 2 Braille numbers.
Taken from http://www.
ridbcrenwickcentre.com/
louisbraille/facts/braille-
punctuation/ (Aug. 12, 2014)

such that the first 10 letters have a specific representation, the next 10 are the same but as the first ones but adding dot 3, and the next letters are the same as the first 10 but adding points 3 and 6, as Fig. 1 shows.

2.2 Numbers in Braille

Numbers 1 to 9 use the same representation as letters "a" to "i". Letter "j" is equivalent to number 0. To distinguish between numbers and letters a prefix before the number is used: points 3, 4, 5 and 6. Figure 2 shows this.

2.3 Music Braille

For music, Louis Braille proposed the same symbols: letter "d" represents the note Do, "e" represents Re, "f" represents Mi, and so on. The duration of the notes is given by the points 3 and 6, which are not used in the first letters of the Braille Alphabet [1]. The notes and their value are shown in Fig. 3.

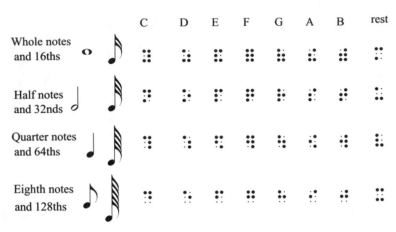

Fig. 3 Taken from the Music Braille Code, 1997 by Braille Authority in North America

2.4 Braille Alternatives

Some other systems for writing and reading for blind people have been used in countries like Spain, USA, Germany, Japan, etc. [3, 8]. Moon system uses big symbols with dots and lines that resemble upper-case letters, making possible that sighted people not trained in this system are able to understand it [17]. The code is still used mostly by people who have lost sensitivity in their fingers due to aging or diseases like diabetes. There is no music notation in this code.

In USA the Modified Braille, or American Braille was developed. It was written with raised dots, using less dots in the symbols for the most used letters, and the bigger number of dots in the symbols for the less used letters, trying to make it more efficient [3, p. 182]. This system was complicated for the blind. It is important to notice that there is a logic in the literary Braille code that makes it easy to remember and to read. In Spain, from 1850 to 1950 many schools used the Abreu system to write and read music. It was designed by Gabriel Abreu, who was a blind musician and knew Braille. It is an 8-dot code, and students who used it said it was comfortable and very clear [2, p. 9]. It added 2 points on the lower part of the Braille box, so there were 4 points used to design a note and other 4 to design the duration. This avoided the use of the same duration signs for two different durations. Also in Spain, in 1855 Pedro Llorens de Llatchos presented another method for reading and writing literature and music. This system used dots and lines with different inclinations, and the signs were understandable for sighted people as they resembled conventional letters. This system was hard to learn, but students that adopted it said it was convenient for them [2, p. 11]. Figure 4 shows a comparison between the Braille, Abreu and Llorens systems.

In 1954 Braille code was officially accepted as the international code for the blind, including the music writing, so other systems were not taught anymore.

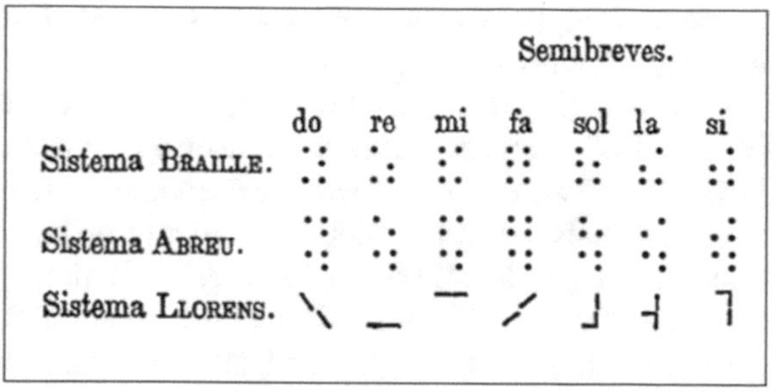

Fig. 4 Comparison between Braille, Abreu and Llorens systems [2, p. 11]

3 Music Braille Problems

In Braille music, there are at least 292 different signs [6, p. 60], and as Braille offers 63 possible dots combinations, it is necessary to use 2, 3 or even 4 boxes to represent some music signs. This leads to one of the major problems of Braille Music Notation: a symbol may change its meaning depending on the context, or the symbols that surround it. The other major problem of this notation is one of the characteristics of Braille itself: it is a linear code. Conventional music notation uses two dimensions, the figure of a note represents a duration but the position of the symbol represents the height of the sound, music notation is written in a graphic space [18, p. 42]. Braille uses only one dimension, so more symbols are needed in order to write the information that position adds to a symbol. Because of all this, Braille music needs a lot of space. Much more pages are needed to transcribe a conventional score. So the use of abbreviations or contractions is sometimes useful in Braille musicography. For example, in a passage with 2 simultaneous voices which are always a third apart, instead of writing each of the notes, only the higher voice is written and another sign is added to establish that the other voice is a third down on the voice written [4, pp. 81–82]. In complicated passages there are several ways to abbreviate music, depending on the characteristic that wants to be emphasized or the one who helps a better memorization of the piece. All this makes the transcription from conventional music notation to Braille Musicography a difficult and tiresome task [10, 12], and sometimes fully automatic transcription is not possible. The study of the notation and interpretation of a Braille score is a tiresome work for blind musicians [4].

4 The Need of a New Musicography for the Blind

In the present stage of this work, the main objective is to present the main problems of Braille musicography in order to show the need of a more efficient system. It should be clear and accessible for the blind musicians and fully automatic transcription of music scores should be possible. In an investigation made for the Open Well-Tempered Clavier project by Robert Douglass, researchers found there were not enough Braille music transcriptions of important works for blind musicians, as they wrote in the webpage: "While online print catalogues list over 8000 scores for Mozart's piano works, searching the Swiss library for the blind shows only 67 scores available" [9]. A lot of efforts are made in the world in order to transcribe scores to Braille. Different softwares are made and revised and other alternatives like Spoken Music are being analyzed and developed in order to make music notation available for the blind people [4]. But there are still great difficulties that are not completely saved by all these efforts. The main difficulties, mentioned in the previous section, make the work hard for transcriptors and automatic transcriptions still need several copy editors in order to assure a good quality Braille score.

5 Proposed Methodology

As the principal objective of this project is to propose a new set of symbols for a clearer and useful musicography for the blind, it is important to take into account several things. First of all, Braille musicography is studied and analyzed so that its virtues and its fails are known and discussed with the blind. The virtues must be kept as much as possible, and its problems should be corrected as well. It is important to notice that the most sensitive zone of our fingers is the fingertip. Two kinds of receptors, Meissner's corpuscles and Merkel's disk, are in charge of tactile acuity given their characteristics [11, pp. 431–437]. There are more of these receptors in the fingertip than in the rest of the hand [11, p. 437]. The area with the greater density is approximately $25\,mm^2$ [5, p. 8]. The density of these receptors decreases drastically from the fingertips to the palm [13, p. 284], delimiting a very sensitive zone as shown in Fig. 5. This allows a clear reading of a Braille symbol [11, p. 435]. Outside this zone, the fast reading of the symbol produces a blurred image, making the reading tiresome and difficult.

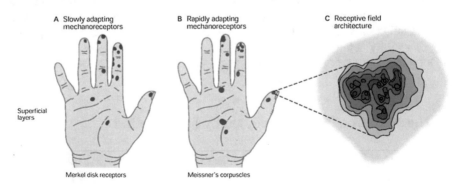

Fig. 5 Merkel disk receptors in Meissner's corpuscles in human hands [11, p. 434]

After the form of a new symbol generator and its characteristics are decided, a set of symbols that are coherent for a music notation for the blind will be chosen. For this, a close collaboration with the blind students of the Faculty of Music of UNAM is being realized. One of the main goals of the new proposal is to enable fully automatic transcriptions of conventional music scores for the blind. During the project this will be held in mind and proved in a prototype for a software to be fully developed in a later stage of the project.

6 Conclusions

Besides the efforts made worldwide, the transcription to music for the blind is still difficult and not fully automatic. As a consequence, blind people have not full access to Braille scores. The proposal of this work is to establish the need of a new code for the notation of music for the blind. This new code, in order to be useful and clear has to obey some principles, some of which are mentioned in this work. In every moment, the opinion and experience of the blind students of the Faculty of Music of UNAM is taken into account, and the virtues of the actual Braille musicography will be kept as much as possible. This problem requires to consider not only the combination of points to make a coherent system of symbols, but the physiology of our hands among other things. It is an aim to offer clear information in less space, so research such as Mazzola's Theory of Gestures [15, 16] and Lobato- Cardoso's study on echolocation [in this volume] may provide new clues in this direction.

References

1. Braille Authority of North America, Braille Music Code. American Printing House for the Blind (1997)
2. Burgos, E.: Las musicografías de Abreu y Llorens: dos sistemas alternativos a la recepción del Braille en España, Integración: revista sobre ceguera y deficiencia visual, vol. 46, pp. 7–12. ONCE Madrid (2005)
3. Burgos, E.: The first Spanish music codes for the blind and their comparison with the american ones. Fontes Artis Musicae **57**(2), 167–185 (2010)
4. Conservatorio di Musica Pollini di Padova: Accessible Music: The State of the Art (2013). www.music4vip.org
5. Detorakis, G., Rougier, N.: Structure of receptive fields in a computational model of area 3b of primary sensory cortex. Front. Comput. Neurosci. **8**(76) (2014)
6. Díaz-Bertevelli, Isabel Cristina: La educación musical de personas con deficiencia visual y la musicografía Braille: De la musicalización a la lectura y la escritura de la partitura en Braille. In: Fillottrani, L.I., Mansilla, A.P. (eds.) Actas de la IX Reunión de SACCoM: Tradición y Diversidad en los aspectos psicológicos, socioculturales y musicológicos de la formación musical, pp. 58–64 (2010)
7. Discapacidad visual. Las personas con discapacidad en México: Una visión censal, Instituto Nacional de Estadística, Geografía e Informática, INEGI, Mexico City (2004)

8. Dixon, J.: Eight-dot Braille. A Position Statement of the Braille Authority of North America (2007)
9. Douglass, R.: The Open well-Tempered Clavier-Bach to Bach (2012). www.kickstarter.com/projects/293573191/open-well-tempered-clavier-bah-to-bach/posts/612949
10. Encelle, B., Jessel, N., Mothe, J., Ralalason, B., Asensio, J.: BMML: braille music markup language. Open Inf. Syst. J. **3**, 123–135 (2009)
11. Gardner, E., Martin, J., Jessell, Th: The bodily senses. In: Kandel, E.R. (ed.) Principles of Neural Science. McGraw-Hill, New York (2000)
12. Homenda, W., Sitarek, T.: Notes on automatic music conversions. In: Kryszkiewicz, M. (ed.) Foundations of Intelligent Systems. Lecture Notes in Computer Science, vol. 6804, pp. 533–542. Springer, Berlin (2011)
13. Johansson, R.S., Vallbo, A.B.: Tactile sensibility in the human hand: relative and absolute densities of four types of mechanoreceptive units in glabrous skin. J. Physiol. **286**(1), 283–300 (1977)
14. MacKenzie, S.C.: World Braille Usage. UNESCO, Paris (1954)
15. Mazzola, G.: Categorical gestures, the diamond conjecture, Lewin's question, and the Hammerklavier Sonata. Jo. Math. Music **3**(1), 31–58 (2009)
16. Mazzola, G.: Musical Performance-A Comprehensive Approach: Theory, Analytical Tools, and Case Studies. Springer, Berlin (2011)
17. Moon Literacy, July 20 (2012). www.moonliteracy.org.uk
18. Wadle, D.C.: Meaningful scribbles: an approach to textual analysis of unconventional musical notations. J. Music Mean. **9**, 38–68 (2010)

Group Theory for Pitch Sequence Representation: From the Obvious to the Emergent Complexity

Emilio Erándu Ceja-Cárdenas

Abstract In the first two sections of this contribution we construct the groups $(S_n, +)$ and $(L(S_n), \circ)$ in order to have an intuitive way to represent musical phrases by their melodic contour. Later we derive an algorithm for composing music using a given number and the group $(L(S_n), \circ)$. Finally we offer a variation of the same algorithm to be able to translate a piece of music in a finite digit number, with analytic and deconstructive aims.

1 Introduction

Fix $n \in \mathbb{N}$. Let S_n be the set whose elements s_j, $j \in \{0, \dots, n-1\}$, are sets of intervals of $\frac{12}{n} j$ semitones, including its octaves; in other words,

$$s_j = \left\{ \frac{12}{n} j + 12m \text{ semitones} \mid m \in \mathbb{Z} \right\}. \tag{1}$$

Fixing n as a divisor of 12 we have the sets S_1, S_2, S_3, S_4, S_6 and S_{12} whose elements are equivalence classes. We shall name elements in S_n using letters in ascending order starting from the letter a.

- $S_1 = \{a = [0]\}$
- $S_2 = \left\{ a = [0], b = \left[\frac{12}{2} \right] \right\}$
- ...
- $S_6 = \left\{ a = [0], b = \left[\frac{12}{6} \right], c = \left[\frac{12}{6} 2 \right], d = \left[\frac{12}{6} 3 \right], e = \left[\frac{12}{6} 4 \right], f = \left[\frac{12}{6} 5 \right] \right\}$

Now we define the operation $+$ as the usual modular arithmetic, that is $[x] + [y] = [x + y]$. E.g. for $a, b, f \in S_6$:

$$b + f = [2 \text{ semitones} + 10 \text{ semitones}] = [12 \text{ semitones}] = a.$$

E.E. Ceja-Cárdenas (✉)
Departamento de Matemáticas, Centro Universitario de Ciencias
Exactas e Ingenierías (CUCEI), Universidad de Guadalajara, Guadalajara, Mexico
e-mail: emilioerandu@gmail.com

© Springer International Publishing AG 2017
G. Pareyon et al. (eds.), *The Musical-Mathematical Mind*,
Computational Music Science, DOI 10.1007/978-3-319-47337-6_7

We see that $(S_n, +)$ is a group with a being the identity element. Now lets define $g : S_n \to \mathbb{Z}_n$, $g(s_j) = [j]$, $j \in \{0, \dots, n - 1\}$, it is clear that g is an isomorphism from S_n to $\mathbb{Z}_n$.

As an example we show the elements in S_4 using middle *do* (i.e. do_4),[1] as reference

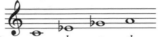

for counting intervals: $\quad a \quad b \quad c \quad d \quad$, where each pitch corresponds to each element in S_4. Since elements in S_n are equivalence classes we have 6 partitioned sets that can be visually represented as subsets of S_{12}, being S_{12} the set of all pitches in the chromatic scale:

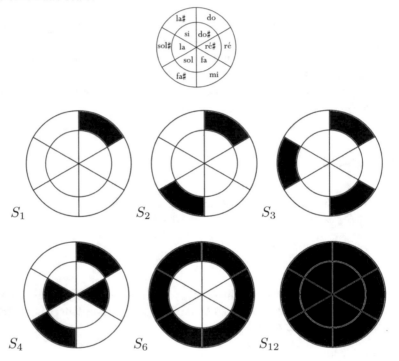

2 $(L(S_n), \circ)$

Let $L(S_n)$ be an infinite set of infinite strings with elements in S_n concatenated in every possible order; also, each string has an infinite string of only a to the right. That is, for S_2, $a\overline{a}$,[2] $b\overline{a}$, $ab\overline{a}$, $babbababa\overline{a}$ are in $L(S_2)$. For convenience we won't write the infinite string of a that goes with every element in $L(S_n)$, this way $babbababa\overline{a}$ will be just $babbabab$, also $a\overline{a}$ will be just a. This way we can represent pitch sequences as elements of $L(S_n)$, that is representing the movement of the melody by sequences

---

[1] We use the do–si pitch nomenclature in order to avoid confusion between letters here used.

[2] We use over line notation to indicate repeating and never ending a.

of musical intervals. For instance, the sequence 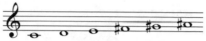 is seen as the element *abab* in $L(S_2)$. We note that having $\bar{a}$ to the right adds nothing to music since it is a, identity element in S_n, concatenated infinitely times and it adds no intervals.

This is how $abbbbb \in L(S_6)$ is seen in a staff:

This is the *whole-tone scale* starting at do_4, the other whole-tone scale can be generated in reference to $do\sharp_4$. Every possible sequence of sounds produced by the use of this scale can be seen as an element of $L(S_6)$. It is trivial to note that every sequence of sounds, as long as it uses some or all of the 12 pitches (disregarding enharmonics) in Western music can be seen in $L(S_{12})$ since the latest set includes all possible sequence of intervals. Also, by fixing any $n \in \mathbb{N}$ and not just divisors of 12 we can extend S_n and later $L(S_n)$ to microtonality. Obviously we may encompass whole-tone scales using the same concepts.

Let's start with a whole-tone scale example. We take a look at Debussy's first two bars of *Prelude No. 2*, *Voiles*, from his First book of *Preludes* for piano [3]:

We can represent the upper melody in reference to do_4 as $effffaf \in L(S_6)$ and the lower melody as $cffffff \in L(S_6)$. Whole-tone elements are present in much of Debussy's repertoire. Just to mention few examples: everything from *Voiles* except 6 bars; the solo between the English horn and the cello at the end of the first movement in *La Mer*, and a number of passages in *Les Images, livre I* for piano solo.

Let $s, \grave{s} \in L(S_n)$, $s = [s_1][s_2] \dots [s_n] \dots$, $\grave{s} = [\grave{s}_1][\grave{s}_2] \dots [\grave{s}_n] \dots$ Now we define the $\circ$ operation as a coordinate-wise addition in the sense of $s \circ \grave{s} = [s_1 + \grave{s}_1][s_2 + \grave{s}_2] \dots [s_n + \grave{s}_n] \dots$ We note that the length of s and $\grave{s}$ does not matter since every element in $L(S_n)$ has $\bar{a}$ to the right; this means there will always be an a to operate. $(L(S_n), \circ)$ is a **group** with $\bar{a}$ being the identity element.[3] In the following example we look at the first beat, bar no. 31 of *Jeux d'eau* for solo piano from Ravel [2]

. Using $(L(S_6), \circ)$ in reference to do_4 the upper melody performed

---

[3]It is important to distinguish between $(L(S_n), \circ)$ and word algebra. We are using concatenated elements in S_n with a coordinate-wise addition which is a fundamentally different operation to the one used in word algebra.

Fig. 1 *Jeux d'eau*, Ravel, bars 31 and 32

Fig. 2 Carrillo's example 25

Fig. 3 Major scale,
$accbcccb \in L(S_{12})$

with the right hand, *la♮ la♮ fa♮ sol♮* can be represented as the *faeb* element in $L(S_6)$, now we arbitrarily select *face* and operate *faeb* ∘ *face* and we obtain *eaaf* which

is the second beat: ℘ed. . Now we represent the upper melody in right hand from bars 31 and 32 of *Jeux d'eau* (Fig. 1) as follows: bar 31, beat 1: *faeb* in reference to do_4; bar 1, beat 2: *faeb* ∘ *face* = *eaaf*; bar 31, beat 3: *eaaf* ∘ *baec* = *faeb*; bar 31, beat 4: *faeb* ∘ *face* = *eaaf*; bar 32, beat 1: *eaaf* ∘ *baac* = *faef*; bar 32, beat 2: *faef* ∘ *faac* = *eaeb*; bar 32, beat 3: *eaeb* ∘ *faaa* = *daeb*; bar 32, beat 4: *caec* in reference to $do♯_4$.

Next we explain the example 25 from Julián Carrillo's treatise *Leyes de metamorfósis musicales* [Music's Metamorphosis Laws][1] using $(L(S_n), ∘)$. Here Carrillo shows a "Major scale metamorphosed to its duple" (Fig. 2).

This is the result of doubling every interval in a major scale: where there was 1 semitone now there is 2 semitones and so on. Using $(L(S_n), ∘)$ we represent every pitch sequence as a sequence of musical intervals. For a Major scale (Fig. 3) that is the element *accbcccb* in $L(S_{12})$:

Now we do *accbcccb* ∘ *accbcccb* = *aeeceeec*. Since ∘ operation is a coordinate-wise addition, the result of operating *accbcccb* to itself is adding every interval in itself (see: Fig. 4).

Fig. 4 *aeeceeec* $\in L(S_{12})$

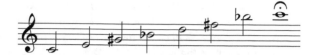

This is the ascending part of the *Major scale metamorphosed to its duple* shown above. Following this process we obtain *aeeceeeclkkkekk* which is the whole example 25. We conclude that a *Metamorphosis to its duple* (according to Carrillo's laws of Metamorphosis) can be seen as an element in $L(S_{12})$ operated to itself.

At the beginning of this exposition we defined n as a divisor of 12 which leaded to 6 different sets, but, as mentioned before, we can extend S_n to microtonality if we choose a different $n \in \mathbb{N}$ to produce an S_n whose elements are additions of any arbitrary division of the octave. Then we use the obtained S_n and expand it to $L(S_n)$ and $(L(S_n), \circ)$. An example is given with fixed $n = 13$:

- $S_{13} = \left\{ a = [0], b = \left[\frac{12}{13}\right], c = \left[\frac{12}{13}2\right], d = \left[\frac{12}{13}3\right], \quad \ldots, \quad l = \left[\frac{12}{13}11\right], m = \left[\frac{12}{13}12\right] \right\}$

3 *Piph Music* for Algorithmic Composition

For a first example on algorithmic composition using number representation, it is convenient to quote one of the first compositions systematically using irrational numbers: π *(A game within the Circle's Constant)*[4] is an awarded composition by Gabriel Pareyon, that uses the first 1000 digits of π in order to produce a solo for the bass flute. This composition associates every chromatic pitch to each digit starting by 0 as *do*, 1 as *do♯* and so on.

In the leftmost part of this example we see the first sound: *re♯* corresponding to 3, then *do♯* corresponding to 1. Afterwards we find the succession 4, 1, 5, 9, 2, 6, 5, 3, 5 where each digit has its defined pitch. We see that for any 1 we will always find a *do♯* while a 9 will always be *la*.

As a creative possibility of $(L(S_n), \circ)$ we present a different algorithm (Fig. 5) capable of reading any given finite number and returning the sequence of pitches (as equivalence classes) in order to compose music:

1. Read first digit $d \neq 0$ and define d instruments.
2. Start with the first instrument, i.e. instrument counter equals 1.
3. Next digit n defines the number of n pitches for the current bar.
4. For each of the next n digits apply[4] $g^{-1} : \mathbb{Z}_n \to S_n$ and consider the corresponding element in $L(S_n)$ for the current bar, e.g. 021 will be $acb \in L(S_n)$.
5. Check if the instrument counter is bigger than the first digit d.
5.1. If not, increase instrument counter and repeat step 3.
5.2. If yes, is this the end of the given number?
5.2.1. If not, repeat step 2.
5.2.2. If yes, end.

The use of digits in this algorithm limits the number of instruments in the score and the number of pitches to a maximum of 9. Also, due to the decimal system there is not much $(L(S_n), \circ)$ interesting options, but this "lack"can be solved using two digits instead of one for each process. Later we will see a different algorithm capable to obtain a finite number from a score. Since by now we do not consider any rhythmic, nor dynamical values, this leads, if waned, to different musical values arising from the same finite number and vice versa.

What results from using the algorithm proposed by Pareyon is different to what results using the $(L(S_n), \circ)$ algorithm. Since we understand every element in $(L(S_n), \circ)$ as a melody that results in adding intervals, it is not obvious to find a pitch with its corresponding digit, but will be easy to understand a whole melody as a sequence of digits.

As a consequent exercise we prepared a music score[5] for two treble and one bass clefs from the number π up to the digit 190 using $(L(S_6), \circ)$ and starting in *do*. Metre was assigned in equal durations $\left(\frac{1}{1}\right)$.

The first digit in π is 3, meaning 3 instruments. Next we find 14, this means 1 pitch, element 4 in $(L(S_6), \circ)$ corresponding to 8 semitones; since we start in *do* the pitch must be *sol♯*. Next there is 15, meaning 1 pitch, element 5 in $(L(S_6), \circ)$, that is *la♯*. Next 10 digits are 9265358979, meaning 9 pitches, element 265358979 in $(L(S_6), \circ)$. Below are the first four bars with a space between bars where every bold digit, the start of a new instrument, assigns how many pitches correspond for current bar: 3. **14159265358979 3238462643383 27950288419 7**16939937510582097494445923.

---

[4] $g^{-1} : \mathbb{Z}_n \to S_n, g^{-1}([j]) = s_j, j \in \{0, 1, 2, \ldots, 9\}$.

[5] An audio sample of this can be listen to at https://soundcloud.com/emilioerandu/pi-in-ls6.

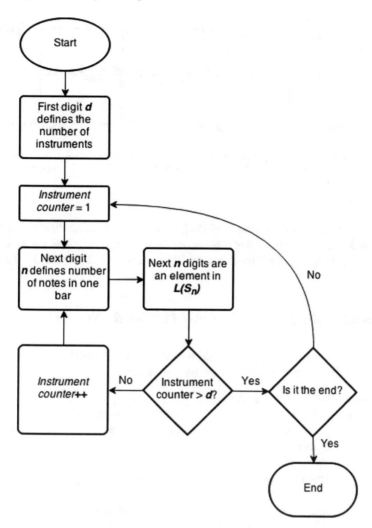

Fig. 5 $(L(S_n), \circ)$ algorithm flowchart

Five bars later there is a triple consecutive digits occurrence in the bass clef: 81284**811**174502:

Using the algorithm with more digits of π we would reach the *Feynman point* 999999 which would result in the addition of the same 9 element in given S_n.

Since we observe that any non-trivially repeated numerical sequence, like π (and typically other irrationals), contains *phrases* (i.e. sequences of ordered digits with their own sequential expressiveness), then we can extend a generalized *Piph Music* as

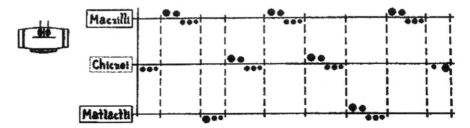

Fig. 6 Pareyon's *Xochicuicatl Cuecuechtli* (2012), excerpt from the manuscript's page 26, with three teponaztlis (wooden, carved log instruments) with the labels Macuilli, Chicuei and Matlactli (i.e. 5, 8, 10)

a branch of Group Theory. We use the term *Piph* after the given example of π as music (*Pi*), containing segments of *musical concatenation* (*ph*rase, therefore making the name *Pi + ph* for any phrasing extracted from irrational numbers segmentation).[6]

4 Translating a Piece of Music into a Single Number

By the reverse usage of the algorithm shown above, we can translate a piece of music into a single finite number. The process we follow is:

1. Number of instruments defines first d digit.
2. Start with the first instrument, i.e. instrument counter equals 1.
3. Count the number of pitches in the current bar and define the next n digit.
4. Next n digits are obtained applying[7] $g : S_n \to \mathbb{Z}_n$ to the corresponding $L(S_n)$ element in the current bar, e.g. *abc* is 012.
5. Increase the instrument counter and check if this is bigger than the first digit d.
5.1. If not, increase instrument counter and repeat step 3.
5.2. If yes, is this the end of the piece of music?
5.2.1. If not, repeat step 2.
5.2.2. If yes, end.

For the last example (Fig. 6) we apply a variation of the proposed algorithm to the instrumental (*teponaztlis*) passage *Macuilli, Chicuei* and *Matlactli* (that is Five, Eight and Ten, in Nahuatl language) in the musical score *Xochicuicatl Cuecuechtli*, also composed by Pareyon [5]:

---

[6]Carrillo's nomenclature is somehow alluded here: we extend the name of π to other irrationals musically useful, as Carrillo employs the name of number 13 (the so called *Sonido 13*) in order to indicate pitch cardinality bigger than the traditional twelve-tone class system.

[7]$g : S_n \to \mathbb{Z}_n, g(s_j) = [j], j \in \{0, \ldots, n-1\}$.

For the *numerical translation* of this excerpt, we use $(L(S_2), \circ)$ with $1, b \in S_2$, being the element that changes between high and low pitch and $0, a \in S_2$ the identity element. Next we numerically represent this example, with a space between bars where every bold digit represents the start of a different instrument: 3 **03**0000 **5**1010**00** **0**03000 **05**101000 **5**1010**00** **05**101000 **00510**100 **5**1010**00** **02**000.

Although this number is "mathematically useless", it may be useful to fulfil a number sequence abstraction, such as the textural-orchestrational pattern, like 0305000030505000500055000020 (i.e. only taking into account bold numbers), or rather in order to abstract the *contrapuntal number* 353555552 as the key number of this segment, in turn able to be treated as a source for musical development from the same source.

Acknowledgements My sincere thanks to Dr. Gabriel Pareyon, who made many helpful suggestions to my manuscript, including the term *Piph Music* and its definition, and allowed me to reproduce his music scores as examples.

This project would have been impossible without the financial support from CUCEI, Universidad de Guadalajara, obtained thanks to Dr. Alfonso M. Hernández Magdaleno.

References

1. Carrillo, J.: Leyes de metamorfósis musicales. Talleres Gráficos de la Nación, México, DF (1949)
2. Demets, E. (ed.): Ravel: Jeux d'eau. Dover Publications, New York (1986)
3. Heinemann, E.G. (ed.): Debussy: Preludes. Henle Verlag, Munich (1986)
4. Pareyon, G.: π (A Game within the Circle's Constant). Klub Muzyki Współczesnej Malwa, Krakow, Poland (2000)
5. Pareyon, G.: Xochicuicatl Cuecuechtli. Fonca, Mexico (2012)

Mazzola's Escher Theorem

Yemile Chávez-Martínez and Emilio Lluis-Puebla

Abstract In this note we give a full proof of Mazzola's Escher Theorem (Mazzola, J Math Music, 3(1):31–58, 2009, [4]). This theorem is needed for the development of the theory that Mazzola seeks to realize, and it helps us to understand better the concept of hypergesture as used in his work (Mazzola, J Math Music 3(1):31–58, 2009, [4], Mazzola, Musical performance-A comprehensive approach: theory, analytical tools, and case studies, 2011, [5], Mazzola and Andreata, J. Math. Music, 1(1):23–4, 2007, [6], Mazzola et al., Musical creativity-strategies and tools in composition and improvisation, 2011, [7]). A *gesture* is a morphism from a digraph into a topological space, and is one of the fundamental blocks in the Mathematical Theory of Performance. A *hypergesture* is a gesture built upon another gesture, describing, in a way, the variation of the latter. The non-trivial fact that the variation of the former gesture, as described by the latter, is given by the *same* hypergesture is essentially the content of the Escher Theorem.

1 Basic Concepts

We review the graph and category theory necessary for fixing notation and deliver the concepts of gesture and hypergesture. The reader already familiar with those, may skip the following paragraphs and proceed directly to Sect. 2.

Definition 1 We consider a **digraph** D as an ordered pair (V_D, A_D), where V_D is a set of vertices and A_D a set of arrows, disjoint from V_D, together with an *incidence function* ψ_D that associates with each arrow of D an ordered pair of vertices (not necessarily distinct) of D. This is $\psi_D : A_D \longrightarrow V_D \times V_D$. Generally speaking, if $\psi_D(a) = (u, v)$, we will call u the *tail* of a and v the *head* of a.

Y. Chávez-Martínez (✉) · E. Lluis-Puebla
Universidad Nacional Autónoma de México, Av Universidad 3000,
Cd. Universitaria, 04510 Coyoacán, D.F., Mexico
e-mail: yemilec@yahoo.com.mx

E. Lluis-Puebla
e-mail: lluispuebla@gmail.com

© Springer International Publishing AG 2017
G. Pareyon et al. (eds.), *The Musical-Mathematical Mind*,
Computational Music Science, DOI 10.1007/978-3-319-47337-6_8

Definition 2 Let D and G be digraphs. A **morphism of digraphs** $f : D \longrightarrow G$ is a pair (ϕ, θ) of functions $\phi : A_D \longrightarrow A_G$ and $\theta : V_D \longrightarrow V_G$, making the following diagram commute:

$$
\begin{array}{ccc}
A_D & \xrightarrow{\phi} & A_G \\
\psi_D \downarrow & \circlearrowleft & \downarrow \psi_G \\
V_D^2 & \xrightarrow{\theta^2} & V_G^2
\end{array}
$$

where $V_G^2 = V_G \times V_G$ and $\theta^2 := (\theta, \theta) : V_G^2 \longrightarrow V_G^2$.

The category **D** of digraphs has as objects the collection $Obj(\mathbf{D})$ of digraphs, and for each pair of digraphs, Γ and Δ, the set $\Gamma @_\mathbf{D} \Delta = \mathbf{D}(\Gamma, \Delta)$ of morphism of digraphs as arrows [1, 3].

The composition of morphisms of digraphs $f = (u, v) \in \Gamma @_\mathbf{D} \Delta$, $g = (w, z) \in \Delta @_\mathbf{D} K$ with Γ, Δ and K digraphs, and where each of the morphisms $u : A_\Gamma \longrightarrow A_\Delta$, $v : V_\Gamma \longrightarrow V_\Delta$, $w : A_\Delta \longrightarrow A_K$ and $z : V_\Delta \longrightarrow V_K$ makes sense, is given by pasting commutative squares. Namely

$$
\begin{array}{ccccc}
A_\Gamma & \xrightarrow{u} & A_\Delta & \xrightarrow{w} & A_K \\
\Gamma \downarrow & & \Delta \downarrow & & \downarrow K \\
V_\Gamma^2 & \xrightarrow{v^2} & V_\Delta^2 & \xrightarrow{z^2} & V_K^2
\end{array}
$$

that is, $g \circ f = (w \circ u, z \circ v) \in \Gamma @_\mathbf{D} K$.

Now consider the set

$$
A_{\overrightarrow{X}} = I @_\mathbf{Top} X := \{c : I \longrightarrow X | c \text{ is a continuous curve}\}
$$

with $X \in \mathbf{Top}$ (the category of topological spaces and continuous functions) and I a fixed closed interval in $\mathbb{R}$ with its canonical orientation [9]. Thus we define $\overrightarrow{X}$ such that $A_{\overrightarrow{X}}$ is the set of its arrows and $V_{\overrightarrow{X}} = X$ that of its vertices. It is clear that $\overrightarrow{X}$ is a digraph.

The digraph $\overrightarrow{X}$ is a very special one, since it is defined *inside* the arbitrary topological space X, and with the concepts above at hand, we may consider the subcollection (of the category **D**) of *spatial digraphs*, **SD**, as follows:

1. $Obj(\mathbf{SD}) = \{\overrightarrow{X} : A_{\overrightarrow{X}} \longrightarrow V_{\overrightarrow{X}}^2 | X \in \mathbf{Top}, \overrightarrow{X}$ the incidence function, where $A_{\overrightarrow{X}}$ are the arrows of a digraph $\overrightarrow{X}$ and $V_{\overrightarrow{X}} = X$ its vertices$\}$.
2. $\mathbf{SD}(\overrightarrow{X}, \overrightarrow{Y}) = \overrightarrow{X} @_\mathbf{SD} \overrightarrow{Y} = \{\overrightarrow{f} : \overrightarrow{X} \longrightarrow \overrightarrow{Y} | \overrightarrow{f}$ is a digraph morphism induced canonically by a continuous function $f : X \longrightarrow Y$ in $\mathbf{Top}\}$.

The aforementioned collection of objects is evidently contained in $Obj(\mathbf{D})$ and in the same way, the collection of arrows for every pair $\overrightarrow{X}$ and $\overrightarrow{Y}$ of spatial digraphs is evidently contained in $\overrightarrow{X} @_{\mathbf{D}} \overrightarrow{Y}$.

The fact that $\mathbf{SD}$ is actually a subcategory of $\mathbf{D}$ is nothing but a straightforward argument, and is left to the reader [8].

2 The Category of Gestures

This section aims at defining the category of gestures.

Definition 3 Let $\Gamma \in Obj(\mathbf{D})$ and $\overrightarrow{X} \in Obj(\mathbf{SD})$ be given objects. A Γ-*gesture* in a topological space X is a morphism $g : \Gamma \longrightarrow \overrightarrow{X}$ between digraphs.

In this case Γ will be called the **skeleton of the gesture,** meanwhile the topological space X will be called the **gesture space,** and the curve defined into X given by g will be called the **body** of the gesture.

Definition 4 Consider $\delta : \Delta \longrightarrow \overrightarrow{X}$ and $\gamma : \Gamma \longrightarrow \overrightarrow{Y}$ two gestures, a **gesture morphism** $\widetilde{f} : \delta \longrightarrow \gamma$ consists of a pair of morphisms $\widetilde{f} := (f, \overrightarrow{h})$, where $f : \Delta \longrightarrow \Gamma$ is a digraph morphism, such that there is a digraph morphism $\overrightarrow{h} : \overrightarrow{X} \longrightarrow \overrightarrow{Y}$, not necessarily continuous, making the following diagram commute:

$$\begin{array}{ccc} \Delta & \xrightarrow{\ \delta\ } & \overrightarrow{X} \\ {\scriptstyle f}\downarrow & & \downarrow{\scriptstyle \overrightarrow{h}} \\ \Gamma & \xrightarrow{\ \gamma\ } & \overrightarrow{Y} \end{array}$$

In particular, note that for *gestures* $\delta : \Delta \longrightarrow \overrightarrow{X}, \gamma : \Gamma \longrightarrow \overrightarrow{Y}$, and $\kappa : K \longrightarrow \overrightarrow{Z}$, and the *morphisms* of gestures $\widetilde{f} : \delta \longrightarrow \gamma$ and $\widetilde{g} : \gamma \longrightarrow \kappa$, such that $\widetilde{f} = (f, \overrightarrow{h})$ and $\widetilde{g} = (g, \overrightarrow{j})$ with $f : \Delta \longrightarrow \Gamma$, $\overrightarrow{h} : \overrightarrow{X} \longrightarrow \overrightarrow{Y}, g : \Gamma \longrightarrow K$ and $\overrightarrow{j} : \overrightarrow{Y} \longrightarrow \overrightarrow{Z}$, the following diagram commutes

$$\begin{array}{ccccc} \Delta & \xrightarrow{\ f\ } & \Gamma & \xrightarrow{\ g\ } & K \\ {\scriptstyle \delta}\downarrow & & \downarrow{\scriptstyle \gamma} & & \downarrow{\scriptstyle \upsilon} \\ \overrightarrow{X} & \xrightarrow{\ \overrightarrow{h}\ } & \overrightarrow{Y} & \xrightarrow{\ \overrightarrow{j}\ } & \overrightarrow{Z} \end{array}$$

that is: $\widetilde{g} \circ \widetilde{f} = (g \circ f, \overrightarrow{j} \circ \overrightarrow{h})$.

If we now consider the collections given by:

1. $Obj(\mathbf{G}) := \{\delta : \Delta \longrightarrow \overrightarrow{X} | \Delta \in Obj(\mathbf{D}), \ \overrightarrow{X} \in Obj(\mathbf{SD})$ and δ a morphism$\}$.
2. $\mathbf{G}(\delta, \gamma) = \delta @_{\mathbf{G}} \gamma := \{\widetilde{f} : \delta \longrightarrow \gamma | \widetilde{f} = (f, \overrightarrow{h})$ are gesture morphisms with $\gamma \circ f = \overrightarrow{h} \circ \delta\}$ (for every pair of gestures δ and γ in $Obj(\mathbf{G})$),

subject to the composition of gestures morphisms with $\widetilde{f} \in \delta @_{\mathbf{G}} \gamma$, $\widetilde{g} \in \gamma @_{\mathbf{G}} \upsilon$, for all δ, γ and υ gestures, as we just mentioned above, then it is clear that we obtain a category $\mathbf{G}$, the category of gestures.

Now if we consider certain gestures as *points* in a space, it is possible to study gestures inside a *gesture space*, which will be called **hypergestures**.

To define them, we need first to know how to make the set of gestures $\Delta @_{\mathbf{D}} \overrightarrow{X}$ into a topological space. This we will show below.

3 Hypergestures with an Approach to Escher's Theorem

First consider the very particular case $\Delta := \uparrow$, that is, a digraph with a single arrow. It is well known how to get a topological space $\uparrow @_{\mathbf{D}} \overrightarrow{X} \cong I @_{\mathbf{Top}} X$ by using the compact-open topology. This, along with the following proposition, is the basis for all that follows.

Proposition 1 *Let Δ be a digraph, then it is the direct limit of a direct system.*

Proof Let $\langle A_\Delta, = \rangle$ be a preordered set. We can give the direct system $\{(\Delta_a)_{a \in A_\Delta},$ $(\varphi_{ab})_{a=b}\}$ where $\Delta_a := \uparrow_a \longrightarrow (t(a), h(a))$ for every $a \in A_\Delta$ and $(\varphi_{ab} : \Delta_a \longrightarrow \Delta_b)_{a=b}$ is a family of isomorphisms of digraphs such that:

$$\varphi_{ab} = (\overline{\varphi}_{ab}, Id),$$

where $\overline{\varphi}_{ab} : \uparrow_a \longrightarrow \uparrow_b$.

Now suppose there is a digraph Γ and a corresponding family of morphisms in $\mathbf{D}$ $(f_\alpha : \Delta_\alpha \longrightarrow \Gamma)_{\alpha \in A_\Delta}$ making the following diagram a commutative one:

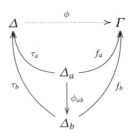

Consider $\phi = f$ such that $f|_{\Delta_a} = f_a$ for each $a \in \Lambda$. Then

$$(\phi \circ \tau_a)(\Delta_a) = \phi(\tau_a(\Delta_a)) = \phi(\Delta_a) = f_a(\Delta_a).$$

Therefore $\Delta \cong \varinjlim_{a \in A_\Delta} \Delta_\alpha.$ ◆

Now that we can regard a digraph as a direct limit, the following results from category theory are at hand. For the proofs, the interested reader may consult [2].

Proposition 2 *If* $\{M_i, \psi_{ji}\}$ *is an inverse system of digraphs, then there is an isomorphism*

$$\omega : \mathbf{D}(\Delta, \varprojlim M_i) \longrightarrow \varprojlim \mathbf{D}(\Delta, M_i)$$

for every digraph Δ. *i.e.,* $\Delta @ \varprojlim M_i \cong \mathbf{D}(\Delta, \varprojlim M_i) \cong \varprojlim \mathbf{D}(\Delta, M_i) \cong \varprojlim(\Delta @ M_i)$.

Proposition 3 *If* $\{M_i, \psi_{ij}\}$ *is a direct system of digraphs, then there is an isomorphism*

$$\theta : \mathbf{D}(\varinjlim M_i, \vec{X}) \longrightarrow \varprojlim \mathbf{D}(M_i, \vec{X})$$

for every digraph $\vec{X}$. *i.e.,* $(\varinjlim M_i) @ X \cong \mathbf{D}(\varinjlim M_i, X) \cong \varprojlim \mathbf{D}(M_i, X) \cong \varprojlim(M_i @ X)$.

Proposition 4 *Let* Δ, Γ *be given digraphs, and* $\{(\uparrow_i)_{i \in A_\Gamma}(\psi_{ij})_{i \le j}\}$ *and* $\{(\uparrow_c)_{c \in A_\Delta}, (\psi_{cd})_{c \le d}\}$ *direct systems of digraphs, then there is an isomorphism*

$$\eta : \varprojlim_{b \in A_\Delta} (\varprojlim_{a \in A_\Gamma}(\uparrow_b @(\uparrow_a @X))) \to \varprojlim_{a \in A_\Gamma} (\varprojlim_{b \in A_\Delta}(\uparrow_b @(\uparrow_a @X)))$$

Proposition 5 *If* Γ, Δ *are digraphs and* X *is a topological space, then there is a canonical homeomorphism*

$$\Gamma \vec{@} \Delta \vec{@} X \cong \Delta \vec{@} \Gamma \vec{@} X$$

This last proposition is nothing but a weaker version of Escher Theorem as the reader will find out in Sect. 4 below.

4   Topological Categories and Mazzola's Escher Theorem

The last ingredient needed for the formulation of Mazzola's Escher Theorem is that of a topological category.

Definition 5 Let K be a category endowed with the property that its set of maps is a topological space, and in which both functions, domain and codomain, and the composition of morphisms as well are continuous.

In this case K is called a **topological category**.

Example 1 The *simplex categoy* ∇ associated with the unit interval I.

In this case the set of maps is $\nabla = \{(x, y)|x, y \in I \text{ and } x \leq y\}$ and the functions domain and codomain are given by $d(x, y) := (x, x)$, $c(x, y) := (y, y)$ respectively. The composition of morphisms is $(x, y) \circ (y, z) = (x, z)$, and the topology on ∇ is the relative topology of the product inherited on $I \times I \subset \mathbb{R} \times \mathbb{R}$.

Definition 6 Let K, L be two topological categories. A **topological functor** $F : K \longrightarrow L$ is a functor which in turn is a continuous function between sets of morphisms.

The definitions above give rise to the category **TopCat** of topological categories, whose objects are topological categories and has as morphisms the topological functors between topological categories. We denote this collection of morphisms by $K \copyright L := \textbf{TopCat}(K, L)$.

Mimicking the construction of a spatial digraph, we may consider two continuous functors, tail and head, respectively by $t, h : \nabla \copyright K \longrightarrow K$.

Now if $\nu : f \longrightarrow g$ is a natural transformation between morphisms (or, which is the same, topological functors) $f, g : \nabla \longrightarrow K$, then $t(\nu) = \nu(0) : f(0) \longrightarrow g(0)$ and $h(\nu) = \nu(1) : f(1) \longrightarrow g(1)$.

The resulting topological diagram of categories and continuous functors is called a **categorical digraph** $\overrightarrow{K}$ of K.

Thus if Γ is a digraph, then the set of morphism $\Gamma @_{\textbf{D}} \overrightarrow{K}$ is the set of digraph morphism in the underlying spatial digraph K. In other words, each morphism assigns an object of K for every vertex in Γ, and a continuous curve (a topological functor) $\nabla \longrightarrow K$ for every arrow of Γ. Then the digraph morphism $g : \Gamma \longrightarrow \overrightarrow{K}$ will be called **a gesture with skeleton in and body in K**.

Proposition 6 *Let $\Gamma @ \overrightarrow{K}$ be the set of gestures with skeleton in Γ and body K, with K a topological category. Then $\Gamma @ \overrightarrow{K}$ is a topological category.*

Proof Recalling that $\Gamma \cong \varinjlim_{a \in A_\Gamma} \Gamma_\alpha$, in particular we have $(\Gamma_a)_{a \in A_\Gamma} \cong (\uparrow_a)_{a \in A_\Gamma}$.

On the other hand, we know that $\uparrow @ \overrightarrow{K} \cong \nabla \copyright K \in \textbf{TopCat}$.
Thus

$$\Gamma @ \overrightarrow{K} \cong (\varinjlim_{a \in A_\Gamma} \Gamma_\alpha) @ \overrightarrow{K} \cong (\varinjlim_{a \in A_\Gamma} \uparrow_a) @ \overrightarrow{K} \cong \varprojlim_{a \in A_\Delta} (\uparrow_a @ \overrightarrow{K}),$$

since each $\uparrow_a @ \overrightarrow{K} \cong \nabla \copyright K$ is a topological category, then $\varprojlim_{a \in A_\Delta} (\uparrow_a @ \overrightarrow{K}) \cong$ $\Gamma @ \overrightarrow{K}$ is also a topological category, because of the properties of inverse limits.

$\blacklozenge$

Theorem (Mazzola's Escher theorem [4]) *Let Γ, Δ be digraphs and K a topological category, then we have a canonical isomorphism of topological categories.*

$$\Gamma \overrightarrow{@} \Delta \overrightarrow{@} K \cong \Delta \overrightarrow{@} \Gamma \overrightarrow{@} K.$$

Proof On one hand, this implies that the space of hypergesture $\Gamma \overset{\rightarrow}{@} \Delta \overset{\rightarrow}{@} K$ is the limit $(\underset{a \in A_\Gamma}{\underrightarrow{\lim}} \Gamma_\alpha) @ (\Delta \overset{\rightarrow}{@} K)$, but in particular we can say that $(\Gamma_a)_{a \in A_\Gamma} \cong (\uparrow_a)_{a \in A_\Gamma}$.

Then

$$\Gamma \overset{\rightarrow}{@} \Delta \overset{\rightarrow}{@} K \cong (\underset{a \in A_\Gamma}{\underrightarrow{\lim}} \uparrow_a) @ (\Delta \overset{\rightarrow}{@} K).$$

Even more, being $(\_ @ (\Delta \overset{\rightarrow}{@} K))$ a contravariant functor which converts direct limits on inverses limits, this must be isomorphic to:

$$(\underset{a \in A_\Gamma}{\underrightarrow{\lim}} \uparrow_a) @ (\Delta \overset{\rightarrow}{@} K) \cong \underset{a \in A_\Gamma}{\underleftarrow{\lim}} (\uparrow_a @ (\Delta \overset{\rightarrow}{@} K)).$$

Proceeding similarly we get:

$$\underset{a \in A_\Gamma}{\underleftarrow{\lim}} (\uparrow_a @ (\Delta \overset{\rightarrow}{@} K)) \cong \underset{a \in A_\Gamma}{\underleftarrow{\lim}} (\uparrow_a @ (\underset{b \in A_\Delta}{\underrightarrow{\lim}} \Delta_b @ K)) \cong \underset{a \in A_\Gamma}{\underleftarrow{\lim}} (\uparrow_a @ (\underset{b \in A_\Delta}{\underrightarrow{\lim}} \uparrow_b @ K)).$$

Because $(\_ @ K)$ is a contravariant functor and converts direct limits in inverse limits

$$\underset{a \in A_\Gamma}{\underleftarrow{\lim}} (\uparrow_a @ (\underset{b \in A_\Delta}{\underrightarrow{\lim}} \uparrow_b @ K)) \cong \underset{a \in A_\Gamma}{\underleftarrow{\lim}} (\uparrow_a @ (\underset{b \in A_\Delta}{\underleftarrow{\lim}} (\uparrow_b @ K)))$$

Thus, since $(\uparrow_a @ \_)$ is a covariant functor preserving inverse limits

$$\underset{a \in A_\Gamma}{\underleftarrow{\lim}} (\uparrow_a @ (\underset{b \in A_\Delta}{\underleftarrow{\lim}} (\uparrow_b @ K))) \cong \underset{a \in A_\Gamma}{\underleftarrow{\lim}} \underset{b \in A_\Delta}{\underleftarrow{\lim}} (\uparrow_a @ (\uparrow_b @ K))) \cong \underset{a \in A_\Gamma}{\underleftarrow{\lim}} \underset{b \in A_\Delta}{\underleftarrow{\lim}} (\uparrow_a @ \uparrow_b @ K)).$$

Then,

$$\underset{a \in A_\Gamma}{\underleftarrow{\lim}} \underset{b \in A_\Delta}{\underleftarrow{\lim}} (\uparrow_a @ (\uparrow_b @ K))) \cong \underset{a \in A_\Gamma}{\underleftarrow{\lim}} \underset{b \in A_\Delta}{\underleftarrow{\lim}} (\uparrow_a @ \uparrow_b @ K)).$$

By proposition 4

$$\underset{a \in A_\Gamma}{\underleftarrow{\lim}} \underset{b \in A_\Delta}{\underleftarrow{\lim}} (\uparrow_a @ \uparrow_b @ K)) \cong \underset{b \in A_\Delta}{\underleftarrow{\lim}} (\underset{a \in A_\Gamma}{\underleftarrow{\lim}} (\uparrow_a @ \uparrow_b @ K)),$$

and considering that $(\uparrow_a @ \uparrow_b) \cong (\uparrow_b @ \uparrow_a) \cong I^2$, then

$$\underset{a \in A_\Gamma}{\underleftarrow{\lim}} \underset{b \in A_\Delta}{\underleftarrow{\lim}} (\uparrow_a @ \uparrow_b @ K)) \cong \underset{b \in A_\Delta}{\underleftarrow{\lim}} (\underset{a \in A_\Gamma}{\underleftarrow{\lim}} (\uparrow_b @ \uparrow_a @ K)).$$

So

$$\underset{b \in A_\Delta}{\underleftarrow{\lim}} (\underset{a \in A_\Gamma}{\underleftarrow{\lim}} (\uparrow_b @ \uparrow_a @ K)) \cong \underset{b \in A_\Delta}{\underleftarrow{\lim}} (\underset{a \in A_\Gamma}{\underleftarrow{\lim}} (\uparrow_b @ (\uparrow_a @ K)).$$

Now, since $(\uparrow_b @\_)$ is a covariant functor preserving inverse limits,

$$\varprojlim_{b \in A_\Delta} (\varinjlim_{a \in A_\Gamma} (\uparrow_b @(\uparrow_a @K)) \cong \varprojlim_{b \in A_\Delta} (\uparrow_b @\varinjlim_{a \in A_\Gamma} (\uparrow_a @K)),$$

and $(\_@K)$ is a contravariant functor which turns direct limits into inverse limits

$$\varprojlim_{b \in A_\Delta} (\uparrow_b @\varinjlim_{a \in A_\Gamma} (\uparrow_a @K)) \cong \varprojlim_{b \in A_\Delta} (\uparrow_b @(\varprojlim_{a \in A_\Gamma} \uparrow_a @K)).$$

Finally, $(\_@(\Gamma \overrightarrow{@} K))$ being a contravariant functor converting direct limits on inverse limits, we have

$$\varprojlim_{b \in A_\Delta} (\uparrow_b @(\varprojlim_{a \in A_\Gamma} \uparrow_a @K)) \cong \varprojlim_{b \in A_\Delta} (\uparrow_b @(\varprojlim_{a \in A_\Gamma} \Gamma_a @K)) \cong \varprojlim_{b \in A_\Delta} (\uparrow_b @(\Gamma \overrightarrow{@} K))$$

$$\cong (\varinjlim_{b \in A_\Delta} \Delta_b @(\Gamma \overrightarrow{@} K)) \cong (\Delta \overrightarrow{@} (\Gamma \overrightarrow{@} K)) \cong \Delta \overrightarrow{@} \Gamma \overrightarrow{@} K.$$

Therefore $\Gamma \overrightarrow{@} \Delta \overrightarrow{@} K \cong \Delta \overrightarrow{@} \Gamma \overrightarrow{@} K$. ⬥

References

1. Bondy, J.A.: Graph Theory, 1st edn. Springer, New York (2008)
2. Chávez, Y.: Teorema de Escher para hipergestos en la Teoría Matemática de la Música. Undergraduate thesis. Universidad Nacional Autonoma de México (2014)
3. Mac Lane, S.: Categories for the Working Mathematician, 2nd edn. Springer, New York (1998)
4. Mazzola, G.: Categorical gestures, the diamond conjecture, Lewin's question, and the Hammerklavier Sonata. J. Math. Music **3**(1), 31–58 (2009)
5. Mazzola, G.: Musical Performance-A Comprehensive Approach: Theory, Analytical Tools, and Case Studies. Springer, Berlin (2011)
6. Mazzola, G., Andreata, M.: Diagrams, gestures and formulae in music. J. Math. Music **1**(1), 23–46 (2007)
7. Mazzola, G., et al.: Musical Creativity-Strategies and Tools in Composition and Improvisation. Springer, Berlin (2011)
8. Rotman, J.J.: An Introduction to Homological Algebra, 2nd edn. Springer Science & Business Media, New York (2009)
9. Schapira, P.: Categories and Homological Algebra. https://webusers.imj-prg.fr/~pierre.schapira/lectnotes/ (2011)

The Mechanics of Tipping Points: A Case of Extreme Elasticity in Expressive Timing

Elaine Chew

Abstract Tipping points are an observable and experienced natural phenomenon that has been invoked metaphorically across various domains external to physics. This article introduces the tipping point analogy for musical timing, and presents three case studies illustrating the concept. Quantitative data from recorded performances presented in score-time graphs support the illustrations. The examples show how musicians employ tipping points in performance, and demonstrate how tipping points play on the listener's expectations to elicit emotion. Tipping points form principal tools for the performer's choreography of expectation; the pervasiveness of tipping points in human experience make them an important strategy also for ensemble coordination.

1 Introduction

Musical timing forms the essence of expressive performance. Expressive timing serves to delineate structures and draw attention to musical features [11]. As in the case of stand-up comedy, the right timing can make the difference between a riveting performance and a lackluster one. As illustration of the importance of musical timing, a simple exercise can show that playing a piece with appropriately shaped timing, albeit with many wrong notes, is preferable to playing all the right notes with broken timing.

Research on expressive timing has centered on aspects of phrasing, which are primarily defined by a rise and fall in local tempo or dynamics. Repp [13] showed that these tempo phrase arcs can be described by quadratic functions; Repp [14] further demonstrated that transitions from one tempo to the next can be modeled by cubic functions. Kinematic approaches to modeling tempo showed that a physical body coming to a stop better approximated ritardandi [5]. Taking the locomotive

E. Chew (✉)
School of Electronic Engineering and Computer Science,
Queen Mary University of London, London, UK
e-mail: elaine.chew@qmul.ac.uk

© Springer International Publishing AG 2017
G. Pareyon et al. (eds.), *The Musical-Mathematical Mind*,
Computational Music Science, DOI 10.1007/978-3-319-47337-6_9

analogy a step further, Chew et al. [2] created a driving interface for the shaping of tempo trajectories.

While much work has focused on the ebb and flow of tempo that mark phrasing, little work addresses gestural forms of timing deviations, which can exhibit far more extreme degrees of elasticity. In 2010, Rajagopal observed that local tempo variations at the start of Gould's 1977 and Pogorelich's 1986 performances of Bach's Saraband (BWV 807) resembled a damped harmonic oscillator, thus suggesting that, beyond modeling beats and meter [8, 12], oscillators can also be used to describe tempo fluctuations.

This article introduces the tipping point analogy for musical timing. A musical tipping point is an extreme distortion of the tempo, a musical hyperbole, which extends well beyond the normal pulse and meter. It can be described as a timeless moment of suspended motion, beyond which a small perturbation will tip the balance and set in motion the return of the pulse. Tipping points vary in magnitude; the largest tipping points are relatively rare over the course of a piece, and form the defining moments of a performance.

The next sections will formally present a definition of tipping points, and three case studies illustrating the concept, followed by discussions on the principles of tipping points and how they work. They incorporate material first presented at the Performance Studies Network Conference 2 in Cambridge (UK) [4] and subsequently developed and presented at the International Congress on Music and Mathematics (ICMM) in Puerto Vallarta (Mexico).[1]

2   Tipping Points: A Definition

Tipping points are experienced and observed in the natural world in which we live. We learn and internalize knowledge about tipping points from a young age. Tipping points constitute not only an experienced pattern of behavior and control, but also a conceptualized one [3]. In physics, it is formally defined as the point beyond which the line through the center of gravity lies outside the base of the object. When the line through the center of gravity crosses this tipping point, the object tips over and falls, hence the name.

Socially, tipping points can refer to the line beyond which one's parents (or friends) get very cross. The term tipping point is first used in 1959 in reference to complex systems for which a tipping point is defined as "the critical point in a situation, process, or system beyond which a significant and often unstoppable effect or change takes place" [9]. In his popular book titled *Tipping Point*, Gladwell refers to the origins of the word in epidemiology, where the term refers to the moment when a virus reaches critical mass and an epidemic takes off, and its applications in criminology, and asks the question: What if everything has a tipping point? [6]

---

[1] The ICMM lecture can be viewed online at https://vimeo.com/112980119.

Tipping points also exist in music. Our focus will be on tipping points in musical timing, which harks back to the basic physics definition of the word.

Music lends itself readily to movement metaphors. With the exception of amorphous music, music generates a pulse that demarcates conceptual units of time. Suppose each conceptualized time unit is a distance, then the time taken to traverse that distance invokes a perception of speed. When less time is taken to travel from one pulse to the next, the music is perceived to be going fast; when more time is taken to traverse the span of a pulse, the music is perceived to be going slowly.

By manipulating the time between pulses, performers can invoke the sensation of acceleration and deceleration; sometimes, the composer notates these speedups and slowdowns in the score (*accelerando*, *ritardando*).

Suppose that experiencing a piece of music is a journey along a path. Then, the performers' moderating of acceleration and deceleration transforms the topology of the path taken: for example, the slight deceleration followed by an acceleration can simulate the perception of easing into a bend in the road then resuming the original speed; a deeper deceleration simulates the perception of traversing a sharper bend.

This link between music and motion is exploited in [2], where the motion metaphor is made concrete through a driving interface. The ESP interface of [2] considers only a two-dimensional path, which fails to capture expressive gestures that are more extreme and require momentum possible only through the addition of vertical motion. With vertical motion, for example when a ball is tossed into the air or when a train reaches the top of a roller coaster, there is a moment when motion stops, when the object is poised at the brink of change, before the (vertical) direction reverses, and motion resumes.

The tipping point analogy in music refers to these moments in time when the movement is perceived to come to a standstill, and is suspended until a (conceptual) tip initiates the return of the pulse. A tipping point can thus be defined as

> *a timeless moment of suspended stillness, of stasis, beyond which a small perturbation will tip the balance and set in motion the inevitable return of the pulse.*

The tipping point is best illustrated by example. The next section presents three case studies of tipping points in various contexts.

3 Three Case Studies

3.1 Case Study I: Puccini's O Mio Babbino Caro

Singers, especially sopranos and tenors, are well known for their ability to execute extravagant timing gestures, such as tipping points. Consider the excerpt of "O mio babbino caro" from *Gianni Schicchi* by Puccini as shown at the bottom of Fig. 1, with the text and translation (from [15]) as shown below:

Mi struggo e mi tormento!   I am anguished and tormented!
O Dio, vorrei morir!   Oh God, I'd like to die!
Babbo, pietà, pietà!   Papa, have pity, have pity!
Babbo, pietà, pietà!   Papa, have pity, have pity!

A video showing the progression of the eighth note lengths as Maria Callas' performance of a part of this excerpt unfolds can be viewed online at https://vimeo.com/
127507105.

Midway through the second line above, "O Dio," the composer writes in a long
note on "Dio" that the singer elongates even further (the first major tipping point
in Callas' performance) to heighten the poignancy of the plea, before the anguished
"vorrei morir." At the next line, "Babbo, pietà, pietà," when "pietà" is desperately
reiterated with an octave leap up, the soprano lingers on the high note, delaying the
expected registral return. There is a dramatic pause at the end of the line (another
major tipping point), before the final "Babbo, pietà, pietà."

Figure 1 shows the eighth note lengths of Maria Callas' performance for the entire
excerpt. The tipping points in Callas' performance are indicated by red cue balls, with
the size of the cue ball reflecting the magnitude of the tipping point. Superimposed on
Callas' performance are plots of lengths of the same eighth notes in performances by
Kathleen Battle and Kiri the Kanawa, showing differences in the narrative strategies
employed by the three performers.

In this case study, tipping points are used to prolong expectation (for example,
by delaying the expected registral return following an upward melodic leap), thus
creating drama, exaggerating emotion cues, and heightening poignancy.

3.2   Case Study II: Strauss' Burleske

A prototypical place for employing or finding tipping points is at the ends of caden-
zas. In the classical concerto, the cadenza, whether improvised or composed, is an
elaboration of the V chord in the V-I progression at an important turning point in the
piece. Figure 2 shows a two-piano arrangement (with the orchestra part in the second
piano) of the cadenza in Strauss' Burleske in D minor for Piano and Orchestra; Fig. 3
shows the tipping points. With D as the tonic, A is the dominant (V). Prominent
octave A's are generously sprinkled throughout the entire cadenza, with the intensity
of the chordal trills and sweeping arpeggios coming to a head at the first tipping point
(indicated by the small cue ball). The tension continues to build as the V has not yet
resolved to the expected I chord. Finally, at the last A, a lone voice in the right hand,
we arrive at the moment of reckoning, of the main tipping point, auguring inevitable
change and release after prolonged suspense.

A video at https://vimeo.com/70618400 shows the bar durations in Chew's performance of the cadenza of Strauss' Burleske synchronized with the audio. Figure 3
shows the bar lengths of Chew's performance annotated with the two tipping points:
a smaller one, and a larger one. Superimposed on Chew's' performance are plots

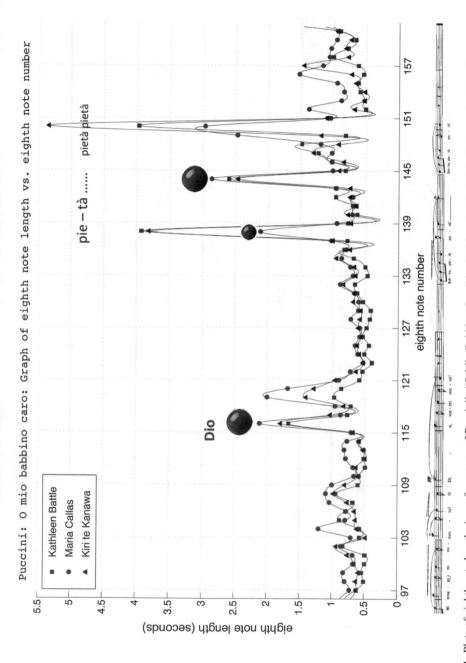

Fig. 1 Plots of eighth note lengths in performances of Puccini's "O Mio Babbino Caro" by Kathleen Battle, Maria Callas, and Kiri te Kanawa, with tipping points in Callas' performance highlighted. Vertical grid lines mark the start of each bar

84   E. Chew

Fig. 2 Score of the cadenza in Strauss' *Burleske* for Piano and Orchestra

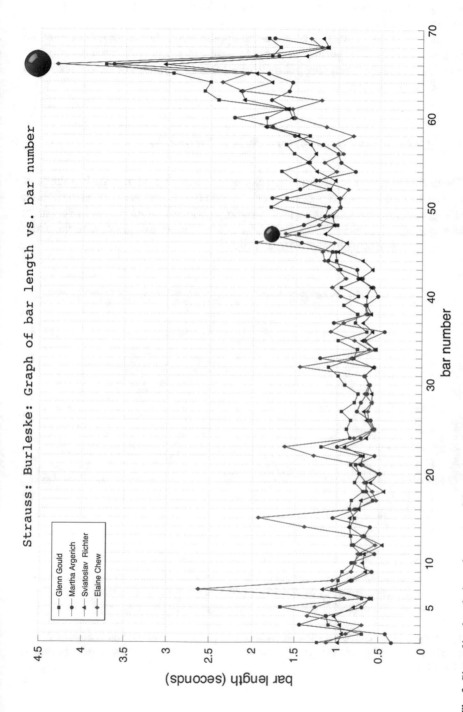

Fig. 3 Plots of bar lengths in performances of the candenza in Strauss' *Burleske* by Glenn Gould, Martha Argerich, Sviatoslav Richter and Elaine Chew, with tipping points in Chew's performance (which concur with Argerich's and Richter's) highlighted. Vertical grid lines mark the start of each bar

of lengths of the same bars in performances by Martha Argerich, Glenn Could, and Sviatoslav Richter.

Here, the main tipping point, a cadential tipping point, signals the release after the prolonged suspense, auguring inevitable change and augmenting the tonal expectations. Additionally, the tipping point also coordinates the return of the orchestra and that of the lyrical theme.

3.3   Case Study III: Kreisler's Schon Rosmarin

The final case study shows an unusual example of a tipping point at the beginning of a piece. Consider the excerpt of Kreisler's "Schön Rosmarin" as shown at the bottom of Fig. 4. Above the score is a graph showing the length of each beat in Kreisler's performance of the excerpt. An animation of a version of this graph with the audio can be viewed at https://vimeo.com/127499857.

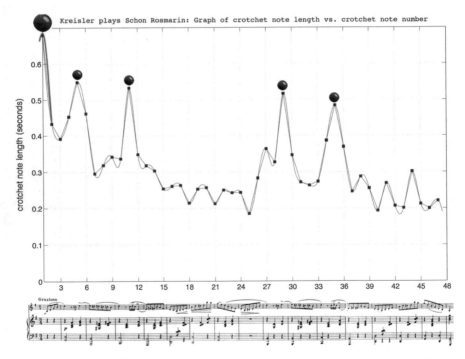

Fig. 4 Plot of beat (crotchet note) lengths in Kreisler's performance of his "Schön Rosmarin", with tipping points highlighted. Vertical grid lines mark the start of each bar (figure reproduced in p. 350 of [1])

As can be seen (and heard), Kreisler begins the piece with a tipping point, before cascading down to the nominal beat length (or tempo), embellishing two melodic target note pairs with small tipping points along the way, in a playful simulation of gravity-defying moves.

4 Discussion and Conclusions

In this paper, I have presented a definition of tipping points in music, in particular, in musical timing. Three case studies illustrated the concept, demonstrating how tipping points can generate expectation (and thus elicit emotion), facilitate and augment cathartic release and coordinate returns, and playfully simulate gravity-defying moves.

In each case, the extreme elasticity of the tipping points are possible because there is full knowledge and maximum expectation of what is to come. At these points of maximum certainty, information is minimized and entropy is low. Because expectation is peaked, and the outcome is fixed, the performer can play freely with time to further pique the listener's expectations.

Musical expectations can be schematic (based on observed patterns) or veridical (based on knowledge of a specific piece, e.g. "Happy Birthday") [7]. Schematic expectations include those pertaining to tonality—such as the tendency to return to a stable state as V needing to resolve to I in a perfect cadence—and melody—such as the ascending leap-descending step and the post-skip reversal tendencies.

Tipping points introduce a piquant element of uncertainty in situations possessing absolute certainty. They delay highly expected outcomes in ways that mix predictability with indeterminacy—for example, the listener does not know when the (time) suspension will tip. Thus, tipping points heighten expectation, increase tension, and elicit emotion.

As noted in [7], Meyer argued in [10] that the emotion content of music arises primarily through the composer's choreography of expectation—by setting up, delaying, thwarting, or delivering on expectations. As shown by the tipping point case studies, the emotion content of music arises through not only the composer's, but also the performer's, choreography of expectation.

References

1. Chew, E.: Playing with the edge: tipping points and the role of tonality. In: McAdams, S., Temperley, D., Rozin, A. (eds.) Milestones in music cognition special issue. Music Perception, vol 33 (no 3), pp. 344–366 (2016)
2. Chew, E., Francois, A.R.J., Liu, J., Yang, L.: ESP: a driving interface for expression synthesis. In: Proceedings of the International Conference on New Instruments for Musical Expression (NIME), pp. 224–227 (2005)

3. DiSessa, A.A.: Systemics of learning for a revised pedagogical agenda. In: Lesh, R. (ed.) Foundations for the Future in Mathematics Education. Lawrence Erlbaum Associates, Mahwah (2007)
4. Chew, E.: The tipping point analogy for musical timing. In: 2nd International Conference of the Performance Studies Network (PSN2), Cambridge, UK (2013)
5. Friberg, A., Sandberg, J.: Does music performance allude to locomotion? a model of final ritardandi derived from measurements of stopping runners. J. Acoust. Soc. Am. **105**(3), 1469–1484 (1999)
6. Gladwell, M.: The Tipping Point: How Little Things Can Make A Big Difference. Time Warner Trade Publishing, New York (2000)
7. Huron, D.: Sweet Anticipation: Music and the Psychology of Expectation. The MIT Press, Cambridge (2006)
8. Large, E.: Resonating to musical rhythm: theory and experiment. In: Grondin, S. (ed.) The Psychology of Time. Emerald, West Yorkshire (2008)
9. Merriam-Webster, Tipping Point. http://www.merriam-webster.com/dictionary/tippingpoint. Accessed 15 Feb 2010
10. Meyer, L.B.: Emotion and Meaning in Music. Chicago University Press, Chicago (1956)
11. Palmer, C., Hutchins, S.: What is Musical Prosody?. The Psychology of Learning and Motivation, vol. 46, pp. 245–278. Elsevier, Amsterdam (2006)
12. Pardo, B.: Tempo tracking with a single oscillator. In: Proceedings of the International Conference on Music Information Retrieval (ISMIR), Universitat Pompeu Fabra, Barcelona (2004)
13. Repp, B.: Diversity and commonality in music performance: an analysis of timing microstructure in Schumann's *Traumerei*, Haskins Laboratories Status Report on Speech Research (SR-111/112), pp. 227–260 (1992)
14. Repp, B.: Expressive timing in a Debussy Prelude: a comparison of student and expert pianists. Musicae Scientiae **1**(2), 257–268 (1997)
15. Wikipedia, O mio babbino caro. http://en.wikipedia.org/wiki/O_mio_babbino_caro. Accessed 15 Feb 2010

Lexicographic Orderings of Modes and Morphisms

David Clampitt

Abstract The context of this paper is the theory of modes of non-degenerate well-formed scales (generalized diatonic or pentatonic scales), within the framework of algebraic combinatorics of words, specifically musical modes encoded as members of the monoid of words in A^* over a two-letter alphabet A, and the monoid of Sturmian morphisms that act on A^*. The paper relates lexicographic orderings of words modes of (non-degenerate) well-formed scales (especially the canonical examples, the diatonic modes) and lexicographic orderings of the special Sturmian morphisms associated with the modes, to the musical scale and circle-of-fifths orderings. These lexicographic orderings are related to Zarlino's 1571 re-ordering of Glarean's 1547 listing of six authentic diatonic modes.

1 Scale Theory Concepts and Notations

Scale theory derives its mathematical character from the fact that musical scales are generally periodic phenomena. Most often the period is the musical octave (associated with the frequency ratio 2:1). One may then identify a given scale with a set of fundamental frequencies f_k, $1 = f_0 < f_1 < \ldots < f_{N-1} < 2$, N a positive integer, and, taking base-2 logarithms, with elements $0 = s_0 < s_1 < \ldots < s_{N-1} < 1$, where $s_k = log_2(f_k)$. We call the elements s_k the *scale steps*, and the differences $(s_j - s_i) \bmod 1$ the *specific intervals*, or the *specific interval sizes*. The specific intervals $(s_{i+1} - s_i) \bmod 1$ for $0 \le i < N$, $i + 1$ reduced modulo N in the case $i = N - 1$, are defined to be the *(specific) step intervals*. The differences mod N between index numbers define the *generic intervals* of the scale. See [1–3] for background. It is shown in [2] that scales in which each non-zero generic interval corresponds to exactly two specific interval sizes are equivalent to non-degenerate well-formed scales. In the present paper we will take that property to be the definition. In particular, the step intervals come in two specific sizes, a and b, with

D. Clampitt (✉)
The Ohio State University, Columbus, OH 43210, USA
e-mail: clampitt.4@osu.edu

© Springer International Publishing AG 2017
G. Pareyon et al. (eds.), *The Musical-Mathematical Mind*,
Computational Music Science, DOI 10.1007/978-3-319-47337-6_10

multiplicities q and p, respectively, coprime with N. The class of well-formed scales may be defined as those admitting a generating specific interval (e.g., perfect fifth) where each specific interval has a constant generic description.

2 From Scales to Modes: Word Theory

Modes of well-formed scales may be represented as strings or words over a two-letter alphabet. In word theory, one considers a finite alphabet A (but here, $A = \{a, b\}$) and the set of words over A, $A^* = \{w = w_1 \ldots w_n | w_i \in A, n \in \mathbb{N}\}$. A^* is a free monoid (semi-group with identity) where the monoid operation is concatenation of words, and one understands that the empty word ε is in A^*, and for all words w in A^*, $\varepsilon w = w = w\varepsilon$. A^* may be extended to become a group if inverses a^{-1}, b^{-1} are adjoined to alphabet A. If $w = uv$ for words in A^*, we say that u and v are *factors* of w. One uses the terms *prefix* and *suffix* as in ordinary usage: in our example, u is a prefix of w. One says that words $w = uv$ and $w' = vu$ are *conjugate* or *conjugates* of each other (an equivalence relation). See Lothaire [4] for an exposition of word theory.

An important set of endomorphisms of the monoid A^* are the following, which map A^* to itself by replacing each letter of $w \in A^*$ by:

$$G(a) = a, G(b) = ab; \tilde{G}(a) = a, \tilde{G}(b) = ba \tag{1}$$

$$D(a) = ba, D(b) = b; \tilde{D}(a) = ab, \tilde{D}(b) = b \tag{2}$$

The set of all compositions of these morphisms forms the monoid St_0 of *special Sturmian* morphisms, under composition of mappings. If $F \in St_0$, F is a morphism by construction: for any words $u, v \in A^*$, $F(uv) = F(u)F(v)$, since in particular for any word $w = w_1 \ldots w_n$ for letters w_i in A, $F(w) = F(w_1)F(w_2)\ldots F(w_n)$ by definition.

The musical application of these morphisms is to produce the authentic diatonic modes (those divided into a perfect fifth and perfect fourth; we omit consideration of the plagal modes here). As discussed in [5], the authentic diatonic modes are produced by applying to the divided root word $a|b$ the morphisms $GGD, \tilde{G}GD, \tilde{G}\tilde{G}D, GG\tilde{D}, \tilde{G}G\tilde{D}, \tilde{G}\tilde{G}\tilde{D}$. For example, $GGD(a|b) = GG(ba|b) = G(aba|ab) = aaba|aab$. The mode rejected by Glarean, the Locrian variety with final B (*hyperaeolius reiectus*) is not reachable via morphisms in St_0; such words are referred to by the word theorists as the "bad conjugates."

The principal tool that mathematical music theory uses to investigate modal varieties of well-formed scales is a refinement of that mathematical duality, obtained by considering the ways each mode gives rise to a pattern of rising perfect fifths and falling perfect fourths (the plain adjoint folding pattern) or a pattern of rising perfect fourths and falling perfect fifths (the twisted adjoint folding pattern). These are defined and discussed in [5], where the plain adjoint is preferred (for music-theoretical reasons) and discussed extensively. Here we will follow Noll and Montiel [6], and consider the twisted adjoint, which from a purely mathematical point of view has many advantages.

The twisted adjoint is defined musically in the following way: consider the pattern of rising perfect fourths and falling perfect fifths, such that all notes lie within the modal octave, i.e., including the modal final and notes of the mode within the octave above it, encoding the pattern of rising and falling intervals as a word over a two-letter alphabet. The twisted adjoint folding pattern for Dorian, assigning x to descending fifth and y to ascending fourth, is the word $xyxyyxy$, as the reader should check. A unique folding pattern is determined for each of the diatonic modes. At this level, no distinction is made between authentic and plagal modes, and a folding pattern may be determined for Locrian (or the bad conjugate generally).

What justifies the use of the term adjoint is that we may lift this duality to the level of the morphisms in St_0, such that for *morphic* words in A^* (those derivable from morphisms in St_0 applied to ab), the twisted adjoint morphism, applied to the root word xy (with definitions analogous to those above), produces the word encoding the corresponding twisted adjoint folding pattern. The mapping of St_0 to itself that accomplishes this, Sturmian involution, replaces each D by $\tilde{D}$ and each $\tilde{D}$ by D, fixes each G and $\tilde{G}$, and reverses the order. The morphism that produces the word $abaa|aba$ corresponding to authentic Dorian mode is $\tilde{G}GD$. Under the twisted adjoint the morphism is thus $\tilde{D}G\tilde{G}$, and $\tilde{D}G\tilde{G}(x|y) = \tilde{D}G(x|yx) = \tilde{D}(x|xyx) = xy|xyyxy$, which corresponds to the word encoding the twisted adjoint folding for Dorian determined by the musical procedure given above. This holds generally for all morphic words. For the scale represented by the bad conjugate, the twisted adjoint folding pattern is also the bad conjugate of its conjugacy class. The folding pattern for Locrian is $yyxyxyx$, which is not the image of a morphism in St_0.

3 Lexicographic Orderings

We are now in a position to take up the musical interpretation of lexicographic orderings of words associated with modes, their twisted adjoint foldings, in relation to lexicographic orderings of the respective morphisms. In the binary alphabets $\{a, b\}$ and $\{x, y\}$ we choose the ordinary alphabetic orderings, $a \prec b$ and $x \prec y$. Because for a given discussion we will fix a pair of Christoffel word conjugacy classes, a conjugacy class of words over $\{a, b\}$ and one over $\{x, y\}$, in the associated morphisms from St_0 we will always have the same patterns of D's and G's, up to location of tildes. We define elements without tildes to lexicographically precede elements with tildes. Since compositions of morphisms act from right to left, as words morphisms will be read right to left, whereas words representing modes or foldings will be read left to right. G and $\tilde{G}$ commute with each other: both $\tilde{G}G$ and $G\tilde{G}$ leave a fixed and $\tilde{G}G(b) = \tilde{G}(ab) = aba$ and $G\tilde{G}(b) = G(ba) = aba$. It follows immediately that they commute over all of A^*. Similarly, D and $\tilde{D}$ commute with each other. Thus, while lexicographically $\tilde{G}G$ precedes $G\tilde{G}$, for example, they are in fact the same mapping.

Here are the two conjugacy classes of words corresponding to the authentic diatonic modes and their twisted adjoint foldings, and the morphisms that produce them as images of the respective two-letter root words, followed by the bad conjugates:

$GGD(a\|b) = aaba\|aab$	$\tilde{D}GG(x\|y) = xy\|xyxyy$	C Ionian
$\tilde{G}GD(a\|b) = abaa\|aba$	$\tilde{D}G\tilde{G}(x\|y) = xy\|xyyxy$	D Dorian
$\tilde{G}\tilde{G}D(a\|b) = baaa\|baa$	$\tilde{D}\tilde{G}\tilde{G}(x\|y) = xy\|yxyxy$	E Phrygian
$GG\tilde{D}(a\|b) = aaab\|aab$	$DGG(x\|y) = yx\|yxyxy$	F Lydian
$\tilde{G}G\tilde{D}(a\|b) = aaba\|aba$	$DG\tilde{G}(x\|y) = yx\|yxyyx$	G Mixolydian
$\tilde{G}\tilde{G}\tilde{D}(a\|b) = abaa\|baa$	$D\tilde{G}\tilde{G}(x\|y) = yx\|yyxyx$	A Aeolian
$baabaaa$	$yyxyxyx$	B Locrian

The first correlation to observe, in the left-hand list, is that the *morphisms* are in lexicographic order, producing words corresponding to authentic modes in step order, beginning with C. The finals corresponding to the morphic words cover Guido's natural hexachord, in ascending step-wise order: C, D, E, F, G, A. The amorphic Locrian on B follows. Turning to the right-hand list, we see that the twisted adjoint folding *words* are in lexicographic order, including the bad conjugate Locrian folding, last in lexicographic order.

Now consider the reverse situation, where the morphisms for the foldings are in lexicographic order, in the left-hand list, with the corresponding twisted adjoint morphisms and resulting words for authentic modes in the right-hand list:

$DGG(x\|y) = yx\|yxyxy$	$GG\tilde{D}(a\|b) = aaab\|aab$	F Lydian
$\tilde{D}GG(x\|y) = xy\|xyxyy$	$GGD(a\|b) = aaba\|aab$	C Ionian
$D\tilde{G}G(x\|y) = yx\|yxyyx$	$\tilde{G}G\tilde{D}(a\|b) = aaba\|aba$	G Mixolydian
$\tilde{D}\tilde{G}G(x\|y) = xy\|xyyxy$	$\tilde{G}GD(a\|b) = abaa\|aba$	D Dorian
$D\tilde{G}\tilde{G}(x\|y) = yx\|yyxyx$	$\tilde{G}\tilde{G}\tilde{D}(a\|b) = abaa\|baa$	A Aeolian
$\tilde{D}\tilde{G}\tilde{G}(x\|y) = xy\|yxyxy$	$\tilde{G}\tilde{G}D(a\|b) = baaa\|baa$	E Phrygian
$yyxyxyx$	$baabaaa$	B Locrian

Arranging the morphisms for the foldings in lexicographic order produces modes in (forward) circle-of-fifths order. Now it is the words for the authentic modes, on the right-hand list, that are in lexicographic order, including the bad conjugate Locrian step-interval pattern, last in lexicographic order.

To say that the words corresponding to the authentic modes ascend in step order as the morphisms increase in lexicographic order is to say that successive words are cyclically permuted by one letter to the right: to go from C Ionian to D Dorian is to conjugate (in the group context) the Ionian step-interval pattern by the step interval a: $(a^{-1})(aabaaab)(a) = abaaaba$. We naturally imagine the location of the final as being at the beginning of the word. Similarly, to say that the words corresponding to the authentic modes ascend in circle-of-fifths order as they increase in lexicographic order is to say that successive words are cyclically permuted by 4, i.e., conjugated by their successive divider prefixes: e.g., $aaab\|aab \rightarrow (aaab)^{-1}(aaabaab)(aaab) = aabaaab$ (F Lydian to C Ionian).

There is nothing special about the alphabet $\{a, b\}$, of course, nor about the particular order of the morphisms, so the folding patterns similarly are cyclically permuted by one letter as one adds tildes, i.e., as the morphisms increase in lexicographic order. What changes is the interpretation: x and y in the folding track the circle-of-fifths backwards from B to B-flat (downwards perfect fifths x and upwards perfect fourths y). To locate the final for each mode we find the the unique location in the given folding pattern of the pair of y's, and see that it moves leftwards from the penultimate letter in the F Lydian folding, until B Locrian is reached:

$$yxyxyxy \rightarrow xyxyx(yy) \rightarrow yxyx(yy)x \rightarrow xyx(yy)xy \rightarrow yx(yy)xyx \rightarrow (yy)xyxyx.$$

Thus, the folding runs backwards through the circle of fifths as we move left to right, while increasing lexicographic order of the morphisms effects moves in the forward circle-of-fifths direction, F $\rightarrow$ C, etc. (This is part of the motivation for *twisted* in the adjoint designation.)

Returning to the first pair of lists, generated by arranging the morphisms for the step-interval patterns in increasing lexicographic order, how are the words for the corresponding twisted adjoint folding patterns cyclically permuted? We saw above that as the morphisms increased in lexicographic order, the corresponding modes ascended by step-intervals, and the words for the folding patterns increased in lexicographic order. The duality between scale order and circle-of-fifths (generating interval) order leads us to guess that the cyclic permutation for the folding patterns should be by 2, length of the divider prefix in the folding pattern words, multiplicity of b in the step-interval patterns, multiplicative inverse of 4 mod 7, where 4 is the multiplicity of y in the folding patterns (and generic length of the generating perfect fifth interval). Indeed, that is the case. As they increase lexicographically they are cyclically permuted by 2, i.e., conjugated successively by their successive divider prefixes: e.g., $xy|xyxyy \rightarrow (xy)^{-1}(xyxyxyy)(xy) = xyxyyxy$ (C Ionian to D Dorian).

The initial words in each list exemplify two important categories in word theory: *Christoffel words* and *standard words*. Christoffel words are images of the two-letter root word under compositions of $\tilde{D}$ and of both varieties of G's (we may call these Christoffel morphisms). Standard words are images of the two-letter root word under compositions of D's and G's (standard morphisms). But our example suggests alternative characterizations in terms of lexicographic order: Christoffel words are least of their conjugacy class in terms of lexicographic order, and standard words are generated by the morphism which is lexicographically least in the class of morphisms that determine the morphic Christoffel conjugates. This characterization of Christoffel words is well known; the characterization of standard words is obvious when one notes that the morphism that determines the standard word is the only one of its class that has no letters with tildes. Note that the Christoffel and standard morphisms map to each other under the twisted adjoint (Sturmian involution).

To summarize the result suggested by the foregoing discussion: increasing lexicographic order of morphisms (starting from the standard morphism) yields successive conjugations by a single-letter prefix, starting from the standard word in the class; the corresponding twisted adjoint morphisms (starting from the Christoffel morphism) yield words in lexicographic order, successive conjugations by their divider prefixes, starting from the Christoffel word in its class. In both cases, the amorphic bad conjugates come last. This covers both directions discussed separately above, because the twisted adjoint on St_0 is an involution. One would like to confirm this for modes of general well-formed scales, i.e., general conjugacy classes of Christoffel words and their associated morphisms.

We appeal to the properties of the Burrows-Wheeler Transform (BWT). This is a general scheme for lossless data compression of words over arbitrary alphabets, but for Christoffel words it has special properties. Following the version of the BWT in [7], BWT(w) results from arranging the conjugates of w in lexicographic order as rows of a matrix and reading the last column as output. In the case of a Christoffel word w or any of its conjugates, by the result in [7] we have $BWT(w) = b^p a^q$. The matrix shows the BWT in the diatonic case:

$$\begin{pmatrix} a\ a\ a\ b\ a\ a\ b \\ a\ a\ b\ a\ a\ a\ b \\ a\ a\ b\ a\ a\ b\ a \\ a\ b\ a\ a\ a\ b\ a \\ a\ b\ a\ a\ b\ a\ a \\ b\ a\ a\ a\ b\ a\ a \\ b\ a\ a\ b\ a\ a\ a \end{pmatrix}$$

The BWT is invertible, so one may also understand it to generate the conjugacy class of Christoffel words of length $N = p + q$ from coprime p and q, respectively the numbers of occurrences of b and a. That is, beginning with an initial column of a cluster of q a's followed by a cluster of p b's, and generating columns of an $N \times N$ matrix by successively rotating upwards by p, one generates rows that are the conjugates in lexicographic order, with the top row the Christoffel word, least in lexicographic order.

One may also determine from the BWT matrix that the rows are rotations by the length of the divider prefix, p^{-1} mod N. Recall from [2] the result that, in the musical interpretation, the generic length of a generating interval is multiplicative inverse of the multiplicity of a step interval. In our example, $p = 2$, and the length of the generating interval and of the divider prefix is 4. In the BWT matrix, the entry in the Nth row that is an isolated b in its column is clearly in column p^{-1} (understanding the matrix as a kind of abacus for performing multiplication by p mod N, the singleton entry marks the jth column such that $pj \equiv 1$ mod N).

We can see that the last row of the BWT matrix represents the bad conjugate. Its prefix of length $j = p^{-1}$ mod N has one more occurrence of b than any of the other divider prefixes, representing the unique specific interval of length j. All the other intervals of length j, corresponding to the divider prefixes, are of the same specific size, i.e., they are the $N - 1$ intervals that generate the scale.

At the level of the words, we have the result that increasing lexicographic order coincides with conjugation by divider prefix, starting from the Christoffel word of the class and concluding with the bad conjugate (circle of fifths order for the words corresponding to the scales, scale order for those corresponding to the foldings). At the level of the morphisms, however, the desired result, that increasing lexicographic order (adding tildes) yields, in the image, conjugation by a single letter, must remain conjectural as of this writing. Here we will see partial results, including those that would prove the case of the usual diatonic scale words, beyond the calculations we have already made.

The very definition of lexicographic order for morphisms needs to be extended because of the commutativities involved: not only that G and $\tilde{G}$ (and their powers) and D and $\tilde{D}$ (and their powers) commute, but that $\tilde{G}\tilde{D}^k G = GD^k\tilde{G}$ and $\tilde{D}\tilde{G}^k D = DG^k\tilde{D}$ (see Lothaire [4]).

Definition 1 Morphisms that are lexicographically (one might say, orthographically) distinct are *equivalent* if they are functionally equal.

It is obvious that equivalence of morphisms under functional equality is an equivalence relation. We define an *inherited lexicographic ordering* of equivalence classes of morphisms, and state a proposition that confirms that this definition makes sense.

Definition 2 Given a standard morphism F of length n and the set of 2^n lexicographically distinct morphisms where the constituents G or D of F are or are not replaced by $\tilde{G}$ or $\tilde{D}$. From each equivalence class, choose the lexicographically lowest and highest elements, L and H. Define the *inherited L-ordering* of the classes to be that induced by the underlying lexicographic ordering, and similarly the *inherited H-ordering*.

The following proposition, that the inherited orderings of the classes coincide, is non-trivial, but is given without proof. The fundamental fact behind it is that in a given equivalence class C_1 of morphisms element L has a left-most constituent $\tilde{G}$ or $\tilde{D}$, and element H has a right-most constituent morphism $\tilde{G}$ or $\tilde{D}$. A contradiction follows if we assume class C_2 to both precede C_1 in inherited L-ordering and succeed C_1 in inherited H-ordering.

Proposition 11 *Given a set of 2^n morphisms as described in Definition 2 above, and the associated equivalence classes of morphisms, the inherited L-ordering and inherited H-ordering coincide. We may therefore define the* inherited lexicographic ordering *of the equivalence classes of morphisms.*

It is evident from the definitions that the point of departure is the standard word of the class, and the next morphism in lexicographic order is obtained by adding a tilde to the left-most element of the composition. Our first result will be that given a special Sturmian morphism F, adding a tilde to the left-most element in the composition, and applying it to ab, conjugates $F(ab)$ by a single letter. We will need to know how to compute in the free group. $D(a) = ba$, so $D(a^{-1}) = a^{-1}b^{-1}$: $\varepsilon = D(\varepsilon) = D(aa^{-1}) = D(a)D(a^{-1}) = baD(a^{-1})$, so $D(a^{-1}) = D(a)^{-1} = a^{-1}b^{-1}$. Trivially, $D(b^{-1}) = b^{-1}$, since $D(b) = b$. By similar computations we have $G(a^{-1}) = a^{-1}$ and $G(b^{-1}) = b^{-1}a^{-1}$; $\tilde{D}(a^{-1}) = b^{-1}a^{-1}$ and $\tilde{D}(b^{-1}) = b^{-1}$; $\tilde{G}(a^{-1}) = a^{-1}$ and $\tilde{G}(b^{-1}) = a^{-1}b^{-1}$.

Lemma 1 Let $x = a, b, a^{-1}$ or b^{-1}. Then $\tilde{G}(x) = a^{-1}G(x)a$; $\tilde{D}(x) = b^{-1}D(x)b$.

Proof $\tilde{G}(b) = ba = (a^{-1}a)ba = a^{-1}(ab)a = a^{-1}G(b)a$; $\tilde{G}(a) = a = (a^{-1}a)a = a^{-1}(a)a = a^{-1}G(a)$. $\tilde{G}(b^{-1}) = a^{-1}b^{-1} = a^{-1}b^{-1}(a^{-1}a) = a^{-1}(b^{-1}a^{-1})a = a^{-1}G(b^{-1})a$; $\tilde{G}(a^{-1}) = a^{-1} = a^{-1}(a^{-1}a) = a^{-1}G(a^{-1})a$. Similarly for $\tilde{D}(x)$.   ◆

Proposition 12 Let $X \in St_0$, and let $X(ab) = w_1 \ldots w_n$, $w_i \in \{a, b\}$. Then $\tilde{G}X(ab) = a^{-1}GX(ab)a$; $\tilde{D}(ab) = b^{-1}DX(ab)b$.

Proof Applying the lemma, we have: $\tilde{G}X(ab) = \tilde{G}(w_1 \ldots w_n) = \tilde{G}(w_1) \ldots \tilde{G}(w_n) = (a^{-1}G(w_1)a)(a^{-1}G(w_2)a) \ldots (a^{-1}G(w_n)a) = a^{-1}(G(w_1) \ldots G(w_n))a = a^{-1}G(w_1 \ldots w_n)a = a^{-1}GX(ab)a$. Similarly for the second case.   ◆

Proposition 13 *(1) If* $Y \in \langle G, D, \tilde{D} \rangle$, *then* $Y\tilde{D}X(ab) = b^{-1}Y'DX(ab)b$, *where* Y' *is* Y *with every* G *replaced by* $\tilde{G}$. *(2) If* $Y \in \langle D, G, \tilde{G} \rangle$, *then* $Y\tilde{G}X(ab) = a^{-1}Y'GX(ab)a$, *where* Y' *is* Y *with every* D *replaced by* $\tilde{D}$.

Proof By induction on the length of Y, applying Proposition 2. Part (1), length of $Y = 1$, show: (i) $G\tilde{D}X(ab) = b^{-1}\tilde{G}DX(ab)b$; (ii) $D\tilde{D}X(ab) = b^{-1}\tilde{D}DX(ab)b$; (iii) $\tilde{D}\tilde{D}X(ab) = b^{-1}\tilde{D}DX(ab)b$.
From $G\tilde{D}X(ab) = G(b^{-1}DX(ab)b) = G(b^{-1})GDX(ab)G(b) = b^{-1}a^{-1}GDX(ab)ab = b^{-1}(a^{-1}GDX(ab)a)b = b^{-1}\tilde{G}DX(ab)b$, i holds. From commutativity of D and $\tilde{D}$ and Proposition 2, ii and iii hold.
Assume the proposition in part (1) for Y of length k. Then for Z of length $k + 1$ we have $Z = GY$ or DY or $\tilde{D}Y$. In the first case, $Z\tilde{D}X(ab) = GY\tilde{D}X(ab) = G(b^{-1}Y'DX(ab)b) = G(b^{-1})GY'DX(ab)G(b) = b^{-1}a^{-1}GY'DX(ab)ab = b^{-1}(a^{-1}GY'DX(ab)a)b = b^{-1}\tilde{G}Y'DX(ab)b = b^{-1}Y''DX(ab)b$, where Y'' is Z with required replacements. Similarly for the other two cases, and for part (2).   ◆

Zarlino's reordering of the modes, when he deposed Dorian from its long-held position as mode 1 and replaced it with Ionian, has of course been vindicated by history, as Ionian leads to our paradigmatic major mode. Noll and Montiel [6] have already argued that Zarlino's decision had mathematical logic behind it. The relations exposed here reinforce that judgment.

References

1. Clough, J., Myerson, G.: Variety and multiplicity in diatonic systems. J. Music Theory **29**, 249–270 (1985)
2. Carey, N., Clampitt, D.: Self-similar pitch structures, their duals, and rhythmic analogues. Perspect. New Music **34**(2), 62–87 (1996)
3. Carey, N., Clampitt, D.: Aspects of well-formed scales. Music Theory Spect. **11**, 187–206 (1989)
4. Lothaire, M.: Combinatorics on Words. Cambridge Mathematical Library. Cambridge University Press, Cambridge (1997)
5. Clampitt, D., Noll, T.: Modes, the height-width duality, and Handschin's tone character. Music Theory Online **17**(1), 1–149 (2011)

6. Noll, T., Montiel, M.: Mathematics and computation in music. In: Yust, J., Wild, J., Burgoyne, J.A. (eds.) Glarean's Dodecachordon Revisited. Lecture Notes in Computer Science, vol. 7937. Springer, Berlin (2013)
7. Mantaci, S., Restivo, A., Sciortino, M.: Burrows-Wheeler transform and Sturmian words. Inf. Process. Lett. **86**, 241–246 (2003)

Music of Quantum Circles

Micho Đurđevich

Abstract We illustrate the basic ideas and principles of quantum geometry, by considering mutually complementary quantum realizations of circles. It is fascinating that such a simple geometrical object as circle, provides a rich illustrative playground for an entire array of purely quantum phenomena. On the other hand, the ancient Pythagorean musical scale, naturally leads to a simple quantum circle. We explore different musical scales, their mathematical generalizations and formalizations, and their possible quantum-geometric foundations. In this conceptual framework, we outline a diagramatical-categorical formulation for a quantum theory of symmetry, and further explore interesting musical and geometrical interconnections.

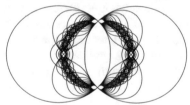

1   Introduction

Quantum geometry maps the ideas and concepts of quantum physics, into the realm of geometrical spaces and their transformations. Quantum spaces, the analogs of atoms, molecules and quantum systems of physics in general, exhibit a nature essentially different from their classical siblings. They are not understandable in terms of points, parts, or local neighbourhoods. In general, these concepts do not apply at all to quantum spaces. The entire fabric of space is considered as the one and indivisible whole.

And there is something profoundly quantum in all of music. A discrete space– the skeleton hosting the partiture, morphs into a true music form, only after being

M. Đurđevich (✉)
Institute of Mathematics, UNAM, Mexico, Mexico
e-mail: micho@matem.unam.mx

© Springer International Publishing AG 2017
G. Pareyon et al. (eds.), *The Musical-Mathematical Mind*,
Computational Music Science, DOI 10.1007/978-3-319-47337-6_11

symbiotically enveloped by a *geometry of sound*. Such a geometry is inherently quantum, as it connects the points of its discrete underlying structure, invalidating the difference between now, then, here and there, thus creating an irreducible continuum for a music piece. Continuous discreteness and discrete continuity. A manifestation of quantum duality and complementarity.

All this inherently promotes *simplicity* in thinking, as we are forced to look for some deeper structure, going far beyond the parts, points, local neighbourhoods... and fragmented classical geometrical views in general. One way of thinking, resonant and transcending the nature of mathematical realms, is the *Harmony Way*: to look at symmetries, the transformational modes of things, and understanding the mathematical creatures in terms of them. Conceptual roots for this are found in the Erlangen Program by Felix Klein, evolving into a whole paradigm, with language and theory of categories.

Circles are children of Simplicity. A principal geometrical realization of an infinite symmetry group. The idea of circle is observed in *repetitions*. Any continuous change, movement, transformation, in which there is something invariant before and after, naturally leads to the idea of circle. In music, such is the concept of *octave*. It promotes a circle representing the geometrical space of abstract tonalities. A more detailed geometrical structure is given by a *musical scale*, interpretable as further 'musical' circular symmetries.

The aim of this essay is to illustrate how these symmetries of the circle, lead to its own projected quantum realizations, and the complementary view of extending the circle into a quantum counterpart. These examples are actually extremely rich in their internal structure, they reflect the spectrum of all principal new phenomena of quantum geometry. In particular quantum circles are interpretable as quantum bundles, in the appropriate sense, unifying a base space parametrized family of classical 'virtual circles'. In the most harmonic scenarios, quantum circles are interpretable as quantum groups as well. We shall briefly talk about a general diagrammatical and categorical formulation of symmetry, which naturally includes our quantum circles and their quantum friends, as well as a variety of all classical structures.

2 Circles, Classical and Quantum

The Pythagorean musical scale invites us to consider the quotients of the classical circle $\bigcirc$ over a free action of the infinite cyclic group of integers $\mathbb{Z}$, generated by a single irrational rotation. The space of equivalence classes has a nice geometro-musical interpretation, as the space of abstract tonality forms within a single octave. In the case of Pythagoreans, we have two principal frequency transformations. The octave itself, given by doubling the frequency $\omega \mapsto 2\omega$, and the *perfect fifth*, given by the shift $\omega \mapsto 3\omega/2$. If we consider the frequency range as covering all positive real numbers $\mathbb{R}^+$ and pass to natural logarithms, then the multiplication becomes addition and frequency range is the whole $\mathbb{R}$. The octave space is given by $\mathbb{R}/\mathbb{Z}\ln(2)$. Within this space, the addition of $\ln(3)$ acts as a symmetry. By transforming $[r] \mapsto \exp(2ir\pi/\ln(2))$ we can identify the octave space with the circle $\bigcirc$ of the unitary complex numbers. In terms of this identification, the Pythagorean perfect fifth becomes a multiplication by $\exp(2i\pi \ln(3)/\ln(2))$ which represents an irrational rotation, by the angle $\varphi = 2\pi \ln(3/2)/\ln(2)$.

Another natural possibility is to play with *rational* rotations. In terms of complex numbers, it corresponds to roots of unity, say primitive solutions of the equation $z^n = 1$ for a given $n \geq 2$. In this case the action of $\mathbb{Z}$ factorizes to the action of the cyclic group of order n on the circle. And the resulting factor space is again a classical circle. So our tonality space is given by a n-fold covering of $\bigcirc$ by $\bigcirc$. Musical scales based on equal temperament provide a realization of such a rational structure, and n is the number of semitones. In terms of the original frequencies, the simplest movement is given by $\omega \mapsto 2^{1/n}\omega$.

However, in the irrational case, there exist infinitely many connectable pitch values, dense in the octave space. In other words, every orbit of the action is dense in the circle $\bigcirc$. The resulting orbit decomposition is *ergodic*, in the sense that there exist no no-trivial decomposition of the circle, into two disjoint measurable sets consisting of whole orbits each. One of them always has measure 0 and hence another is of the normalized measure one. To put it differently, there exist no no-trivial measure theory on the orbit space Q. It exhibits a kind of intrinsic *wholeness*. And if there is no non-trivial set measuring in Q, then there is simply no hope to build, in the spirit of classical geometry, any meaningful higher-level theory.

So Q is consisting of points, however the points are behaving quite wildly, and there is no any effective and operational separability between them. On the other hand, Q is *natural*, as it is directly *constructed* form a classical object–the circle $\bigcirc$.

Exactly the same kind of phenomenon we encounter in studying certain *aperiodic tilings* of the Euclidian plane.

A paradigmatic example is given by the space of isomorphism classes of Penrose tilings. There exist (uncountably) infinitely many classes, however every two tilings are indistinguishable by comparing their finite regions. Every finite region of one tiling, is encountered infinitely many times faithfully echoed in any other tiling.

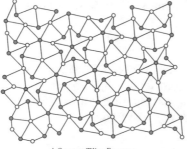

A Quantum Tiling Fragment

This space can also be described as the quotient space of the full binary sequences space $\{-, +\}^{\mathbb{N}}$ which is the same as the Cantor triadic set, by an equivalence relation identifying sequences which coincide on a complement of a finite subset of $\mathbb{N}$.

One possibility to deal with such quantum points, is to construct a non-commutative C*-algebra B, which captures the space Q in terms of equivalence classes of its irreducible representations. Such an approach is presented in detail in [1]. A central idea is that the associated algebra reflects not only the space Q as such, but also the way of its construction, from a classical equivalence relation carrier space ($\bigcirc$ or $\{-, +\}^{\mathbb{N}}$ in our examples). Another, related but inequivalent approach, is to apply the theory of quantum principal bundles developed in [2, 3], and consider non-trivial geometric (necessarily quantum, as in music) structures on discrete and extremely disconnected classical spaces and groups.

Let us outline a hybrid re-formulation, applicable to every discrete group G freely acting (say, on the right) on a compact topological space X. The Pythagorean example is given by $G = \mathbb{Z}$ and $X = \bigcirc$, with the described irrational twist defining the action, while the aperiodic tilings space correspond to the Cantor triadic space $X = \{-, +\}^{\mathbb{N}}$ of binary sequences and G is its countable subgroup consisting of sequences stabilizing at $+$ from some moment.

The freeness of the action allows us to introduce a natural bijective transformation $X \times G \leftrightarrow R \hookrightarrow X \times X$, where R is the equivalence relation in $X \times X$ associated to the group action. The correspondence is given by $(x, g) \leftrightarrow (x, xg)$.[1] The building blocks for our C*-algebra are continuous complex functions ψ with compact support on $X \times G$. This means that $\psi(x, g)$ is possibly different from 0, only for finitely many values of $g \in G$. We can introduce a 'support weight' n_ψ, the number of such levels. On the vector space B of these functions, we can introduce a non-commutative convolution-type product

$$(\psi \cdot \varphi)(x, g) = \sum_{h \in G} \psi(x, h)\varphi(xh, h^{-1}g) \tag{1}$$

and it is easy to see that this product is associative. Furthermore, the formula

$$\psi^*(x, g) = \overline{\psi}(xg, g^{-1}) \tag{2}$$

defines an antilinear involution $*: B \to B$. It is easy to check that also

$$(\psi \cdot \varphi)^* = \varphi^* \cdot \psi^* \tag{3}$$

for every $\varphi, \psi \in B$, in other words $*$ is antimultiplicative. So we have a nice *-algebra B. We can equip this *-algebra with a C*-type norm, by considering a family of Hilbert space representations, defined as follows. For every orbit $\omega = [x] \in Q$, let $H_\omega = \ell^2(\omega)$. The formula

---

[1] For an arbitrary action, we can surjectively map $X \times G$ to R via $(x, g) \mapsto (x, xg)$, but the correspondence will be in addition injective (that is, bijective) iff the action is free.

$$(\psi u)(x) = \sum_{g \in G} \psi(x, g)u(xg) \qquad x \in \omega = [x] \tag{4}$$

defines an action $B \times \mathcal{H}_\omega \to \mathcal{H}_\omega$. This action turns out to be a *-representation D_ω of B by bounded operators, because of

$$(\psi \cdot \varphi)u = \psi(\varphi u) \qquad \langle \psi u, v \rangle = \langle u, \psi^* v \rangle \tag{5}$$

where $\psi, \varphi \in B$ and $u, v \in H_\omega$, which directly follow from our definitions of basic objects. The operators are indeed bounded, and moreover we have

$$|D_\omega(\psi)| \le \sqrt{n_\psi \max_{x \in X} \left\{ \sum_{g \in G} |\psi(x, g)|^2 \right\}} \le n_\psi |\psi|_\infty \quad \text{for every } \psi \in B.$$

Every representation D_ω is *irreducible*. For a given orbit ω the representation will be *faithful*, if and only if $\overline{\omega} = X$. Different orbits (points of Q) give rise to mutually *non-equivalent* irreducible representations of B.

The constructed algebra can also be viewed as the cross-product $B = C(X) \ltimes G$ of $C(X)$ by the action of the discrete group G. Both $C(X)$ and the group *-algebra $\mathcal{A}$ of G are unital *-subalgebras of B, in a natural way. As a vector space we can write $B = C(X) \otimes \mathcal{A}$. Indeed, the functions on X can be interpreted as those functions φ of B such that $\varphi(x, g) = 0$ for all $g \ne e$. And the elements g of G are interpretable as functions such that

$$g(x, h) = \delta_{g,h} = \begin{cases} 1 & g = h \\ 0 & g \ne h \end{cases}$$

To complete the construction, let us observe that the representations D_ω distinguish the elements of B, and taking the supremum of the above operator norms, we obtain a C*-norm $|, |$ on B. Finally, by completing B with respect to this norm, we obtain a C*-algebra $B = \overline{B}$. By construction, every D_ω extends by continuity to B.

In the above example of a quantum tilings space, the group G is discrete, countable and given by binary sequences of $\{+, -\}$ stabilizing at $+$. It is exactly the Pontryagin dual of the Cantor triadic set $X = \{+, -\}^{\mathbb{N}}$ of all binary sequences, viewed as a compact topological group. And symmetrically, X is the dual of G. Every point $\omega \in Q$ of the tilings space (a tilings class) is represented by a unique irreducible representation D_ω. So, although the space Q is totally unfriendly to functions (there are no no-trivial measurable complex valued functions on it), it is equipped with a natural Hilbert space bundle, and is quite rich in operator-valued functions, acting pointwise—in fact every $b \in B$ induces such a function. It is worth mentioning that B is simple, as a C*-algebra.

And in our irrational rotation case, represented by a unitary complex number z, the algebra B allows a simple definition in terms of generators and relations. We have $G = \mathbb{Z}$, the additive group of integers, and $X = \bigcirc$, the circle. The symmetry group is the dual group to the circle. The algebra $C(\bigcirc)$ is generated by a single unitary element U, coming from the canonical embedding, as unitary complex numbers, of $\bigcirc$ into $\mathbb{C}$. The algebra $\mathcal{A}$ is also generated by a single unitary element V, corresponding to our principal rotation (the number one, as the generator of G). These unitaries then satisfy the following commutation relation

$$VU = zUV \tag{6}$$

which completely defines the C*-algebra B. In other words, we obtain a quantum torus space. For irrational numbers z, the algebra B is simple. If z is rational ($\Leftrightarrow$ root of one), then B turns out to be (strongly) Morita equivalent to the classical algebra $C(\bigcirc)$.

So the rational rotations give us classical circle as the tonality classes space. And irrational rotations produce quantum objects. It is interesting to observe that from a purely geometrical perspective, the quantum behaviour is *the generic one*. Indeed, although the rational and irrational unitary complex numbers are intertwined, both being everywhere dense in the circle, the roots of unity constitute a countable and therefore negligible, subset. With probability one, $\bigcirc$ will choose an infinite covering mode, and cast a quantum shadow.

A kind of complementary approach to the described quantization of the circle, is to consider quantum extensions of the commutative algebra of the classical circle. In this complementary picture, the classical circle is *embedded* into a quantum counterpart. As an illustrative example, consider a short exact sequence

$$0 \to K(H) \hookrightarrow B \twoheadrightarrow C(\bigcirc) \to 0$$

where $K(H)$ are compact operators in the infinite dimensional Hilbert space $H = \ell^2(\mathbb{N})$ and B is the C*-algebra generated by the unilateral shift operator $U \colon H \to H$. The classical circle $\bigcirc$ appears as the classical part of the quantum object given by the non-commutative C*-algebra B, which in turn provides a base for constructing a quantum hyperbolic plane, possessing a full classical symmetry group $\mathrm{SU}(1, 1)/\{-1, 1\}$. Indeed, the rule

$$U^* \mapsto \frac{aU^* + b}{\bar{b}U^* + \bar{a}} \qquad \mathrm{SU}(1, 1) = \left\{ \begin{pmatrix} a & b \\ \bar{b} & \bar{a} \end{pmatrix} \,\middle|\, |a|^2 - |b|^2 = 1 \right\}$$

consistently defines an action of $\mathrm{SU}(1, 1)/\{1, -1\}$ by *-automorphisms of B. The adjoint shift operator U^* corresponds to the complex variable z restricted to the unit complex disc, in the classical Poincaré model of the hyperbolic plane. The algebra B can be also described in abstract terms, as the C*-algebra generated by a symbol U and a single relation $U^*U = 1$, forgetting about the counterpart $UU^* = 1$ which would force the structure into the classical circle. The classical circle is also interpretable as the 'horizon heaven' for such a quantum space.

Another interesting interpretation of the same C*-algebra, is via quantum principal bundles. Here we would need to adopt an extended formulation [3, 4], allowing non-standard 'tensor products' to represent *collectivity configurations* on a given quantum space.

Let us consider a map $F: B \to B \otimes A$ where $A = C(\bigcirc)$, defining the canonical action of the classical circle on B. In terms of transformations associated to the single elements of the circle (viewed as unitary complex numbers z) this action takes the form

$$U \mapsto zU, \qquad U^* \mapsto \bar{z}U^* \tag{7}$$

If we now consider a free C*-algebraic product $B \star B$ of B with itself, the map F, acting on the second factor and combined with the canonical inclusion $B \hookrightarrow B \otimes A$ acting on the first factor, naturally extends by multiplicativity, to a unital *-homomorphism $F: B \star B \to B \otimes A$. It is easy to see that the extended map is surjective, in other words, the action F is *free*, and we have a circular quantum principal bundle. The algebra B is the closure of $\mathcal{B}$, which has a natural linear base consisting of $p_{kl} = U^k U^{*l}$, where $k, l \in \mathbb{N}$.

The bundle algebra B is generated by a single element U, a kind of circular coordinate. The base space C*-algebra, consisting of all invariant elements, is a closure of the *-algebra $\mathcal{V}$, which is a linear span of elements of the form $p_k = p_{kk} = U^k U^{*k}$ where $k \in \mathbb{N}$. These are mutually commuting descending orthogonal projections, and hence the base space is a classical extremely disconnected topological space M, naturally viewable as

$$M \leftrightarrow \mathbb{N} \cup \{\infty\}$$

where, in terms of the identification of points of M with characters $\varkappa: \mathcal{V} \to \mathbb{C}$, the point ∞ corresponds to a character $\varkappa_\infty$ evaluating to one, in all these projectors, while $l \leftrightarrow \varkappa_l$, so that

$$\varkappa_l(p_k) = \begin{cases} 1 & \text{for } l \geq k \\ 0 & \text{for } l < k \end{cases}$$

By taking the inverses $l \mapsto 1/(1+l)$ and $1/\infty = 0$, we can identify $M = \{0, 1, 1/2, 1/3, \cdots\}$. The geometrical picture is that we have circular object, unifying infinitely many circular 'oscillating modes'. The limiting oscillating mode is the classical mode (corresponding to the projectability of the bundle to the classical circle itself, represented by the above mentioned short exact sequence). All other modes are purely quantum 'virtual modes', we can not distinguish separate fibers over the classical points labeling these modes. The entire structure is a unified and irreducible quantum circle.

This can be taken as a starting point, towards its natural generalization, defining quantum circles as circular quantum principal bundles, such that the bundle algebra B is generated by a single element U, satisfying the property (7) and such that the action F is free (the freeness condition being a central property for general quantum principal bundles). In this more general setting, the freeness condition can be viewed as the invertibility of

$$\varrho = U^*U + UU^* \tag{8}$$

in the base space algebra $\mathcal{V} \leftrightarrow M$. The oscillating modes base space M, represented by the F-invariant elements of $\mathcal{B}$, in general, will be quantum ($\leftrightarrow$ noncommutativity of the algebra $\mathcal{V}$).

3   Universal Harmony Partiture

When we liberate the symmetries from shells of a secondary mathematical concreteness, then we come to their true essence. The harmony for the sake of harmony. A possible development of such a harmony partiture,[2] will be sketched now. We shall construct an elegant *diagrammatic category* Δ. Symmetry objects of a given mathematical realm, manifested by a category $\mathcal{C}$, are then simply the appropriate functors from Δ to $\mathcal{C}$.

Our category is constructed from 2 fundamental symbolic morphisms.

We shall imagine that a 'time' flows vertically and downwards. The first morphism corresponds to the composition movement, of two symmetry transformations. The second morphism can be interpreted as the duplication movement, that is, the diagonal map. Equivalently, at the dual level of algebras describing spaces in terms of their 'observational properties', the first morphism correspond then to the diagonal map (the product-induced map) and the second morphism is the coproduct, the dual version of the product. See [5] for more details.

Our first assumption is that

This can be interpreted as a rule to move from one diagram to another. Whenever we find a motif containing one of these four trees, we can replace it with its mirrored counterpart.[3] The second basic assumption is that the following two mutually symmetric combinations

---

[2] We do prefer the term "partiture" because of its semiotic and etymological contents (related to It. Sp. "partitura", Fr. "partition").

[3] The same interpretation applies to all equalities derived in this diagrammatic theory.

of our principal morphisms are *invertible* in Δ.

The morphisms in our category are thus represented by diagrams constructed using the primary morphisms and the rules described. The objects of our diagrammatic world are simply—natural numbers $1, 2, 3, \cdots$. Every diagram is a morphism between its entry points number, and its exit points number. The composition of morphisms is thus simply the operation of vertically connecting the diagrams. Another basic operation that we can perform with the diagrams is putting them horizontally side-by-side, which defines a monoidal structure on Δ (with $+$ being the monoidal composition of numbers-objects). In particular the identity morphism on the number-object n, is given by n parallel vertical lines.

It turns out that there naturally appear some very special 'circular' morphisms, indexed by integers $k \in \mathbb{Z}$. They satisfy the following

interesting convolution property for every $i, j \in \mathbb{Z}$. The circular morphism indexed by zero corresponds to the 'neutral transformation'. The one associated to $k = 1$ turns out to be the identity morphism of the object-number one. And the morphism indexed by $k = -1$ is a personification of the inverse transformation–the antipode.

It turns out that our quantum circles provide very interesting examples for realizations of this diagrammatic symmetry category. One class of such examples is constructed by considering *-algebras where U is invertible. Then of course U^* is invertible, too, with $(U^*)^{-1} = (U^{-1})^*$. The freeness condition (8) is automatically satisfied. If we consider a free construction, then we can construct higher-order collectivity algebras B_n where $n \geq 2$, as n-fold free products of B with itself. The expression

$$\phi(U) = U_{[1]}U_{[2]} \tag{9}$$

defines a coproduct *-homomorphism $\phi \colon B \to B_2$. The dual morphism is simply the standard product-like *-homomorphism $m \colon B_2 \to B$.[4]

In the case when B is not free, but say generated by some relations, these relations must be propagated to the higher-order collectivity algebras, via the representatives of diagrams, and we should pass to the factor-algebras. Only in some very special cases, the construction will lead to a non-trivial quantum group structure. One particularly nice case is given by relations of the form $U^m = 1$, which leads to a *cyclic quantum group*. In any case, additional geometrical considerations should complete our interpretation of the group being the circle. For instance, considering the spectrum of U and requiring that it be within the classical circle of unitary

---

[4]Here the square-bracketed indexes simply refer to the associated copy of the initial algebra within the corresponding higher order collectivity algebra.

complex numbers (the above mentioned relations automatically imply this spectrum property).

The whole non-commutativity of the algebra comes from non-commuting U and U^*. If they commute, then we are back in classical geometry, at the 'one-particle' level (and in particular $U^* = U^{-1}$). However, even in this case there is a highly non-commutative world of higher-order collectivity algebras B_n. This can be used to capture the geometry of rotations, like those appearing in classical Pythagorean octave versus perfect fifth considerations. Specifically, we can construct the algebras B_n by requiring that all non-trivial commutators

$$U_{[i]}U_{[j]}U^*_{[i]}U^*_{[j]} = z_{ij} \qquad i \neq j \tag{10}$$

be central elements of these algebras. In this sense the algebras B_n are interpretable as quantum versions of n-dimensional tori.

An elegant general construction of quantum groups is given by a *matrization procedure*. We start with an algebra defined by certain generators and relations. Then we substitute every generator, by a $n \times n$ matrix of new generators. If we apply this procedure to the classical circle (with universal higher-order collectivity algebras), we obtain the universal unitary quantum matrix groups $QU(n)$. Indeed, if U is converted into a $n \times n$ matrix, then by the construction in the new quantum group, its entries satisfy

$$\phi(U_{ij}) = \sum_{k=1}^{n}(U_{ik})_{[1]}(U_{kj})_{[2]} \tag{11}$$

In this sense, the music of quantum circles, covers all possible quantum symmetries, describable by unitary matrices.

4 Concluding Remarks

In the introduction, we mused about continuous discreteness and discrete continuity of music. Quantum geometry provides a number of ways to view this duality and complementarity.

For example, we can consider non-trivial differential calculi over classical discrete spaces. These calculi in a way unify the points, mutually totally separated when viewed within the classical perspective. In [6] a quantum calculus is used to provide a quantum geometrical framework for certain non-local operators generalizing partial derivatives and naturally associated to Coxeter groups in Euclidean spaces.

Another interesting phenomenon in resonance with this, comes from a geometric interpretation of matrix algebras $M_n(\mathbb{C})$ for $n \geq 2$. The automorphism groups of these algebras are quite rich, given by $U(n)/U(1) = SU(n)/\mathbb{Z}_n$, unitary matrices factorized over unitary scalars (all automorphisms are inner, of the form $a \mapsto UaU^*$ with $U \in \mathrm{U}(n)$). On the other hand, the algebra $\mathrm{M}_n(\mathbb{C})$ is finitely-dimensional, a kind of quantization of $n \times n$ finite space (where symmetries are just permutations). The quantum space Q corresponding to $\mathrm{M}_n(\mathbb{C})$ is completely 'pointless', as there are no characters on the matrix algebras (they are all simple). If $n = 2$, then the space is interpretable as a quantum sphere. Indeed, the automorphism group is $\mathrm{SO}(3) = \mathrm{SU}(2)/\{1, -1\}$ and pure states ($\Leftrightarrow$ one-dimensional projectors) are naturally viewed as points of a 2-dimensional classical sphere.

This allows a natural generalization in terms of coherent states, and in such a way we can connect the stochastic approach to quantum geometry [11] with the non-commutative C*-algebraic interpretations.

And yet another beautiful example of a quantum unification between discrete and continuous: the existence of non-trivial orthogonal projections in quantum torus C*-algebras. In the irrational case, as the one associated to the Pythagorean musical scale, the torus space exhibits a total wholeness, no divisibility at all (the C*-algebra is simple). However these projectors can be used to construct a surjective map from the quantum torus space, over the extremely disconnected Cantor triadic set.

5 Further Reading

About categories: We refer to treatises [7], as a beautiful conceptual introduction to categories, and [9], as an extensive and inspirative categoric-theoretic foundational framework for music. For a detailed historical overview and a deep discussion of the concept of point, and development of a pioneering self-contained quantum geometry paradigm, see [11]. Another pioneering (C*-algebraic and categorically friendly) approach to quantum geometry is given by [13]. The quantum circle representation on the title page was inspired by the artworks of [10], and realized in Maxima (http://maxima.sourceforge.net). A charming introduction to C*-algebras and K-theory is [12]. The book contains a detailed analysis of diverse fascinating examples of C*-algebras, including irrational rotation/quantum torus algebras. A classical treatise on tessellations, including regular tilings, and also aperiodic tilings and in particular Penrose tilings, is [8].

Acknowledgements I am very indebted to Dr. Gabriel Pareyon, for extending me a warm invitation to participate in this Conference, various interesting and fruitful conversations, and unlimited enthusiasm which contributed to creating a unique scientific, academic and artistic atmosphere in Puerto Vallarta.

References

1. Connes, A.: Noncommutative Geometry. Academic Press, Dublin (1994)

2. Đurđevich, M.: Geometry of quantum principal bundles. Part I, Commun. Math. Phys. **175**(3), 457–521 (1996). Part II–extended version, Rev. Math. Phys. **9**(5), 531–607 (1997). Part III–structure of calculi and around, Alg. Groups Geom. **27**, 247–336 (2010)
3. Đurđevich, M.: Geometry of Quantum Principal Bundles 4. In preparation
4. Đurđevich, M.: Categorical Frameworks for Quantum Spaces, Groups and Bundles ∼ Geometric Quantum Groups as Realizations of Diagrammatic Symmetry Category ∼ Quantum Riemann Surfaces. Alg. Groups Geom. **33**, 3 (2016)
5. Đurđevich, M.: Diagrammatic formulation of multibraided quantum groups. Contemp. Math. **318**, 97–106 (2003)
6. Đurđevich, M., Sontz, S.: Dunkl operators as covariant derivatives in a quantum principal bundle. SIGMA **9**(040), 29 (2013)
7. Goldblatt, R.: Topoi-The Categorical Analysis of Logic. Elsevier Science, Amsterdam (1984)
8. Grünbaum, B., Shephard, G.C.: Tilings and Patterns. Freeman & Company, San Francisco (1987)
9. Mazzola, G.: Collaborators: The Topos of Music: Geometric Logic of Concepts, Theory, and Performance. Birkhauser, Basel (2002)
10. Neubäcker, P.: The Vibrating String, (Harmonics and the Glass Bead Game). Harmonik & Glasperlenspiel, Beiträge (1993). http://www.harmonik.de
11. Prugovečki, E.: Quantum Geometry-A Framework for Quantum General Relativity. Kluwer Academic Publishers, Dordrecht (1992)
12. Wegge-Olsen, N.E.: K-Theory and C*-Algebras. Oxford Science Publications, Oxford (1993)
13. Woronowicz, S.L.: Pseudospaces pseudogroups & pontriagin duality. Proceedings of the International Conference of Mathematical Physics, Lausanne: Lecture Notes in Physics **116**, 407–412 (1979)

Partitiogram, Mnet, Vnet and Tnet: Embedded Abstractions Inside Compositional Games

Pauxy Gentil-Nunes

Abstract This paper integrates a broad research about the pragmatic modelling of compositional process, and some mathematical abstractions that arises from the relations between textural configurations. As the available choices for textural organization are limited, it is possible to provide a global map of all possible configurations for a given number of sources (exhaustive taxonomy) and assess all the kinship and metrics between them (topology). The graphic called *Partitiogram*, in fact, constitutes this phase space, where three basic nets of parsimonious relations are drawn: *mnet*, *vnet* and *tnet*. Each net deals with a different kind of textural transformation. This framework is part of Partitional Analysis (PA) — an original proposal of mediation between mathematical abstractions derived from the Theory of Integer Partitions and compositional theories and practices. The main goal of the theory is the study of compositional games. It has been used in the pedagogy of composition and in the creation of new pieces.

1 Partitional Analysis

Partitional Analysis (therefore, PA; [6, 8]), is an original approach concerning textural analysis based on an approximation of the Theory of Integer Partitions (therefore, TIP; [1, 2, 5]) and compositional theories and practices.

In TIP, *partition* refers to the various representations of an integer by the sum of integers. The summands are generically called *parts*. For example, the number four has five partitions: $(1 + 1 + 1 + 1), (1 + 1 + 2), (1 + 3), (2 + 2)$ and (4). These representations can be written in abbreviated form, where the bases show the different parts and the exponents show the multiplicity of each of them: $(1^4), (1^2 2), (13), (2^2)$ and (4), for instance.

PA is inspired by Wallace Berry's work [3] about texture. PA can be thought, in fact, as an extension and development of some of his conceptions and suggestions.

P. Gentil-Nunes (✉)
Federal University of Rio de Janeiro, Rua do Passeio, 98 - Centro, Rio de Janeiro, RJ, Brazil
e-mail: pauxygnunes@musica.ufrj.br

© Springer International Publishing AG 2017
G. Pareyon et al. (eds.), *The Musical-Mathematical Mind*,
Computational Music Science, DOI 10.1007/978-3-319-47337-6_12

Berry proposes the assessment of concurrent parts of a musical plot, concerning mainly rhythmic and melodic profiles. This reading generates a representation of the sucession of textural settings in the form of stacked numbers that mirror the vertical organization of the vocal or instrumental parts. The absolute number of concurrent elements or sources at each time, called *density-number*, generates the so called *quantitative curve*. On the other side, the evaluation of the relations of independence and interdependence between the parts generates the *qualitative curve*. These curves are the basis of Berry's concept of textural *progression* and *recession* (Fig. 1).

The point of departure of PA lies in the detailed consideration of the pairwise relations between agents of the musical plot, which in fact ground the process of definition of the parts. These binary relations are categorized in collaboration and contraposition types, according to a given criterion (congruence between time points and duration, belonging to a line inside a melody, proximity of timbre or orchestral group, spatial location in the stage, and so on). The counting of the two types of binary relations leads, respectively, to the establishment of the agglomeration and dispersion indices *(a, d)*.

The total number of relations *(T)* of a given textural configuration is evaluated by the pairwise combination of its elements (Eq. 1, where *n* is the number of elements).

$$T = C_{(n,2)} = \frac{n(n-1)}{2} \tag{1}$$

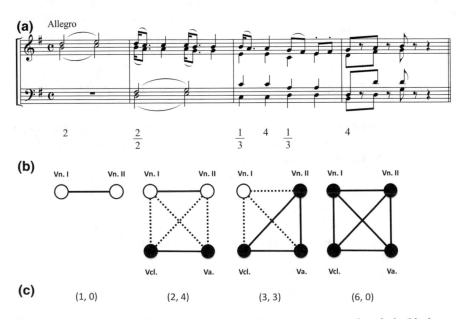

Fig. 1 W.A. Mozart, *Eine Kleine Nacthmusik*, K. 525, excerpt: **a** textural analysis (blocks are formed from similar attacks and durations); **b** binary relations: collaborations (*full lines*) and contrapositions (*dotted lines*), in each of the four configurations used in the excerpt — (2), (2²), (1 3) and (4); and **c** the *agglomeration* and *dispersion* indices *(a, d)* for each configuration

The agglomeration index *(a)* is calculated by the sum of all pairwise combinations of each part (Eq. 2, where p is the number of parts and T_i is the integer related to each part.

$$a = \sum_{i=1}^{p} C_{(T_i,2)} \qquad (2)$$

The dispersion index *(d)* is the simple difference between T and a.

Plotting of one index against the other forms a phase space, called *partitiogram*, where the dynamic movement between textural progressions (under the chosen criterion) is registered as a trajectory (Fig. 2a). This outline highlights the used partitions and the frequency of the used paths, working as a mapping of the involved textural modes. Geometric patterns that emerge from this process can be used to compare parts or sections of a piece, to the development of typologies or in the creation of gestures or compositional designs.

The partitiogram constitutes too an exhaustive taxonomy of all textural configurations and available movements inside the field (number of sources and criteria). It shows also the kinship of its elements, evidenced by the metrified distances between the locations of partitions. In this regard, the partitiogram can be considered as a textural *Tonnetz*.

The *indexogram* is the temporal counterpart of the partitiogram, where the two indices are mirrored vertically against a horizontal axis representing time points or beats (Fig. 2b). The polygonal graphic structures that result from the textural progressions (called *bubbles*) are then available to visual analysis and can bring important informations about musical form and gestures [4].

The structure of the partitiogram has an intimate affinity with Young's Lattice, a partially ordered set formed by all integer partitions, related by inclusion relationships, and ranked according to their sums. *Partitional Young's Lattice (PYL)* is an

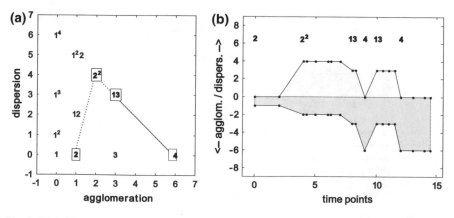

Fig. 2 W.A. Mozart, *Eine Kleine Nacthmusik*, K. 525, excerpt: **a** partitiogram; **b** indexogram

enriched version of this structure with classification of the relations according with their qualities and with the indication of the *(a, d)* indices for every node [6]. The internal structure of partitiogram is in fact a metrified version of a Young's lattice (Fig. 3).

2   Textural Nets

Inside Partitions Analysis (PA), the relations between adjacent partitions are qualified according to the nature of the specific progression. Three basic parsimonious operators emerge from this procedure. Each one can be preceded by a positive or negative sign, according to its eventual progressive or recessive quality (Fig. 3):

1. Resizing (*m*), involving unitary change in the size of one part (tapering or fattening). For example, $1^2 \preceq 12 \preceq 2^2 \preceq 23$.[1]
2. Revariance (*v*), referring to the addition or subtraction of a unitary part (changing in diversity of content inside partition). For instance, $1 \preceq 1^2 \preceq 1^3$;
3. Transference (*t*), when both operations (*m* and *v*) come into play simultaneously, but with opposite signs, in a complementary way and with steady density-number. In this case, the progressive nature of the movement points towards dispersion, following analytical and compositional traditions. For example, $4 \preceq 13 \preceq 22 \preceq 1^2 2$

The formalization of textural relations allows the manipulation of textural progressions through canonic compositional operations, like transposition, inversion,

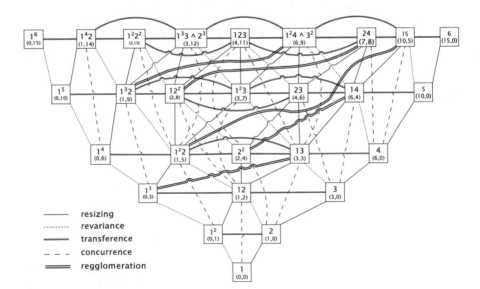

Fig. 3 Partitional Young's Lattice (PYL) for $n = 6$

[1] The symbol $\preceq$ means "is precedent".

retrogression or any other variational or derivational abstract device. The operators can also be combined and accumulated. The partial order of PYL gives eventually more than one result for a given operation, which can be very convenient for creative purposes.

Simple operators form networks of parsimonious relationships, each one with specific profile and structure. Plotting of the relations in computational applications leads to three functions: *mnet*, *vnet* and *tnet*, which basically draw the requested networks for a given number of agents or sources. The superposition of the three basic networks can represent the global field of actions available for the composer and can give rise to new applications, like typologies of compositional procedures, styles, and fingerprints (Fig. 4). All the functions were developed inside Matlab environment and are integrated in the software *Partitions* for Windows [7, 9].

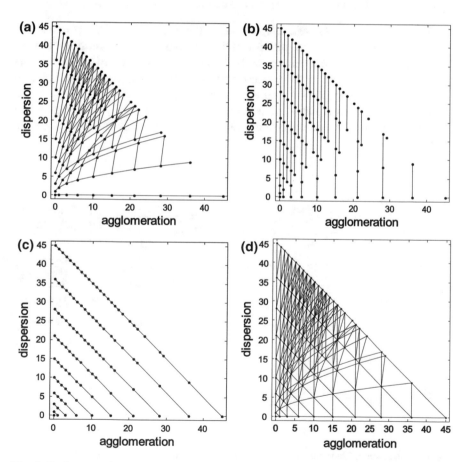

Fig. 4 Basic textural nets: *mnet* **a**, *vnet* **b**, *tnet* **c** and the global set **d** for $n = 10$. Each point represents a partition or group of partitions sharing the same *(a, d)* indices

3   Homology Between Textural Fields

As the partitiogram can be applied under different criteria, it leads PA to exceed the scope of Berry's textural analysis, constituting an abstraction that can be used as a unifying element between heterogeneous compositional and analytical practices. In other words, it allows the establishment of perfect homologies between texture, melody, timbre, spatial distribution, form and the like, always under the field of verticality.

Three main examples are presented, using the same progression to generate different musical surfaces, according to the chosen criteria. The arbitrary progression is < (1), (2), (12), (2²), (12), (1²2), (12²), (32), (5) >. This sequence generates a vector of successive operators, namely < *(+m)*, *(+v)*, *(+m)*, *(-m)*, *(+v)*, *(+m)*, *(-t)*, *(-2t)* > (for a clearer understanding, this structure is illustrated in Figs. 5, 6 and 7)

1. *Rhythmic or Textural Partitioning* considers the coincidence of attack points and durations, leading to the constitution of lines, blocks and textural entities (Fig. 5).
2. *Linear or Melodic Partitioning* can be thought as a sub-level of Rhythmic Partitioning, where lines inside a monophonic part are dismembered, according to the Schenkerian concepts of conjunction and disjunction (Fig. 6).
3. *Channel or Events Partitioning* considers the units defined by the composer as significant from an external criterion, in this case the timbre and type of kinetic profile (Fig. 7).

Fig. 5 Application of Rhythmic or Textural Partitioning to the chosen progression (partitions and operators)

Fig. 6 Application of Linear or Melodic Partitioning to the chosen progression (partitions and operators)

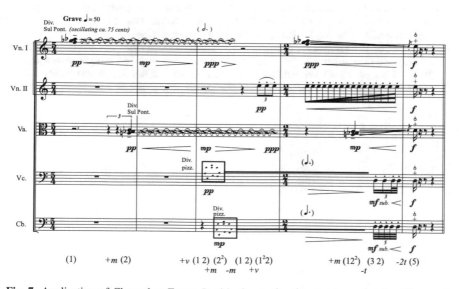

Fig. 7 Application of Channel or Events Partitioning to the chosen progression (partitions and operators)

4 Conclusions

In this paper, the Partitional Analysis framework was briefly presented, with focus on textural nets and its applications. The study of the topologic structure and properties of partitiogram are still in progress and the applications are being implemented in computer programs developed by the MusMat Research Group. The possibilities opened by PA are numerous and available to composers, theorists and analysts interested in the production of knowledge about the musical texture.

References

1. Andrews, G.E.: The Theory of Partitions. Cambridge University Press, Cambridge (1984)
2. Andrews, G.E., Eriksson, K.: Integer Partitions. Cambridge University Press, Cambridge (2004)
3. Berry, W.: Structural Functions in Music. Dover, New York (1976)
4. Codeço, A.: Movimento de derivação gestual textural a partir de dados da Análise Particional [Model Of Gestural Textural Derivation From Partitional Analysis]. Anais do XXIV Congresso da ANPPOM. UNESP, São Paulo (2014)
5. Leonhard, E.: Introduction to Analysis of the Infinite. Springer, New York (1748)
6. Gentil-Nunes, P.: Análise Particional: uma Mediação entre Composição Musical e a Teoria das Partições [Partitional Analysis: a Mediation between Musical Composition and the Theory of Integer Partitions]. Federal University of Rio de Janeiro, Rio de Janeiro (2009)
7. Gentil-Nunes, P., Moreira, D.: Partitions (2014). www.musmat.org
8. Gentil-Nunes, P., Carvalho, A.: Densidade e Linearidade na Configuração de Texturas Musicais [Density and Linearity in the Configuration of Musical Textures]. IV Colóquio de Pesquisa do PPGM-UFRJ, pp. 40–49. Federal University of Rio de Janeiro, Rio de Janeiro (2003)
9. Musmat Research Group: MusMat Research Group site. www.musmat.org

Algebraic Combinatorics on Modes

Franck Jedrzejewski

Abstract In the mid-1970s, Alain Louvier worked out microtonal scales called *modes of progressive transposition* and used them in many musical works. These modes have similar properties to major modes and are related to diatonicism. Some of them were known by Ivan Wyschnegradsky and Georgy Rimsky–Korsakov, the grand-son of Nikolai. Deep scales are well known in diatonic theory, and are special cases of these modes. However, their algebraic structure is not known. Although the diatonic theories have been developed by many musicologists, such as Agmon, Balzano, Carey, Clampitt, Noll, Zweifel and others, many questions remain open. In this paper, we describe some studies on microtonality, published over the last century, and we review what is known and what remains to understand in this field, in both theoretical and compositional aspects. In the first section, we study the modes called by Alain Louvier "*imperfect modes*", a special case of modes of progressive transposition. He used them in several important works as *Le Clavecin non tempéré* (1973), *Canto di Natale* (1976) and *Anneaux de lumière* (1983), written in the 24 tone equal temperament. The mathematical study would be to find a criterion to easily determine all the modes of progressive transposition in any equal temperament, and in particular to determine all deep scales. In the second section, we study the enumeration of the modes of limited transposition, also known as *Messiaen modes* in any equal temperament. In the last section, we present another classification of the modes related to the plactic monoid. Along this paper, we question what, if they exist, the microtonal diatonicism and the microtonal modality could be.

1 Introduction

At the end of the 19th century, microtones generated new interest. John Foulds wrote a string quartet (today lost) and some composers discovered some impressive non-Western Tunings and Balinese gamelan performance, especially at the

F. Jedrzejewski (✉)
French Atomic Energy Commission (Commissariat à l'énergie atomique et aux énergies alternatives), F-91191 Gif-sur-Yvette, France
e-mail: franckjed@gmail.com

© Springer International Publishing AG 2017
G. Pareyon et al. (eds.), *The Musical-Mathematical Mind*,
Computational Music Science, DOI 10.1007/978-3-319-47337-6_13

Exposition Universelle of 1889 in Paris. In the 1910s and 1920s, microtones received attention from Richard Stein, Jörg Mager, Willy Möllendorf, Julian Carrillo, Aloïs Haba, Ivan Wyschnegradsky, Mordecai Sandberg and many others. In Soviet Russia, Georgy Rimsky–Korsakov and some pupils of the Conservatoire founded the *Circle of quarter-tone music*. Georgy wrote a *Basis of the musical quarter-tone system* and some pieces have been performed in the Red-corner of the Leningrad Conservatory [1].

The program of the first concert of quarter-tone music in Russia (13 April 1925) included some works sent by Wyschnegradsky, Möllendorf and Mager, *Melody* and *Poem* by Malakhovsky, *Etude*, *Prelude* for two pianos and *Poem* for harp, harmonium and piano by Georgy Rimsky–Korsakov, and a lecture by Malakhovsky. This concert was followed by a second one (15 May, 1925) with some new pieces by Wyschnegradsky, a new *Prelude* for two pianos, two french horn and harmonium by Georgy Rimsky–Korsakov and *Two sketches* by Alexander Kenel. Later, Georgy gave some lecture at Moscow where the interest in microtones had also grown.

In 1927, Aloïs Hába wrote *Neue Harmonielehre des diatonischen, chromatischen, viertel, drittel, sechstel und zwölftel-Tonsystems* [8]. He tried to construct instruments that would be able to play microtones. At the same time, Ivan Wyschnegradsky, who had emigrated to France with his family, wrote *La Journée de l'existence*, began to experiment microtones and spent his time writing his theoretical treatise *La Loi de la pansonorité*, at least three versions until 1953 [15, 16]. In Mexico, Augusto Novaro wrote *Sistema Natural Base del Natural Aproximado* (1931) and Julian Carrillo discovered *El sonido 13* (1921). Later, Harry Partch wrote *Genesis of a Music* (1949) and Adriaan Fokker promoted the 31 tone equal temperament and founded what is now called *Stichting Huyghens-Fokker*. Basics of microtonal system were raised and theoretical research would become increasingly important over the years.

In this paper, I would like to give an overview of some theoretical aspects of micro-tonal music, starting with little known theory of "modes of progressive transposition" invented by Alain Louvier. Alain Louvier (born in 1945) is a French composer who was director of the Conservatoire National Supérieur de Paris (1986–1991) and whose music, far ahead of his time, often inspired by Olivier Messiaen, is based on original theoretical concepts and for some pieces, on microtones. Particular cases of these modes are the well-known *deep scales*. As we will see later, these scales reproduce an important property of major scales in the microtonal field and are related to the concept of tonality, and hence, to the concept of diatonicism. The combination of these two concepts, that are tonality and diatonicism, may be related to the presence of modes of limited transposition, at least that is what is suggested by the role played by the tritone in classical music. That is why we study the Messiaen modes of limited transposition in the microtonal field. Finaly, we review how to classify modes through the plactic relations. The title of this paper mimics the title of the book by Lothaire [13], the pseudonym of a group of mathematicians, because we are convinced that the combinatorics on words is the most suitable mathematical tool for studying microtonal modes and scales.

2 Progressive Transposition Scales

In the 12 tone system, all major scales have the same interval sequence 2212221. The circular permutations of this sequence are the usual church modes (Dorian 2122212, Phrygian 1222122, etc.). The transposition a fifth higher (or a fourth lower) of C major leads to a new major scale which has only one different pitch with C major. The scale C major $\{0, 2, 4, 5, 7, 9, 11\}$ a fifth up is G major $\{7, 9, 11, 0, 2, 4, 6\}$. The two sets differ only by one pitch (F $= 5$ in C major and F$\sharp = 6$ in G major). This property continues if the transposition is repeated and leads to the definition of a progressive transposition scale [14].

In the N tone equal temperament (N-tet for short), a *scale of progressive transposition* $L = \{a_1, \ldots, a_k\}$ is a set of k pitches such that each transposition at v steps higher leads to a set $M = \{a_1 + v \mod N, \ldots, a_k + v \mod N\}$ such that L and M differ only by one pitch. The number v is called the *transposition index*. The scales are identified by their interval sequence, which is a word on an alphabet $\mathcal{A}$. By definition, the alphabet of 2-scales has 2 letters, and the alphabet of 3-scales has 3 letters. By convention, scales have at least 5 pitches. The chromatic mode of k notes is always a mode of progressive transposition with transposition index equals 1. The scale is said *trivial* if $v = 1$. A non trivial scale is *tame* if the transposition index v is coprime with N. Otherwise, cycles could appear in the transposed scales (e.g. For $N = 12$, the scale $\{0, 2, 4, 8\}$ cycles for $v = 4$).

In the 12 tone system, a simple computation shows that there are only three 2-scales of 7 notes: the major scale 2212221 with transposition index $v = 5$, the pseudo whole scale 1122222 with transposition index $v = 2$ and the chromatic scale 1111116 with $v = 1$. It is remarkable that the only tame scale of 7 notes is the major scale. With 3 letters, there are only 5 scales of 7 notes: three 3-scales with transposition index equal to 6 (1113123, 1113213, 1122114) and two 3-scales with $v = 4$ (1121313, 1131312), but no tame scale.

In the 24 tone system, the number of scales of progressive transposition with 2 or 3 letters ranges from 1 to 180. Some scales come from the 12 tone system. For example, 1221222 is a scale of progressive transposition in the 12-tet. It follows that 2442444 (multiplying each interval by 2) is also a scale of progressive transposition in the quarter-tone system. The scale 4334343 of 7 notes ($C, D, E \flat, F, G, A \flat, B \flat$), used by Alain Louvier in *Aria, récit et carillon* (*Le Clavecin non tempéré no 1*) is like a major scale with modal degrees (E, A and B) lower by a quarter-tone (see Fig. 1).

If we now look at scales with 12 notes, there are only one 2-scale of 12 notes (the chromatic scale) and only one 3-scale 414131131131 having a transposition index of 5. With 13 notes, there are only three 2-scales: the chromatic scale, the pseudo-whole tone scale 1122222222222 and the generalized diatonic scale 1222221222222. But if the alphabet has three letters, there are 180 scales. Most of them have index of transposition of 12 (or 8). Only one has a transposition index equal to 7 (2133121312131) and only one has transposition index equal to 5 (4131131131131) This is a part of what Messiaen called the *charm of impossibilities*.

Fig. 1 A. Louvier, mode of progressive transposition

3 Deep Scales

A *deep scale* is a scale containing each interval class a unique number of times. It follows that the deep scale property ensures that there is a different number of common tones associated with each transposition level with one exception. In the 12-tet, the *C* major scale is deep as each interval class appears only ones. Following the circle of fifths, the C major transposed one step up or down yields *G* major or *F* major which have 6 common tones with C major. The same scale, C major, transposed two steps up or down leads to *D* major or Bb major which have 5 common tones with *C* major. The process goes one until the transposition by a tritone which leads to *F*♯ major which have 2 common tones with *C* major, the same as the transposition by 5 steps (*B* major or Db major). The exception is that there are two common tones in *F*♯ major and *C* major rather than only one tritone in the major scale. This inconsistency, said T. Johnson [12], may be accounted by observing that the tritone is counted only once and could be counted for 2 if we consider enharmonic intervals (*F-B* and *B-E*♯), suggested by the circle of fifths. It follows that all deep scales have the property that one transposition differs only by one note, and hence are also progressive transposition scales.

A deep scale is trivial if its transposition index is 1 or 2. In the N-tet, the chromatique scale $11 \ldots 1n$, where the number of 1 is $N - n$ with $n = \lceil N/2 \rceil$ or $n = \lceil N/2 \rceil + 1$ (the symbol $\lceil x \rceil$ is the ceiling function, the smallest integer not less that x) is trivial as its transposition index is 1. These chromatic scales are not maximally even [6]. If the transposition index is 2, the scales $122 \ldots 22$ with 2 repeated $(N - 1)/2$ times and the scales $22 \ldots 223$ with 2 repeated $(N - 3)/2$ times are trivial deep scales. Unlike the previous one, there are maximally even.

If we now look at non trivial deep 2-scales, the following 2-scales with interval sequence ($k \geq 3, n \geq 1$)

$$\left(1^{k-2}k\right)^n 1^{k-3}k \quad \text{with } N = (2k-2)n + 2k - 3$$
$$1^{k-1}k\left(1^{k-2}k\right)^n \quad \text{with } N = (2k-2)n + 2k - 1$$

and the 2-scales

$$1^{n-1}n1^{n-1}n1^{n-2}n \quad \text{with } N = 6n - 4$$

are deep 2-scales. The symbol 1^n means that 1 is repeated n times. The first formula gives the scales $(13)^n 3$ for $k = 3$ and $N = 4n + 3$, and the scales $(114)^n 14$ for $k = 4$, $N = 6n + 5$, etc. The second formula means $113(13)^n$ for $k = 3$, $1114(114)^n$ for $k = 4$, etc. And the last formula leads to the scales 12122 for $N = 8$, 11311313 for $N = 13$, etc. All these scales are not maximally even.

The main result of this section is the following. The only non trivial deep 2-scales maximally even are the scales

$$12^{n-1}12^n$$

for $N = 4n$ and $n \geq 1$ is an integer. In the 12-tet, this scale is the major scale, and in the 24-tet, it is what Wyschnegradsky called *diatonicised chromatism*. We will see that these scales are archetypes of generalized diatonic scales.

The next step is to determine a criteria for finding deep scales on an alphabet of 3 letters, for example $A = \{1, 2, 3\}$, and to report hierarchy of deep scales such that $12^n 32^n$ or $12^{n+1}12^n 32^n$ where $n \geq 1$ is an integer. The hope is that an hypothetical procedural way to build deep scales could be described by a generative grammar. For example, the six scales

$$r = (12)^n 13^n, \qquad s = (12)^n 13^{n+1}, \qquad t = (12)^n 13^{n+2}$$
$$R = (12)^{n-1}13^{n-1}, \qquad S = (12)^{n-1}13^n, \qquad T = (12)^{n-1}13^{n+1}$$

are deep 3-scales (with $n \geq 1$ or 2), but it is also the case for 12 scales issued by some concatenation of two words (rs, sr, rS, Sr, RS, SR, st, ts, sT, Ts, ST and TS), 33 scales of 3 concatenated words (rsT, rsr, sst, etc.), 40 scales of 4 concatenated words ($rsTr$, $rsTT$, $ssrs$, $SSRS$, etc.), 95 scales of 5 concatenated words ($rsTrs$, $rsTrr$, $sssst$, $TsTrs$, etc.), and so on.

To determine if a scale M is deep or if M is a progressive transposition scale in the N-tet, it is sufficient to look at the subwords of the interval sequence. Let $\omega_1\omega_2 \ldots \omega_n$ be an interval sequence where ω_k are letters of the alphabet $\mathcal{A}$, the subwords ω of $\omega_1\omega_2 \ldots \omega_{n-1}$ (the last letter is dropped) counted with multiplicity determine the number of notes that are in common between a scale and its transpositions in the following way. Consider the function

$$S : \omega \longrightarrow \min(|\omega|, N - |\omega|)$$

where $|\omega|$ is the sum of all letters of ω, namely $|\omega| = \sum \omega_j$ and consider for all interval classes $m \in \{1, 2, \ldots, [N/2]\}$ the function

$$\chi : m \longrightarrow \mathrm{card} S^{-1}(m)$$

If the point $[N/2]$ is reached, then the scale M is a scale of progressive transposition. And if all values of χ are different, then M is deep. The number of notes that are in common between M and the transposed scale $T_r(M)$ is $\chi(r)$ if r is an interval class $r \in \{1, 2, \ldots, [N/2]\}$, except if N is even and $r = [N/2]$ in which case the number of notes is $2\chi(r)$.

For example, if the scale $M = 112222$ for $N = 10$ is considered as an algebraic word on the alphabet $\{1, 2\}$, the subwords of 11222 counted with multiplicity are $\{1, 1, 2, 2, 2, 11, 12, 22, 22, 112, 122, 222, 1122, 1222, 11222\}$. Taking the length S of each subword modulo N and counting the number χ of words for each interval class leads to the table:

m	1 2 3 4 5
$\chi(m)$	2 5 2 5 1

Since $5 = N/2$ is reached for $m = 2$, the scale M is a scale of progressive transposition. But it is not a deep scale since $\chi(1) = \chi(3)$.

4   Microtonal Diatonic Scales

Trying to better understand what a diatonic scale is, we first review some results on well-formed scales [4]. Following the book of Timothy Johnson [12], we distinguish two kind of distances. The scales are arranged on a circle and all distance are measured clockwise. The c distance between two notes is the number of *steps* between these two notes, and the d distance between two notes is the number of *notes* between the two notes $+1$.

In a given N-tone equal temperament, a *well-formed scale* of k notes is a collection formed by repeatedly adding a constant interval (called the *generator*) around the chromatic circle until a complete k notes scale is formed, such that a single d distance corresponds to the c distance of the generator. The scale can be represented mathematically by the set $mj \bmod N$ for some consecutive values j. The scale is *degenerated* if m-step presentation circles, that is to say if the last note plus m is equal to the first note. Example: $N = 12, m = 4$, the set $\{0, 4, 8\}$ is a degenerated well-formed scale. The major scale $P = \{0, 2, 4, 5, 7, 9, 11\}$ is well-formed with $(m = 5, k = 7, N = 12)$, and has presentation:

$$11 \xrightarrow{5} 4 \xrightarrow{5} 9 \xrightarrow{5} 2 \xrightarrow{5} 7 \xrightarrow{5} 0 \xrightarrow{5} 5$$

Each c distance corresponds to a single d distance. The scale is built by taking a note each p notes (here $p = 3$),

$$\underline{0}, 2, 4, \underline{5}, 7, 9, \underline{11}, 0, 2, \underline{4}, 5, 7, \underline{9}, 11, 0, \underline{2}, 4, 5, \underline{7}, 9, 11, \underline{0}$$

A collection is *maximally even* if for each d distance, there are only one or two c distances, and if there are two c distances for a particular d distance, then the d distances are consecutive numbers. Thus to determine if some scale is maximally even, just count the number of steps between all pairs of notes. For example, the collection $A = \{0, 3, 6, 10\}$ is not maximally even since the pairs $(0, 3)$, $(6, 10)$ and $(10, 0)$ have three d distances: 3, 4 and 2.

A collection A has Myhill's property if A has exactly two c distances for every d distances. For example, the major scale has Myhill's property. But the harmonic minor scale $U = \{0, 2, 3, 5, 7, 8, 11\}$ does not have Myhill's property. For two consecutive notes, there are three interval qualities, namely (B-C, C-D and A flat-B). The scale U is not maximally even.

Let N be the degree of the chromatic universe (N-tet), and k be the number of notes of the scale A. Usually, a well-formed scale A is said to be a diatonic scale if A is maximally even. But it has been shown that if A is maximally even, then

$$N = 2(k - 1) \quad \text{and} \quad N \equiv 0 \bmod 4$$

It follows that the definition of diatonicity is not suited for all chromatic universe. Several theories have emerged, see for example [5, 7]. Eytan Agmon [2, 3] found two kinds of diatonic scales depending on the parity of N. In his theory, the diatonic scales are

$$2^{(N-1)/2}1, \quad \text{if } N \text{ is odd}$$
$$2^{(N/2-3)}12^{(N/2-4)}1, \quad \text{if } N \text{ is even}$$

All these scales are maximally even and well-formed. In [11], another definition of generalized diatonic scales was given. It is equivalent to the following. A scale A is a *generalized diatonic scale or a microdiatonic scale* if A is a scale of progressive transposition, well-formed, maximally even, built on the alphabet $\{1, 2\}$ (which corresponds to the white and black keys of the keyboard), and such that there is no two black keys side by side, and the number of the white keys minus the number of black keys is positive, minimal and different of 1. For $N \equiv 0 \bmod 4$, this definition corresponds to Agmon definition and to the *diatonicized chromatic scale* used by Ivan Wyschnegradsky and its generalization. But for $N = 13$, this definition leads to the scale 22122121 which is different from Agmon's diatonic scale. However this scale is the same as the one designed by Erv Wilson (as we can see on its keyboard plan). In the 24-tet, the *diatonicized chromatic scale* is a scale of 13 notes, constructed by two connected heptachords. Wyschnegrasky used this scale in his *24 Preludes op. 22* and *Premier Fragment Symphonique, op. 23* (see Fig. 2).

Fig. 2 Ivan wyschnegradsky, prelude op. 22 no. 4

Furthermore, this scale can be generalized to N-tone equal temperament in the following way. Let k be an integer $k \geq 2$. The set

$$W_k = \{0, 1, 3, 5 \ldots, 2k + 1, 2k + 2, 2k + 4, \ldots, 4k + 2\}$$

is the generalized diatonic scale with $|W_k| = 2k + 3$ notes, and generator $m = 2k + 1$ in the N-tone equal temperament with

$$N = m + |W_k| = 4k + 4$$

The scale W_k can be interpreted as a generalized major scale since W_k contains one chord of limited transposition $\{0, 2k + 2\}$. This covers the case for $N = 20, 28, 32,$ etc.

Another way to consider diatonicity is to introduce deep scales and to change the alphabet. For example, we can define *triolic diatonic scale* as a deep well-formed scale built on the alphabet $\{1, 3\}$. Two black keys can be side by side. In fact, the scales do not exist for all values of N (namely, $N = 12, 20, 24,$ etc.). It can be shown for $n \geq 1$ integer, the triolic diatonic scales are $113(13)^n$ for $N = 4n + 5$, $(13)^n 133$ for $N = 4n + 7$ and $113(13)^{n-1} 113(13)^n$ for $N = 8n + 6$.

The *quadriolic diatonic scales* are well-formed, deep, built on the alphabet $\{1, 4\}$. There are of the form: $(114)^n 14$ if $N = 6n + 5$, $1114(114)^n$ if $N = 6n + 7$ and $1114(114)^{n-1} 1114(114)^n$ if $N = 12n + 8$.

In the same way, *quintolic diatonic scales* are defined on the alphabet $\{1, 5\}$, and *sextolic diatonic scales* on the alphabet $\{1, 6\}$. More generally, for an integer $p \geq 3$, the *p-olic diatonic scales* are well-formed, deep scales built on the alphabet $\{1, p\}$ and are of the form:

$$(1^{p-2}p)^n 1^{p-3}p \qquad\qquad \text{if } N = (2p-2)n + 2p - 3$$
$$1^{p-1}p(1^{p-2}p)^n \qquad\qquad \text{if } N = (2p-2)n + 2p - 1$$
$$1^{p-1}p(1^{p-2}p)^{n-1}1^{p-1}p(1^{p-2}p)^n \quad \text{if } N = 4(p-1)n + 2p$$

Monotonic deep diatonic scales are deep scales defined on the alphabet $\{1, p\}$, where p is smallest as possible. These scales generalized Agmon diatonic scales (when N is odd and $N \equiv 0 \bmod 4$), but are not always maximally even. For example, 111115 for $N = 10$ and 11311313 for $N = 14$ are deep, but not maximally even.

The depth criterion is not always essential. In his *Manuel d'harmonie à quarts de ton*, Wyschnegradsky considers the *tridecatone quasi diatonic scale* $1^3 2^5 12^5$ ($N = 24$) which is a progressive transposition scale, but not a deep scale. It follows that replacing deep by progressive transposition in the definition of *p-olic diatonic scales* leads to the definition of *p-olic quasi diatonic scales*. A *p-olic quasi diatonic scale* is a well-formed, maximally even, progressive transposition scale defined on the alphabet $\{1, p\}$, with index transposition different of 1 and coprime with N. The alphabet could also be extended to $\{p, q\}$, for example to define *hemiolic diatonic scale* on the alphabet $\{2, 3\}$. In the 48-tet, the scales $54(544)^3$ with transposition index $v = 13$, and $(344)^4 4$ with transposition index $v = 11$ are maximally even and have progressive transpositions.

Furthermore, the study of diatonic scales on an alphabet of 3 letters has to be done. *Tritonic diatonic scales* are deep scales on 3 letters, and tritonic quasi diatonic scales are progressive transposition scales on 3 letters. For example, the scales $(12)^n 13^n$ if $N = 6n + 1$ and $12^n 32^n 32^n$ if $N = 6n + 7$ are tritonic diatonic scales. In the 24-tet, the tritonic diatonic scales 213311312131 or 4131131131131 play a structural role as the one played by chromatic diatonic scales. As we can see, the concept of diatonicity is far from being well understood.

5   Microtonal Modes of Limited Transposition

Another problem that I studied some years ago, is the enumeration of modes of limited transposition. Modes of limited transposition (MLT) in the 12-tet are well-known since Olivier Messiaen used them in many compositions. But it is a rather difficult question to give a way to construct MLT or to enumerate them in a given N-tone equal temperament. I first studied modes of limited transposition in quarter-tone system and found 381 modes. Later, with F. Ballon, we give a complete answer [10]. The number of modes of limited transposition is given by the following formulas:

$$L_n = P_n + \frac{2}{n}(K_n - M_n)$$

where P_n is the total number of collections (φ is the Euler totient function)

$$P_n = \frac{1}{n} \sum_{d|n} \varphi\left(\frac{n}{d}\right)$$

M_n is the number of modes $M_n = 2^{n-1}$, and K_n is given by

$$K_n = \frac{1}{2} \sum_{k=1}^{r} (-1)^{k+1} \sum_{i_1=1}^{r-k+1} \sum_{i_2=i_1+1}^{r-k+2} \cdots \sum_{i_k=i_{k-1}+1}^{r} 2^{\frac{n}{p_{i_1} p_{i_2} \cdots p_{i_k}}}$$

where n is decomposed in prime factors $n = p_1^{k_1} p_2^{k_2} \ldots p_r^{k_r}$, avec $p_i > 1$, $k_i > 0$, $(i = 1, \ldots, r)$ et $r > 0$. For a N-tone equal temperament, usual microtonal universes have some very large numbers of modes of limited transposition as shown in the following table.

Tones	N	MLT	Chords
1/2	12	16	351
1/3	18	69	14601
1/4	24	381	699251
1/5	30	2300	35792567
1/6	36	14939	1908881899
1/8	48	703331	5864062367251
1/12	72	1909580799	65588423374144427519
1/16	96	5864196582931	8252933595235898782053586451

It must not be supposed that these modes are pure abstraction. In the 24-tet, Georgy Rimsky–Korsakov used the scale 33333333. More recently, Alain Louvier wrote *Prelude et Fugue no. 2* (1978) (*Le clavecin non tempéré no 2*) in the 18 tone system. In this work, Louvier used a mode of limited transposition of interval sequence: 111311131113. In the same way, he used in *Prelude et Fugue no 3* (1973) (*Le clavecin non tempéré no 3*) the mode of limited transposition 111117111117 of the 24-tet. Today, more and more composers are interested in microtonality and its new concepts [9].

6 Plactic Modes Classification

As modes are relatively large, the goal of this section is to classify them. There are many ways do to so. Here I classify them using what is called by mathematicians *plactic relations*. Modes are identify by their interval structures, or more abstractly by letters *a,b,c*, etc. For example, 4334343 is coded *baababa*.

The plactic monoid [13] over some totally order alphabet $A = \{a, b, c, \ldots\}$ with $a < b < c < \ldots$ is the monoid whose generators are the letters of the alphabet verifying the *Knuth congruence relations*

$$\begin{cases} bca \equiv bac \text{ whenever } a < b \leq c \\ acb \equiv cab \text{ whenever } a \leq b < c \end{cases}$$

Namely, if we have two letters in the alphabet $A = \{a, b\}$ with $a < b$, Knuth relations reduce to:

$$bab \equiv bba, \quad aba \equiv baa$$

and for an alphabet with three letters $A = \{a, b, c\}$ with $a < b < c$, Knuth relations are the two relations:

$$bca \equiv bac, \quad acb \equiv cab$$

Platic classes can be represented by a graph. The vertices of this graph are the words corresponding to the modes. Two modes are connected by an edge if and only if their respective words are related by Knuth relations. A *plactic modal class* is a non trivial graph (non linear graph with more than 5 vertices), otherwise the class is a *linear plactic class*. Most of the graphs have less than five vertices. In the plactic classification, non-trivial plactic classes are interesting as they show how one move from a mode to the other when switching two notes.

In the 12-tet, the 14-modes class of some heptatonic modes is composed of church modes and some karnatic modes (see Fig. 3). In the 24-tet, the class of 14 modes (with $a = 1, b = 2$) in the 12-tet remains the same class in the 24-tet (with $a = 2, b = 4$). The dual class (reverse each word and change the name of the letters) of 14 modes has two new implementations ($a = 2, b = 7$ and $a = 3, b = 4$). The heptatonic mode 4334343 used by Alain Louvier in its *Clavecin non tempéré* belongs to this class.

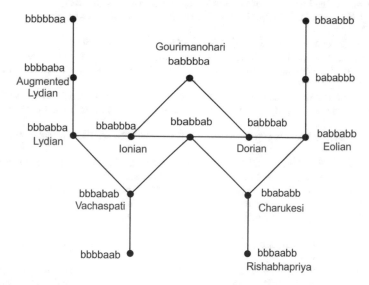

Fig. 3 Heptatonic modes class

7 Conclusion

Since the use of microtones is nowadays a standard in contemporary music, some composers like Alain Bancquart, Warren Burt, Pascale Criton, Dean Drummond, Georg–Friedrich Haas, Ben Johnston, Bernhard Lang, Michaël Levinas, Joe Maneri, Jean–Étienne Marie, Laurent Martin, Bruce Mather, Pauline Oliveros, Gérard Pape, François Paris, Enno Poppe, Alberto Posadas, Henri Pousseur, Horatiu Radulescu, Johnny Reinhard, Franz Richter Herf, Marc Sabat, Ezra Sims, Martin Smolka, Manfred Stahnke, Karlheinz Stockhausen, James Tenney, Lasse Thoressen, Toby Twining, Samuel Vriezen, Julia Werntz and many others, have shown different approaches in their use of microtones [9]. Today, new microtonal investigations require further studies in microtonality. From the first paper of Georgy Rimsky–Korsakov to the one of Alain Louvier in 1997, and some more recent papers, the investigation of microtonal modes is a great way for the understanding of diatonicity.

References

1. Ader, L.: Microtonal storm and stress. Georgy Rimsky-Korsakov and Quarter-Tone music in 1920s Soviet Russia. Tempo **63**, 27–44 (2009)
2. Agmon, E.: A mathematical model of the diatonic system. J. Music Theory **33**(1), 1–25 (1989)
3. Agmon, E.: The Languages of Western Tonality. Springer, Berlin (2013)
4. Carey, N., Clampitt, D.: Aspects of well-formed scales. Music Theory Spect. **11**(2), 187–206 (1989)
5. Clough, J.: Aspects of diatonic sets. J. Music Theory **23**, 45–61 (1979)
6. Clough, J., Douthett, J.: Maximally even sets. J. Music Theory **35**, 93–173 (1991)
7. Gould, M.: Balzano and zweifel: another look at generalized diatonic scales. Perspect. New Music **38**(2), 88–105 (2000)
8. Haba, A.: Neue Harmonielehre des diatonischen, chromatischen, viertel, drittel, sechstel und zwölftel-Tonsystems, Leipzig, 1927; herausgegeben von Horst-Peter Hesse, Books on Demand (2007)
9. Jedrzejewski, F.: Dictionnaire des musiques microtonales, 2nd edn. L'Harmattan, Paris (2003). (2014)
10. Jedrzejewski, F.: Mathematical Theory of Music. Editions IRCAM/Delatour, Sampzon (2006)
11. Jedrzejewski, F.: Generalized diatonic scales. J. Math. Music **2**(1), 21–36 (2008)
12. Johnson, T.: Foundations of Diatonic Theory: A Mathematically Based Approach to Music Fundamentals. Key College Publishing, Emeryville (2003)
13. Lothaire, M.: Algebraic Combinatorics on Words. Cambridge University Press, Cambridge (2002)
14. Louvier, A.: Recherche et classification des modes dans les tempéraments égaux. Musurgia **4**(3), 119–131 (1997)
15. Wyschnegradsky, I.: La loi de la pansonorité (version de 1953), Texte établi et annoté par Franck Jedrzejewski avec la collaboration de Pascale Criton. Postface de Franck Jedrzejewski, Genève, Editions Contrechamps (1996)
16. Wyschnegradsky, I.: Une philosophie dialectique de l'art musical La loi de la pansonorité, Version de 1936. Texte inédit, établi et annoté par Jedrzejewski, F. Paris, L'Harmattan (2005)

Proportion, Perception, Speculation: Relationship Between Numbers and Music in the Construction of a Contemporary Pythagoreanism

Juan Sebastián Lach Lau

Abstract This investigation is a departure point for understanding what Pythagoreanism can mean today, how can harmony be conceived at several time scales and what might a hierarchical model of form together with an algebra of perception entail for music composition. The study of qualitative aspects of music through mathematics is made by taking James Tenney's theory of musical form together with Alain Badiou's 'objective phenomenology' in order to imagine new ways of composing music.

1 Qualitative Numbers

The tradition commonly ascribed to as Pythagoreanism can refer to various doctrines, groups of people, disciplines and genealogies of research lead by common problems. Of special interest for contemporary musical and harmonic research is the relationship between perceptual qualities and numbers. From this standpoint, the tradition that bears this name does not begin in ancient Greece nor is it limited to a single culture or lineage, going back as far as we know through Egypt, Mesopotamia, India, China, passing also through Native American cultures as well as going forward through Semitic cultures, Scholastic philosophy and further on to mainstream modernity and involving musicians, philosophers, mathematicians and other kinds of inventors and eccentric characters not limited, as is commonly portrayed, to a single gender.

It is interesting, both as inspiration and point of departure, to think what a renewed Pythagoreanism might involve. Music composition might seem like a natural terrain for this to happen, a position that can preserve the speculative hallmark of this lineage due to its synthetic and artistic objectives, while also dealing with some of the problems that are commonly associated with this stance. Firstly, there is no interest in the mystical or sectarian facets typified by the transmigration of souls and the so-called school of *akousmatikoi*. There is also the misconstrued image of Pythagorean

J.S. Lach Lau (✉)
Conservatorio de las Rosas, Santiago Tapia 334, 58000 Morelia, CP, Mexico
e-mail: lachjs@gmail.com

© Springer International Publishing AG 2017
G. Pareyon et al. (eds.), *The Musical-Mathematical Mind*,
Computational Music Science, DOI 10.1007/978-3-319-47337-6_14

movements as expressing eschatological mentalities of intellectual (male) elites, a sociological prejudice which need not be an issue in this context. The main difficulty, however, is the danger of overextending claims based on contingent aspects of mathematical models, leading to unwarranted conclusions. The topics must therefore be approached carefully and with a pinch of skepticism to avoid extravagant speculation, while at the same time taking the ideas at the heart of scientific Pythagoreanism seriously. This aspect is embodied by the other Pythagorean school, that of the *mathematikoi*, epitomized by Archytas of Tarentum and particularly by the science of harmonics, that branch of philosophy that showed that sensible (audible) nature can be cognized through mathematical means ('means' acquiring here a double sense of 'procedure' as well as mathematical partitions and averages) [1].

In particular, to contemporize harmonics involves confronting a problem that afflicts Pythagoreanism from the inside, namely, the status of the disproportions that resist incorporation into mathematical notation. As illustrated by the discovery of the incommensurability of irrational numbers, there are aspects of reality which resist incorporation into quantitative units or ciphers, exceeding laws and explanations, be they audible but not recordable phenomena, mathematical 'monsters', logical paradoxes, etc.[1] Exceptions have always been part of mathematics and still today apparently trivial problems from arithmetic quickly lead to the frontiers of knowledge. I don't believe creativity or the lived aspect of musical experience can be mastered by mathematical models, nor is that the main reason for delving into the subject matter. In contemporary mathematics numbers are far less central or foundational than abstract structures, processes and articulations between spaces, so the conception of quantity and the problem of the 'unspeakable' may acquire new perspectives. Furthermore, there is no desire to tame or integrate these residues into ever more complex mathematics (although that happens in certain cases) but to acknowledge them as irreducible marks of contingency, as productive gaps, without that implying the collapse of the orientational role of the whole edifice, which need not be totalizing in its ambit(ion).

The transcription between the sensible and intelligible realms carries a fascination that persists, after centuries of thought and progress in science, in the process of abstraction, the measurement of qualities, the ability to go beyond what is intuitive and the aesthetic implications of formalization. Instead of the Aristotelian definition of Pythagoreanism as the principle that everything is governed by number, i.e., being *is* number, our focus is on the principle 'number is a bridge between matter and psyche' [3]. Other features are also relevant for today's harmonics such as micro-macro relationships or the string as a model, an important invariant throughout Pythagoreanism which provides links between the continuous and the discrete, between mathematics, physics and music, as well as with theories of perception. The spirit behind the proposal of a rehabilitation of Pythagoreanism necessarily commits us to a rethinking what sonorous number can mean in music and the openings this might bring to composition.

---

[1]For an account of these problems and their history see [2].

2   Harmonic Duality

Harmonic materials in music have two independent but intertwined aspects operating simultaneously: proportionality and pitch-distance. The former relates to intervallic 'characters', while the latter concerns features such as 'high', 'low', 'bright', 'dull', and corresponds to timbre, 'not timbre in the sense of spectrum, but timbre in the sense of regular pitch perception, coloured' [4]. The proportional facet comprises relations between whole numbers, concerns fundamental pitches and does not take timbre and register into account. The timbral aspect, on the other hand, involves register, spectral constitution and its main qualitative effect is sensory dissonance. Harmonicity, which does not always coincide with consonance, is proportionality's main perceptual quality.

These two aspects are entangled and their prominence and balance differs according to musical styles, performance practices, tunings, timbres and so forth.[2] They have a parallel in the division between mathematical and empirical schools in Greek harmonics as well as with the dichotomy between the discrete (arithmetic proportions, the Pythagorean approach) and the continuous (geometric pitch-distance, the Aristoxenian approach). This relationship between music and mathematics goes in two directions, as when integer harmonic means are discovered as solutions to musical problems, and, conversely, when properties of mathematical objects arise in sensory qualities, as it happens, for example, when prime numbers are understood as generating the fundamental types of harmonic intervals.

The link between numbers and sounds occurs in connection with measurement. Proportionality is indirect, having features that lie beyond the senses, pertaining to the intellect and mediated by experimental apparatuses (such as the monochord). It is empirical more than merely sensory, as is the case with pitch distance, where the relation is directly phenomenal. I agree with Michael Pisaro when he states that "perception tends to make a continuum out of the world: (our ears) are better at finding continuity than at finding fissures" [7]. Information from the world is 'folded' by perception into a qualitative immediate continuum which can be understood in terms of log morphisms that compress extensive physical sound quanta into intensive qualia (with different transfer functions for sound parameters such as pitch, intensity, timbre, time integration and so on). Proportionality can be modeled in the multidimensional harmonic space of Euler tonal lattices consisting of discrete nodes that map to linear pitch space in such ways that what is near in harmonic space does not coincide with what is near in pitch-distance space.[3] In terms of numerical structures, proportions are enfoldings of $\mathbb{N}$ into $\mathbb{Q}$, while timbre is a folding of the multidimensional space of waves inside $\mathbb{R}$. The log morphisms mediating these foldings switch the algebra of intervals from multiplicative ratios to additive distances, exchanging the identity element from 1 to 0. They wrap diverse multiples into single magnitudes.

---

[2]For more details on this entanglement and its compositional uses see [5]. The seminal research on the two facets of harmony initially took shape in the fascinating book [6].

[3]An octave or a fifth, for instance, lie close to a given pitch in harmonic space, while in 12 tone equal temperament, a semitone, which is harmonically relatively far, would be the nearest interval.

This harmonic dichotomy can also be seen as a question of orthogonality. Fourier analysis relates multidimensionality and multiple degrees of freedom to linear information: proportionality as multi-dimensionality is projected into linear pitch-height. Harmonic space projects into pitch space in the same way as a waveform, which can be depicted as a surface on the $\mathbb{C}$ plane in terms of vectors, phases, poles, etc., folds into a single pitch with timbre. Timbre space operates in an intuitionist way, with no excluded third and no order relation. By way of the Fourier transform, n-dimensional orthogonality is projected into a linear scalar number (the fundamental) with a timbral qualia (represented by the coefficients of its partials).

A question arises: can this duality that belongs to intervallic perception be extended past the timescale of immediate harmony and beyond the attribute of pitch?

3   Time Scales

The present interest in mapping out connections and analogies between mathematics and perceptual fields in music follows composer James Tenney's theory of musical form as a nested hierarchy of polyphonies, with a discrete/continuous polarity within each level: a morphological, continuous, facet of audible parametric contours in time, and a structural, discontinuous aspect of relations between parts and between parts and wholes (see [8, 9]). A mereology of perceptual fields, of interlocking objects, where at each scale there is an assimilation of differences under determined identities which are different from those at other scales. The underpinnings for realizing this synthesis are prompted by philosopher Alain Badiou's *objective phenomenology*, a particular philosophical reading of set-theoretical and categorical ideas (see [10, 11]), where set theory provides the *noumenal* material for the construction of phenomenal *logics* which govern differentiations into orders and degrees of intensities.

Tenney approached the question of form and content through these temporal scales, where forms at a given level become the content at the next higher one. In terms of sets, the elements of a level are composed from the powerset of elements at its next lower domain, *urelements* which become the (quantitative-multiple) forms that coalesce into (qualitative-unitary) matter at a higher scale. The *logic* at a particular level determines the nuances, qualities and thresholds of appearing by means of an *appearance* function that attributes to each pair of elements of the underlying constituting set, an element of a set whose elements represent degrees of relation. Each level operates through a specific *protocol of differentiation* that captures the multiples that appear in music through a network of differences and identities.

Perceptual constraints make the relative scale of each temporal domain have a specific quality to it, yielding three main strata, notwithstanding the fact that each piece of music produces its own context with any number of extra intermediate levels. These main qualities of levels correspond to the micro, meso and macro scales. There is a strong *Gestalt* of the clang at the micro to meso level, its qualities resulting from the contracting of vibrations into a single qualitative state of timbre, color, note, etc.

The meso level of sequences involves comparative memory, and the macro level of piece constitutes the architectural form of a work.

Levels are inaccessible among each other, finitude is relative to a model (in reference to Skolem's *finitization of infinity*). Urelements are always finite with respect to their powerset, forming a chain which in principle goes both up and down indefinitely. Material is extensional, perception intensional at the next higher strata, which marks the horizon of inaccessibility, the potential infinity from the point of view interior to this level, outside of which it is intensionally accessed. What is quantitative becomes qualitative 'at the limit', when a finite, countable set is seen from the 'point of view' of its inaccessible powerset.

What do these levels suggest from the different perspectives of music, psychology and mathematics?

4 Levels of Perception

The following table shows some categories that help us understand music in terms of levels. Between content and context lies the principal playground for music. There are the scales of elements, clangs, sequences and piece according to Tenney's view, next to which we can see musical notions and some continuous and discrete qualitative kinds parallel to them. The highest level of context has to do both with the general atmosphere and the space of the situation that surrounds a piece, as well as to the style and genre of groups of pieces. Analogous to atmosphere is some kind of macro structure associated with long term pieces as well as to sociological aspects of music making. Below follow form, morphology, profile and timbre in continuous formations, as well as architecture, structure, pattern and proportion in discreteness. These categories are not fixed, as it is difficult to imagine sufficiently general concepts that adapt to all kinds of music, which after sufficient reflection can be replaced by other categories that can better trace them. This table should be seen as a starting point for further research, both theoretical and musical (Fig. 1).

The last two columns show psychological processes corresponding to each level and next to it are their comparable relevant mathematical fields: in the first level we find the tools of harmony (harmonic arithmetic, means, harmonic space and metrics, etc.) that can also relate to networks of proportionality in durations; there is also the tools of logarithmic pitch with its compressions, expansions and rotations as well as the 'smooth space' of non metric durations. To give an example of how these discontinuous/continuous polarities rely on each other while being epistemically independent, think how the same continuous morphological profile can convey very different perceptual information depending on which subjacent discrete pitch grids are used to realize them.

The next level involves combinatorics (groups, graphs, knots, etc.) of patterns of units integrated from the lowest level. The corresponding continuous field could

scale	level	music	quality continuous/discrete		psyche	math
micro	**content**				subliminal stimulation	
	element	material	timbre	proportion	sensory perception	$Q \leftrightarrow R$, harmonic space, Fourier
meso	**clang**	rhythm / texture	profile	pattern	proprioception / echoic memory	combinatorics / math morphology
	sequence	method / technique	morphology	structure	working memory	processes / networks
macro	**piece**	aesthetic model / narrative	form / drama	unity / architecture	episodic memory	interactions / schemes
	context	space/place styles/genres	atmosphere	global/social structure	'autonoetic' memory / social memory	

Fig. 1 Table of time scales according to different perspectives. The *level* column contains Tenney's categories, the next one general musical ones, followed by qualitative musical aspects (divided into continuous/discrete), psychological processes and mathematical structures that could pertain to each level

relate to the mathematics of morphology and the continuous mutability of profiles.[4] At the next higher level, the fields we suggest have to do with setting up processes and networks of relations, interactions of already given musical forms that might be studied through categories and morphisms, comprehending, at a higher level of abstraction, relations among the structures and parametric spaces that have been presented at lower levels. This can go further into the macro levels of drama and narrative organization (both directional and non directional) in consonance with the overall aesthetic models that pertain to atmosphere and style.[5]

Statistical measures at each level can also explain and be useful to generate the distribution of musical structures. Tenney proposes the idea of *ergodic* form to explain non directional morphologies while directed processes can be grasped in terms of parametric densities and limits.

Musical material should be understood as a hypothesis, not just as an inert 'stuff' manipulated arbitrarily by forces of thought, but also having a say in the process of creation, getting to kick back and impose constraints. Abstraction is a back and forth process where thought, imagination and matter (however immaterial musical and sonorous materials might seem) influence each other. Matter is not pre-constituted but must be understood as information is gathered while it is manipulated, thereby making up the material during the process: its constructibility becomes isomorphic

---

[4]In the wake of Tenney, two of his colleagues have proceeded to study each of these aspects in turn through morphological [12] and structural metrics [13].

[5]Some suggestions in this direction are pursued in [14].

to its understanding. This approach to creativity is experimental in the sense that it requires an empirical intervention and is not reduced to pure speculation or the imposition of abstract ideas on passive, maleable material. Art can be considered a form of engineering,[6] and the approach we take by means of an algebraic phenomenology of sorts means that art can also reverse-engineer the conditions of its reception and production.

5 Objective Phenomenology

Tenney's theory comes from phenomenology and Badiou sets up a phenomenology based on the theory of locales, providing tools for complementing and going beyond the *Gestalt* principles of similarity, continuation, closure, proximity as well as the figure/ground dichotomy. Phenomenology, as the theory of appearing and objectivity, concerns relations between qualitative differences and an order structure that provides the unity through which a manifold is unified as an object. When a thing (a set) is localized in a world, this is because the elements of the set are inscribed within a distribution of degrees over all the differences that appear in this world, the *disposition of the infinite hues of a concrete world* ([16], p. 51). An object is a multiple associated with an evaluation of the identities and differences immanent to it. There are many types of orders and, consequently, many possibilities for the logical organization of worlds. Two worlds with the same things can be absolutely different from each other because their transcendental gradings are different. There is always in a world a certain number of limits to appearing's intensity.

Lets give a short summary of Badiou's theory[7]:

- Beings can be determined in their abstract form by the pure multiple of set theory, as the infinite composition of elements with a metaphysical stopping point at the void set.
- Elements compose localized entities in relation to each other within situations or 'worlds'.
- Beings can appear in different situations while being ontologically the same; a multiple co-belongs in general to many worlds.
- The appearance of an entity implies differences with itself and with other entities through degrees of gradation: a relational network.
- A transcendental is the operator set that allows giving meaning to the 'more or less' of identities and differences in a determinate world.
- The scale of evaluation of appearing depends on the situation. These degrees do not depend on any subject or consciousness. There is no privileged frame of reference (listener, performer, composer, for instance). The conditions for experience of a subject are not pre-given, there is no subjective receptivity nor constitution because

---

[6]For more on this interesting topic of abstraction and material, see [15].

[7]It is beyond the scope of this article to provide a comprehensive account of Badiou's theory. For more information see [10, 11, 16].

the transcendental is an intrinsic constitution of being as it belongs to the same world as the appearances.

- Transcendentals are local, there are many of them, 'difference is differentiated'.

The algebraic structure of a world, corresponding to a Heyting algebra,[8] is the following:

- The set A that ontologically subtends a situation, the 'material' set.
- A transcendental T, the set of degrees of appearance.
- Degrees of identity $\mathbf{Id}(\alpha, \beta) = p; \alpha, \beta \in A, p \in T$.
- Order relation, $\leq$, obeying reflexive, transitive and antisymmetric axioms.
- Minimal degree, μ.
- Conjunction, $\cap$.
- Envelope, $\sum B$; regions defined by a predicate over intensities of subsets B of A.
- $\cap$ is distributive with regard to $\sum$.
- Dependence of degrees, $p \Rightarrow q$; the degree of connection between two intensities.
- Every degree admits a reverse and $p \cap \neg p = \mu$.
- Maximal degree, $M = \neg \mu$.
- The phenomenon of a relative to A:
  $\Phi(a/A) = \{a, [\mathbf{Id}(a, x_1), \mathbf{Id}(a, x_2), \ldots, \mathbf{Id}(a, x_\alpha), \ldots]/x_\alpha \in A\}$; the set of a and the degrees of appearing of all x's which co-appear with a in A.
- Degrees of existence, $\mathbf{E}x = \mathbf{Id}(x, x)$; the extent to which x appears in a world.
- Phenomenal components, $\pi(x) = p$; identity function with respect to a fixed degree.
- Atoms of appearing (phenomenal identities).
- Objects, $(A, \mathbf{Id})$, a support set together with a transcendental indexing.

  - Localizations, $a\lceil p = \pi(x) \cap p$; a local decomposition from the spectrum of intensities.
  - Compatibility, $a \ddagger b$, $a\lceil \mathbf{E}b = b\lceil \mathbf{E}a$; atomic equality through reciprocal localization on existences: a and b are compatible if they are in 'the same zone of existence'.

- Proper inexistent of an object, $\varnothing_A$; $a \in A$ inexists if $\mathbf{E}a = \mu$. An element a of an object is said to be its inexistent if its being is attested but its existence is not. Every object admits of one (and only one) inexistent.
- Transcendental functor, $\mathbf{F}A(p) = \{x/x \in A \text{ and } \mathbf{E}x = p\}$.

There is a lot of work to do in order to think through what this structure can mean musically, both from the point of view of understanding perceptual cues as well as to generate new ones.

Maybe the transcendental functor, which marks the territories or the 'retroaction of appearing on being' ([11], p. 221), associating to every element p of T that part of A composed of x's such that $\mathbf{E}x = p$, could be interpreted as a connection from

---

[8]To further our investigations we might turn away from Badiou towards more mathematically oriented literature. Also, Badiou's theory of change and the Event is not directly relevant to our purposes (although it is not incompatible either).

the percepts to the conditions of production of those percepts, between listener and producer, referring the phenomenal realm back to the material facts of music, an operation common in music composition. On the other hand, the inexistent, which is local and 'testifies, in the sphere of appearance, for the contingency of being-there' (*ibid*, p. 324), i.e., is present materially but does not appear, is a measure for what can happen to a world, and can thus become pivotal in delineating the unexpected, contingent and open, so this approach is compatible with musical indeterminacy.[9]

Especially interesting is the positing of atoms of appearance by way of language propositions which function conceptually at a high level of aesthetic abstraction, also showing how this model can be adapted to simple settings which do not need insinuate any explicit use of mathematics. Atoms can also be defined to posit appearances by way of arbitrary functions, sampled material, descriptions rather than definitions, both real and fictional (recordings, data for sonification, random distributions, patterns, algorithms in general, translations from other media or disciplines, and so on). Following Zalamea [17], the former kind can be said to be *eidal* in mode, while the latter are *quiddital*. Any degree of mixture between eidal and quiddital modes is possible.

It is not immediately obvious what the reverse of a particular sound may be. It has to do with absence but not necessarily with silence. In a sense it is all that begins when a thing (a sound) 'ends', i.e., its form (and in sound this implies abstracting time spatially along with other attributes). It is what contains the sound, the region of the world that envelops it. There are many possible concrete ways in which reverses can happen in music, for example in remainder sonorities, as in Alvin Lucier's *Slices* for cello and orchestra (2007), where a large orchestral cluster is punctuated note by note with 'holes' by the solo cello.

Envelopes, conjunctions, dependencies, phenomenal components, atoms, compatibilities and inexistents have to be imagined for music in general as well as for specific pieces.

6 Analysis and Synthesis

Badiou's theory does not explicitly deal with mereologies and here the meeting with Tenney might prove fruitful. Atoms and appearance functions can be 'plugged in' to each temporal scale, either as perceptual transfer functions or as the extension of concepts, properties or metaphors the give rise to gestalt-like forms over a ground. They can be defined both from the top-down and from the bottom-up, in any order and at any level, engendering hierarchies whose structure can change over time. These functions determine what musical variables are to be taken into account and the arbitrariness of it means that they do not have to be tied to traditional musical parameters, can appear from a variety of perspectives (listener, performer, author, situation, etc.) and incorporate multi-modal media (other 'senses': visual, performa-

---

[9]I'm interested in establishing collaborations both with musimathicians as well as with mathemusicians in order to find out what can be made of these ideas.

tive, theatrical, etc.). Crossing the natural/cultural distinction, musical experiences can be launched that have political consequences in that they are driving forces for a subject which is not presupposed.

It is a question of dual eidal ascents and quiddital descents: the expansion and dilation of 'the entanglement between thought and matter, the intelligible and the sensible' ([15], p. 19) (synthesis), as well as the contraction of thought to a point (what is it like to 'be' a grain of sound?) (analysis).[10] The creative process becomes a coming and going between real and ideal as diverse forms of transit between multiple material and conceptual strata, where the notion of parameter becomes enlarged to a perceptual field, a phase space or manifold with intrinsic properties (curvature, orientability, symmetries, connectivities boundaries, etc.).

Acknowledging and intervening in the interweaving between the continuous and the discrete can impart more depth and dimensionality to these parameter fields. Inner periodicities encode within degrees of intensity 'hues' that stand out from other saliences; contours can have qualities that go beyond the up/down and long/short dimensions of morphology, adding breadth to spatial relationships and alluding to the vertical from within the horizontal.

As much as this algebra permits thinking music and perception in terms of mathematical structures, it can also be used inversely to imagine what these structures might imply as applied to music: to imagine transcendentals with arbitrary limits and conditions of individuation and appearance going against the grain of intuition. This is the speculative aspect, where instead of modeling nature and music mathematically, new musical thresholds can be imagined out of the mathematical structures that prescribe intersections involving quanta and qualia indiscriminately.

Finally, there is also the perspective from the totality, where the reciprocal actions between levels and their morphologies, in loops between qualities and time scales, take place to produce something which is more than the sum of the parts: a harmony.

References

1. Barker, A.: Harmonics in Classical Greece. Cambridge University Press, Cambridge (2007)
2. Heller-Roazen, D.: The Fifth Hammer. Pythagoras and the Disharmony of the World, Zone Books, New York
3. Watkins, M.: Prime evolution (interview). Collapse **1**(1), 93–189 (2006)
4. Gilmore, B.: Clarence Barlow interviewed by Bob Gilmore, Amsterdam, 1st August 2007, Paris Atlantic Magazine (2015). www.paristransatlantic.com/magazine/interviews/barlow.html
5. Lach, J.S.: Harmonic Duality. From interval ratios and pitch distance to spectra and sensory dissonance, Leiden University (2012)
6. Barlow, C.: Bus Journey to Parametron. Feedback Papers, Cologne (1981)
7. Pisaro, M.: Continuum Unbound (notes to CD). Gravity Wave, Jersey City (2014)
8. Tenney, J.: Form in 20th century music. In: Vinton, J. (ed.) Dictionary of Contemporary Music. E.P. Dutton, New York (1974)

---

[10]"As the thought experiment is fully immersed within the material system, it permits the abstracting force to assume material behaviors and new generative schema otherwise unavailable to an isolated account of thought trapped in naive intuitions of itself."[15], p. 24.

 9. Tenney, J.: Meta-Hodos and META Meta-Hodos. Frog Peak Music, Lebanon (1988)
10. Badiou, A.: Mathematics of the transcendental. Bloomsbury Academic, New York (2014)
11. Badiou, A.: Logics of Worlds: Being and Event. Bloomsbury Academic, New York (2014)
12. Polansky, L.: Morphological metrics. J. New Music Res. **25**(4), 289–368 (1986)
13. Winter, M.: Structural Metrics: An Epistemology. University of California, Santa Barbara (2010)
14. Vriezen, S.: Action Time, The Ear Reader (2014)
15. Negarestani, R.: Torture Concrete. Sequence Press, New York (2014)
16. Badiou, A.: Second Manifesto for Philosophy, Polity (2011)
17. Zalamea, F.: Filosofía Sintética de las Matemáticas Contemporáneas, Editorial Universidad Nacional de Colombia, Bogotá (2009)

Topos Echóchromas Hórou (The Place of the Tone of Space). On the Relationship Between Geometry, Sound and Auditory Cognition

Jaime Alonso Lobato-Cardoso

Abstract Based on the spatial composition method proposed by the author and its application in the piece, *Materia Oscura* (This work was premiered at MediaLab-Prado, Madrid, Spain on October 11, 2013.), some geometric representations that allow the description and documentation of the relation between cognition, sound and space are proposed. The purpose of developing this analysis is to establish a formal precedent for future studies related to different perceptual skills with which we abstract three-dimensional space information through our ears. The applications of these studies range from artistic creation to the development of educational tools for music and mathematics.

1 Introduction

Since the beginning of my career, I began doing spatialization experiments with chamber music groups. Then, for accuracy reasons, I changed the musicians for speakers setups and the scores for computer programming to continue my research.

This approach to sound spatialization allowed me to include in my compositional language concepts like position in space, trajectory or symmetry. Superficially, these notions are thought to be more linked to the visual world than to the sonic one, so I pay attention to several ways in which humans can interact with space through listening. Eventually, I came across the concept of *echolocation* that is a perceptual skill that, through self-production of sound, can offer information to the listener on the three-dimensional qualities of space and the objects within it.

Then, several questions on the artistic level arose, like what nature a piece in which the listener would actively interact through sound production should have, what I would be interested in transmitting with this work and whether echolocation would be the best way to represent it, or what the best place would be to present to the public.

J.A. Lobato-Cardoso (✉)
SEMIMUTICAS-IIMAS, UNAM, Circuito Escolar 3000, Ciudad Universitaria,
Coyoacán, D.F., Mexico
e-mail: jaimelobatocardoso@gmail.com

© Springer International Publishing AG 2017
G. Pareyon et al. (eds.), *The Musical-Mathematical Mind*,
Computational Music Science, DOI 10.1007/978-3-319-47337-6_15

The current article is the result of the need to develop methodological and conceptual tools for shaping pieces based on the features of the space where the listener is. The main goal of this paper is to propose geometric representations or coordinate systems that allow us to describe the relationship between cognition, sound and space based on the spatial composition method and to document its application in my work, as a composer, as well as to open new lines of research for my creative activity.

2 Spatial Composition Method

After finding an article about the physical characteristics of the best self-produced sounds for echolocation I got in contact with the researchers who wrote it and I carried out a research residency with them. There, Juan Antonio Martinez-Rojas introduced me to a technique derived from echolocation that is under experimental development called *evanescent perception*. Upon realising that echolocation is not the only ability available to interact with the space, I opened the investigation to a method that would allow me to integrate all these skills and not only to describe echolocation.

In a traditional method of composition, the musical parameters with which a composer works to structure a piece may be: pitch or notes (melody/harmony), temporal articulation (rhythm) and tone or harmonic content (instrumentation). In analogy, I chose three parameters for my system: binaurality, echolocation and evanescent perception (remote vibroception) (See Fig. 1).

2.1 Binaurality

Binaurality refers to any hearing performed simultaneously with the two ears. This capability given by the own human physiology allows us to interact with the space

Fig. 1 Intersection of aural perception-cognition

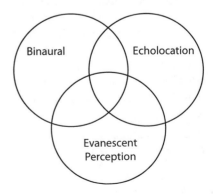

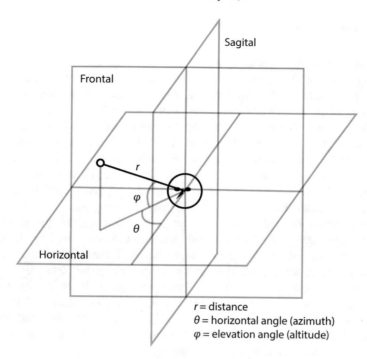

Fig. 2 Spherical coordinates of an auditory system

due to the fact that our brain can interpret, spatially, the differences between the sounds that reach us. In other words, the fact that we have two ears placed in two different locations allows us to perceive things from two slightly different listening positions. These interaural differences allow us to locate sonorous objects in space.

To locate a sonorous event in space, we can use spherical coordinates due to the perceptual scope of our auditory system and due to the fact that the position is always related to a point of origin, that is, a listener. The position is given by the distance, r, and two angles; one, a horizontal plane or azimuth, θ, the degree zero, and the other one that rises on this plane, the altitude, φ. (See Fig. 2).

This position will be interpreted by the spectromorphological variable of the sound in each ear. The first one is the Interaural Level Difference ILD, that occurs due to the fact that the head separates the ears and acts like a screen, therefore, if the sonorous event happens to the left of the listener, the sound that reaches the right ear, filtered by the head, will have a smaller amplitude. The second one is the Interaural Time Difference ITD that also depends on the angle of positioning of a source.

There is an important monaural indication, expressed by the spectral differences of a sonorous event in relation to its position. This transformation is due, mainly, to the filtering action of the pinna in relation to the sonorous source.

2.2 Echolocation

Echolocation is a perceptual skill possessed by some mammals and birds, linked to the ability to obtain information about the three-dimensionality of the space in which they are immersed as well as information of the eminent bodies found within. This is done through processing and interpretation of sound reflections (reverberation and echo). To be able to develop this sense, the listener needs to produce sound in order to make a comparison between the spectromorphology[1] of the direct sound and the reflections of this same sound that reaches the ears with a time delay and a spectral difference. Sound is a wave spread through the air in three dimensions. This allows the sound produced by the listener to bounce off the objects around so the listener can have at least two different sounds to compare. (See Fig. 3).

The best sounds for echolocation are claps, palatal clicks and the English phoneme ʃ, and depending on the way they are produced, they can be more or less directional.

This is an innate sense in humans and we use it all the time unconsciously for orientation and balance. An ability or special condition, like blindness, is not needed to develop it. We all have the physiological composition needed to develop it. One of the clearest examples of our capability of abstracting three-dimensional and metric characteristics of space is how easily we are able to notice the difference between the reverberation of a cathedral and a small room. Echolocation, like binaurality is a cognitive operation consisting on comparing sound.

2.3 Evanescent Perception

Juan Antonio Martínez Rojas who works for the Department of Theory of Signal and Communications at the University of Alcala, develops a line of research on applied physics to biomedical and cognitive sciences, so he got interested in developing a research on the physical characteristics of different organic signals (sound produced

Fig. 3 The difference between mouth-ear and mouth-wall-ear distance can be interpreted by the brain and help us to find silent objects in space

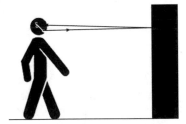

---

[1]Denis Smalley developed the term *spectromorphology* in 1995 as "tools for describing and analysing listening experience. The two parts of the term refer to the interaction between sound spectra (spectro-) and the ways they change and are shaped through time (-morphology). The espectro- cannot exist without the -morphology and vice versa: something has to be shaped, and shape must have sonic content" [4].

by the listener) for human echolocation [2]. During his research he realized that: "When natural palatal clicks or certain body vibrations are structured as precise rhythmical patterns, a new sensing modality emerges which does not depend on echo hearing. This new perception, remote or evanescent vibroception, is similar to a vibrotactile exploration of objects, but without direct touch. This sensory ability can be quickly developed through a relatively easy training technique... Analysis of experimental data strongly suggests that remote vibroception can be interpreted as a complex combination of acoustic tunneling of infrasound vibrations near the main resonance of the human body, vibrotactile perception without hearing, and both neural and cognitive biofeedback." [1]. This mode of perception is still in development and research.

3   Materia Oscura (Dark Matter)

The title of this piece refers to the concept of astrophysics used to name the hypothetical matter that does not emit enough electromagnetic radiation to be detected with current technology, but whose existence can be inferred from gravitational phenomena that it causes to the visible matter. This is in analogy to the possibility of perceiving and interpreting acoustic differences depending on certain characteristics of space that cannot be seen, like the mass or density of objects around us. The piece attempts to transform an urban space into a sound sculpture, but not by adding items to those that are already part of it or by modifying the existing ones, but by intervening the cognition of the audience with auscultation dynamics of the space chosen. The proposition is to perform this cognitive intervention by different listening exercises starting from the three parameters of the method.

First of all, a short lecture about sound and its physical parameters is provided, such as frequency, amplitude, tone, etc. So that a common language can be established among the audience in a way that they can describe their experience during the intervention, in case they are not familiar with sound technicalities and its nature.

In regard to binaurality, the audience performs listening exercises designed by the composer Murray Schafer [3] and they are invited to have the chance to spatialize some tracks in a surrounding speaker setup. Then, we can have two different situations to locate sonorous sources, in a controlled environment (quotidian) and in a non-controlled one (electroacoustic). In the non-controlled one, we perceive the sonorous sources thanks to the binaural process (See Fig. 4), in the controlled one, we can create the illusion that there are virtual sonorous sources due to fact that it's possible to produce several spectromorphologicaly identical sounds at the same time, in this way the ITD mechanism can be deceived. By controlling the amplitude of each sound we can deceive the ILD and even animate the virtual position in space. (See Fig. 5).

The following coordinate system intends to describe some situations of electroacoustic spatialization in a controlled context: the first two ordinates are polar and they refer to the sound spectromorphology, the first distance, r, which represents the amount of amplitude, the second is an angle, θ, that represents the tonal similarity,

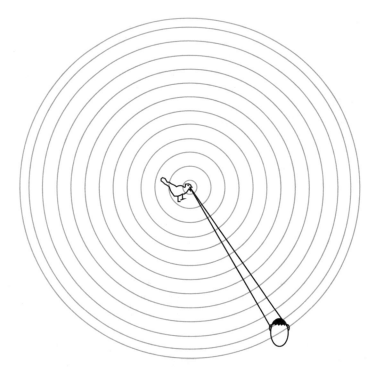

Fig. 4 In this sample the bird is located to the left of the listener, so the sound will arrive first to the left ear and with more amplitude as well

from 0 to 180° the transformation trend is towards the difference, from 180 to 360° the transformation trend is towards the similarity. The third ordinate is Cartesian and it represents the location in the physical space of the virtual sound source, ranging from −1 to 1. The last one is an index from, 0 to 1 that represents the way in which the brain interprets the positions of sounds, if the index is 0, then the audience computes the interaural differences with the binaural classical skills. If the index is 1, it means that the sounds are identical, spectrally speaking, and the binaural system can be deceived to perceive a virtual sonorous source. Three scenarios in stereo setup are shown in the example. (See Fig. 6).

For echolocation, two proposals are presented. The first one consists in listening to the filtering capabilities of the space chosen. The formal and materic qualities of each space allow us to have the same effect on the sound than on the pinna. A speaker is placed to spread a white noise approximation (WNA) with sufficient amplitude to be heard at all points and the audience is asked to freely walk around the room to start discovering an architecture hidden from view but that coexists with the visual one thanks to the way in which the space filters sound. The proposal of coordinates, in this case, consists in combining, first, two Cartesian coordinates to find specific locations or trajectories in the physical space. Secondly, we have a barycentric coordinate system within a polygon with the number of sides by which one wants to divide

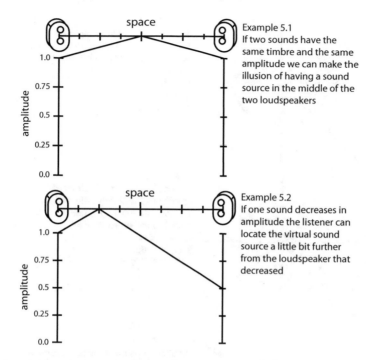

Example 5.1
If two sounds have the same timbre and the same amplitude we can make the illusion of having a sound source in the middle of the two loudspeakers

Example 5.2
If one sound decreases in amplitude the listener can locate the virtual sound source a little bit further from the loudspeaker that decreased

Fig. 5 Sound source deplacement by correlation of spatial perception

the sound spectrum of the WNA. The energy changes between each frequency band shall be represented by the distance between a point and the center of each side of the polygon.

Secondly, an exercise to understand the difference between echo, reverb and standing wave is proposed, as well as the concept of spectral harmony that refers to the frequencies that are reinforced or canceled in reference to a specific point in the physical space. For this geometric representation, two ordinates will be combined. The first one, the width of a line that describes the trajectory recorded and the second one, a representation of the spectrum of reflections that we want to document. The two coordinate systems applied to a specific space are shown in the example (see Fig. 7).

These two examples allow us to find two cases in the perception of space with echolocation. One in which we process information about the space around us. The second one, where we focus on the objects that are in this space, affecting the total configuration of space, but which are surrounding us. Depending on the kind of sound that we use for echolocation we are be able to get more information from one case or the other.

As for the evanescent perception, due to extension limitations, we only mention that spherical coordinates can be used to show the relationship between space of

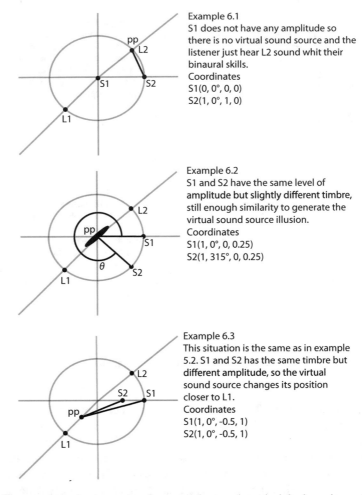

Example 6.1
S1 does not have any amplitude so
there is no virtual sound source and the
listener just hear L2 sound whit their
binaural skills.
Coordinates
S1(0, 0°, 0, 0)
S2(1, 0°, 1, 0)

Example 6.2
S1 and S2 have the same level of
amplitude but slightly different timbre,
still enough similarity to generate the
virtual sound source illusion.
Coordinates
S1(1, 0°, 0, 0.25)
S2(1, 315°, 0, 0.25)

Example 6.3
This situation is the same as in example
5.2. S1 and S2 has the same timbre but
different amplitude, so the virtual
sound source changes its position
closer to L1.
Coordinates
S1(1, 0°, -0.5, 1)
S2(1, 0°, -0.5, 1)

Fig. 6 Three scenarios in stereo setup, from spatial perception principle shown in previous figure

resonance of the listener and the rhythmic changes in the rhythmic patterns that allow to structure this sense. (See Fig. 8).

4   Conclusion

Thanks to the description of these perceptual phenomena, through coordinate systems, we can better understand their nature and infer certain differences and similarities based on the study of their geometries and topologies. This first paper intends to be an introduction to further studies resulted from this theoretical and artistic expe-

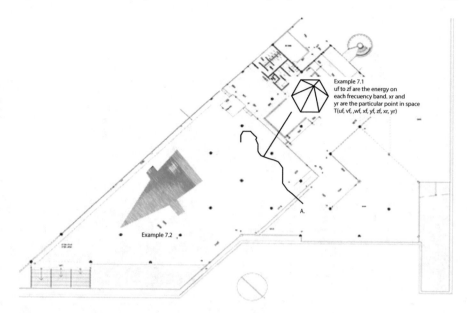

Fig. 7 First example from my presentation of my piece *Dark Matter* at museum Paço Das Artes, Saõ Paulo, Brasil, May 2014

Fig. 8 Second example from *Dark Matter* at Paço Das Artes, 2014 (aural-spatial design)

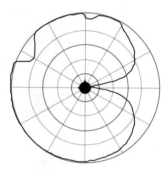

rience. Finally, the importance of art as a generator of meaning and how it can help in various areas as a tool for knowledge creation becomes evident.

References

1. Martínez-Rojas, J.: Remote Vibroception: Can Humans Sense Through Acoustic Tunneling?
2. Martínez-Rojas, J., et al.: Physical analysis of several organic signals for human echolocation: oral vacumm pulses. en Acta Acustica United with Acustica **95**(2), 325–330 (2009)
3. Schafer, R.M.: A Sound Education. Arcana Editions, Indian River (1992)
4. Smalley, D.: Spectromorphology: explaining sound-shapes. Organized Sound

Models and Algorithms for Music Generated by Physiological Processes

Jaime Alonso Lobato-Cardoso and Pablo Padilla-Longoria

> *This science [mathematics] is the easiest. This is clearly proved by the fact that mathematics is not beyond the intellectual grasp of anyone. For the people at large and those wholly illiterate know how to draw figures and compute and sing, all of which are mathematical operations.*
>
> Roger Bacon, c.1265

Abstract Generative art emphasizes processes. On the other hand, mathematical models are used to understand underlying biological, economic, physical or social phenomena (among others). The outcome of such models can be considered as processes in their own right. For instance, physiological processes give rise to a wide variety of signals which can, in turn, be detected by changes in pressure, temperature, electrical potential and so on. When measured and converted with an appropriate transducer, they constitute the raw material which algorithms and models may translate into sound. In this paper we explore a mathematical model of the human circulatory system based on differential equations. We then use this model as a generator of melodic and rhythmic structures in a compositional multimedia context.

1 Introduction

Generative art is focused on processes and not only on the outcome. This is probably the reason why an algorithmic approach in creative disciplines has become more and more a subject of interest. On the other hand, the possibility of controlling in a very precise way the context and parameters with which such algorithms are

J. Alonso Lobato-Cardoso (✉)
SEMIMUTICAS-IIMAS, UNAM, Cd. Universitaria, Cto. Escolar 3000,
Coyoacán, D.F., Mexico
e-mail: jaimelobatocardoso@gmail.com

P. Padilla-Longoria
IIMAS, UNAM, Cd. Universitaria, Cto. Escolar 3000, Coyoacán, D.F., Mexico
e-mail: pabpad@gmail.com

© Springer International Publishing AG 2017
G. Pareyon et al. (eds.), *The Musical-Mathematical Mind*,
Computational Music Science, DOI 10.1007/978-3-319-47337-6_16

153

being implemented provides flexibility in manipulating the basic material an artist has at his disposal. Here is not the appropriate place to discuss in detail the trends and most representative positions in this matter (for such a discussion, including an introduction about the circulatory system as an endosymbiotic context for music, see [2]). However, we would like to discuss here what in our opinion might constitute a significant and interesting line of creative work. More specifically, we would like to discuss the role mathematical models might have as process generators in their own right. It is rather natural to think that the resulting description of a real process by means of a mathematical model, in the form of a simulation for instance, can be used in the same way as other algorithms.

In recent years the use of biosensors in the context of multimedia art has attracted lots of attention. Rather than obtaining data or information in general from direct measurements, we propose to explore the use of mathematical models of physiological processes as sources of data for creating compositional tools complementing or interacting with information obtained by means of such sensors. In this paper we limit ourselves to a relatively simple mathematical model of the human circulatory system, which is given by ordinary differential equations. Then we solve these equations numerically obtaining two time series, for the circulatory pressure and for the blood flow respectively. Later on we transfer these data generated in MATLAB to Supercollider, in order to manipulate them and construct rhythmic and melodic structures. We also automatically generate a score with this material. We would like to point out that what we are presenting here is not a final work or piece of electroacoustic music, but rather a compositional tool that can be used to generate sound material in different compositional and sonification contexts.

2   The Model

Dynamical systems models have traditionally been used to study and understand natural processes. In particular, they constitute a suitable methodological framework to deal with the evolution of systems with time. Quite often the model leads to a system of differential equations. In a physiological model this is rather complicated and involves not only the dynamical understanding of the variables, but also stability, robustness, control and feedback aspects. In order to simplify our presentation we consider a model for the blood pressure and blood flow in the human circulatory system. It is derived and explained in detail in [1], Sects. 1.11 and 1.12. Here we write down the corresponding equations

$$C_{sa}\dot{P}_{sa} = Q_{A_0} - P_{sa}/R_s, \tag{1}$$

where $P_{sa}(t)$ is the arterial pressure, C_{sa} is the systemic arterial compliance, Q_{A_0} is the outflow from the left ventricle through the aortic valve into the systemic arterial tree and is a given periodic function of time, each period being a heart beat (see Fig. 1). Finally R_s is the systemic resistance.

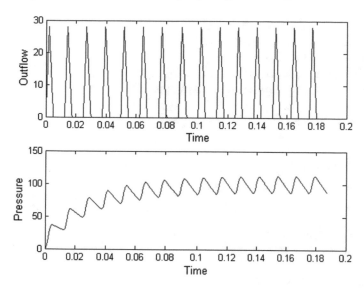

Fig. 1 Typical simulation

3 Numerical Implementation

The previous equation is solved by the Euler's method. Let us observe that in other applications a more accurate implementation—for instance the use of Runge–Kutta methods—would be desirable, but for the present purposes our choice suffices. For the sake of completeness we include the MATLAB code below. It is essentially the same as that presented in the above mentioned reference and the modifications are related to the interface and sending data to Supercollider.

```
%filename: sa.m
clear all %clear all variables
clf       %clear all figures
global T TS TMAX QMAX;
global Rs Csa dt;
in_sa %initialization
for klok=1:klokmax
    t=klok*dt;
    QAo=QAo_now(t);
    Psa=Psa_new(Psa,QAo), %new Psa overwrites old
    %Store values in arrays for future plotting:
    t_plot(klok)=t;
    QAo_plot(klok)=QAo;
    Psa_plot(klok)=Psa;
end
```

```
%Plot results in one figure
%with QAo(t) in upper frame
%and Psa(t) in lower frame
subplot(2,1,1), plot(t_plot,QAo_plot)
subplot(2,1,2), plot(t_plot,Psa_plot)

function Psa=Psa_new(Psa_old, QAo)
%filename Psa_new.m

global Rs Csa dt;
Psa=(Psa_old+dt*QAo/Csa)/(1+dt/(Rs*Csa));
end

function Q=QAo_now(t)
%filename: QAo_now.m

global T TS TMAX QMAX;
tc=rem(t,T); %time elapsed since
% the beginning of the current cycle
%rem(t,T) is the remainder when t is divided by T
if (tc<TS)
    %SYSTOLE:
    if (tc<TMAX)
        %BEFORE TIME OF MAXIMUM FLOW:
        Q=QMAX*tc/TMAX;
    else
        %AFTER TIME OF PEAK FLOW:
        Q=QMAX*(TS-tc)/(TS-TMAX);
    end
else
    %DIASTOLE:
    Q=0;
end


end


% filename: in_sa.m (initialization for the script sa)
T=0.0125      %Duration of the heartbeat (minutes)
TS=0.0050     %Duration of the syastole  (minutes)
TMAX=0.0020   %Time at which flow is max (minutes)
QMAX=28.0     %flow through aortic valve (liters/minute)
Rs=17.86      %Systemic resistance (mmHg/(liter/minute))
Csa=0.00175   %Systemic arterial compliance (liters/(mmHg))
```

```
%This value of Csa is approximate and will need adjustment
%to make the blood pressure be 120/80.
dt=0.01*T          %Time step duration (minutes)
%This choice implies 100 timesteps per cardiac cycle.
klokmax=15*T/dt    %Total number of timesteps
%This choice implies simulation of 15 cardia cycles.
Psa=0
%Any intial value is OK here.
%Initialize arrays to store data for plotting:
t_plot=zeros(1,klokmax);
QAo_plot=zeros(1,klokmax);
Psa_plot=zeros(1,klokmax);
Psa_plot_red=zeros(1,klokmax);
```

4 Generation of Musical Structures

This numerical implementation in Matlab gives us as a result 1500 numbers describing a circulatory process. We send these numbers from Matlab to Supercollider via User Datagram Protocol (UDP) and keep them in a matrix. Then we apply mathematical and logical operations to obtain different melodic contours. In fact we use only a few specific transformations, but we could have applied any function to the elements of the array. Below we only give a few of these functions as typical examples:

Mapping using a linear transformation: The original data range 0.12450207–110.27082 is transformed into an audible range in MIDI numbers 60–72, using a subset of this array by

Selection of elements:

(a) Selecting elements whose indices lie within a pre-established range, for example from 100 to 250, 150 elements of the array, which constitute 10.
(b) Discard elements with even or odd indices (50).
(c) Discard a percentage of elements of the array randomly.

Tuning:

(d) Consider only the integer part of each element in the matrix.
(e) Select some ranges taking into account the decimal part, which enables us to generate a microtonal pitch profile.

Transformation range:

(f) Allow only a certain number repeated values.

Representing the contrapuntal derivations:

(g) Melodic inversion, retrograde, retrograde inversion, augmentation, diminution.

Articulation:

(h) Take some items as rests.

5 Compositional Application

This implementation is a compositional tool applied to the creation of melodic contours but it also can be used to structure a piece more focused on timbral developments or on spacialization parameters. The way in which we map the matrix to musical or sonic values is going to suggest more adequate musical forms relating the material generated by the system. Here is the corresponding code in Supercollider which receives the data generated by the model and computed using MATLAB.

```
~laLista=[];
(
~lis = OSCresponder(n, "/test", {|...msg|
~dato = msg[2][1].postln;
~laLista = ~laLista.add(~dato);
if(~laLista.size == 1500, {
~lista10 = ~laLista.collect({arg item, i; (item / 5) + 60});
~lista11 = ~lista10.collect({arg item, i; item.asInteger});
~itembuf = 0;
~lista12 = ~lista11.collect({arg item, i;
if(item == ~itembuf, {item = 0}, {item; ~itembuf = item;});
});
~lista13 = ~lista12.reject({arg item, i; item == 0});
~lista14 = ~lista13.collect({arg item, i;
if(i.odd, {~sel = [0,1].choose; if(~sel == 1, {\r}, {item})}, {item});

});
p = Pdef(\parti,
Pbind(
\chan,   0,
\midinote, Pseq(~lista14, inf),
    \dur, Pwrand([0.25, Pn(0.125, 2)], #[0.8, 0.1], inf),
    \legato, sin(Ptime(inf) * 0.5).linexp(-1, 1, 1/3, 3),
\amp, Pwrand([1, 0.5, 0.25, 0.125], [0.1, 0.8, 0.05, 0.05], inf)
)
).play;
m = SimpleMIDIFile( "~/Sangre/Midis/Prueba06.mid" );
m.init1( 2, 120, "4/4" );
m.fromPattern( p );
m.write;
});
}
).add;
)
```

6 Conclusions

In this paper we have presented a compositional platform based on mathematical models. The solutions of the dynamical systems and their corresponding simulations have been used to generate different musical structures. We hope it may be useful as a starting point example in the development of musical compositions in which real time interaction and setting of the context via adjustment of the parameters of the model, can be included as an important part of the compositional process.

References

1. Hoppensteadt, F.C., Peskin, Ch.S.: Modeling and Simulation in Medicine and the Life Sciences. Texts in App. Math. Springer, New York (2004)
2. Pareyon, G.: On Musical Self-Similarity. The International Semiotics Institute, Imatra-Helsinki (2011)

Music, Expectation, and Information Theory

D. Gareth Loy

*Music conveys general forms of feelings, related to specific ones
as algebraic expressions are related to arithmetic [expressions].*
Susanne Langer [5]

Abstract What makes a musical work successful? In Darwinian terms, music is successful if listeners attend to it, repeatedly, for then it can live on. However, attention is fleeting: successful music holds listeners' interest by manipulating their expectations using deception and confirmation. The ratio of the rate at which listeners follow music to the rate at which music unfolds is a predictor for musical success. This paper informally presents a theory of musical interest, based on some ideas from music theory, cognitive psychology, and information theory.

1 When is Music Successful?

Music is successful in Darwinian terms if we are repeatedly willing to hear it. Successful music is so because it cultivates and sustains listeners' interest. In no other way can music live on. Music's most powerful attractant is our *curiosity*. By exploiting it, successful music lives to be heard another day.

We are curious when we want to learn about something. When the discovery process is going well, the learner is *engaged*. This is more than simple attraction. If the discovery process continues commensurate with the rate of new information received, we can sustain our curiosity if we wish. The same is true of music: we can, if we wish, remain engaged if the rate at which we follow the music is commensurate with the rate at which it unfolds. However, if the rates are unmatched, and our minds outrace or fall behind the music, we lose interest.

D.G. Loy (✉)
Gareth Inc., POB 151185, San Rafael, CA 94915, USA
e-mail: dgl@garethinc.com

© Springer International Publishing AG 2017
G. Pareyon et al. (eds.), *The Musical-Mathematical Mind*,
Computational Music Science, DOI 10.1007/978-3-319-47337-6_17

161

If our minds race ahead—figuring out where the music is going faster than it gets there—we risk boredom, and loose interest. After all, it is redundant, and a waste of time, for us to know what will happen *and* to bother hearing it out. On the other hand, if we fall behind because the music outstrips our ability to keep up, we grow frustrated and lose interest.

Interest in music is closely tied to the rate at which we can make sense of what we hear. To follow music means to be able to orient oneself, to understand what has been heard, and to have a prediction, or an expectation, of where it is going. If our understanding increases commensurate with the rate of musical information then we believe ourselves to be in possession of enough knowledge to remain current as it unfolds, and to have some confidence that we can anticipate forthcoming musical events. Some degree of such confidence is required for interest to persist. But it is the *vulnerability* of this confidence that successful music exploits.

Even for very simple music, we form and evaluate large numbers of mostly unconscious predictions as we listen to music. The key to engaging listeners is to satisfy some expectations while frustrating others as the music unfolds. This is the art of *entertainment*.

Example of musical expectation Figure 1 shows an elementary motive of four notes sequenced up repeatedly by diatonic steps.

Suppose you were hearing it played for the first time. By the end of measure 2, you'd probably have heard the repeated motive. You might think, "I bet the music is sequencing a four note motive up diatonically." If, as in the third measure, the music meets your expectation, your prediction is confirmed [4]. You feel a fleeting sense of satisfaction... *and, curiously, the music starts to lose your interest* because no sooner is the pattern you've predicted realized than it ceases to be interesting: because there is little to no new information to digest, it's a waste of time to continue listening.

If the musical pattern continues unvarying into the fourth measure as shown, a new sensation, *boredom* may arise. Interest is allergic to deadeningly predictable patterns. Music dies when listeners don't care to hear it. But suppose instead the music veers off as shown in Fig. 2.

Here, after exactly 2.5 repetitions of the four-note motive, the music switches from horizontal to vertical motion—from melodic sequencing to a dominant-tonic (V–I)

Fig. 1 Elementary sequenced motive

Fig. 2 Elementary sequenced motive with cadence

cadence. The listener, having already heard two repetitions of the motive, expects the pattern to continue and is surprised by its interruption. Surprise is invoked by the introduction of an asymmetry that violates the listener's expectations, and the surprise serves to fetch the listener's interest back, thereby *entertaining* the listener.

Music requires a degree of structural ambiguity to gain and maintain interest. The structure of successful music must continually mutate to sustain listeners' engagement, i.e., to entertain listeners. Violating motivic regularity is but one way to accomplish this.

How is it that we were able to predict the evolution of the musical motive in Fig. 2 even before we'd heard it all the way through? This suggests we carry models—schema—of what we expect, which we apply to fathom novel circumstances. Schemas describe patterns of thought that organize and categorize our experiences, and express the relationships among them.

Aristoxenus said,

> Musical cognition implies the simultaneous recognition of a permanent and a changeable element... for the apprehension of music depends upon those two faculties, sense perception and memory; for we must perceive the sound that is present, and remember that which is past. In no other way can we follow the phenomenon of music. – Aristoxenus [1]

How indeed can we follow music unless we can compare the sound that arises to what we expected to hear? Leonard Meyer said,

> Emotion or affect is aroused when a tendency to respond is arrested or inhibited... What a musical stimulus or a series of stimuli indicates... [is] not extramusical concepts and objects but other musical events which are about to happen... Embodied musical meaning is, in short, a product of expectation. – Leonard Meyer [6]

Representational momentum When comparing what we hear in the present to our expectations from the past, we experience varying degrees of confirmation and surprise, much as, when following a ball in flight, we may experience confirmation if it hits its mark, and surprise if it is suddenly deflected. Freyd and Finke discovered that,

> Under appropriate conditions an observer's memory for the final position of an abruptly halted object is distorted in the direction of the represented motion, much as a physical object continues along its path of motion because of inertia [2].

The authors termed this phenomenon *representational momentum* [3].

We can adapt the concept for musical purposes by reference to Fig. 2, where the repetitive motivic sequence sets up representational momentum in the listener's mind in the form of a belief that the pattern will continue. The surprise elicited when the cadence breaks the pattern is analogous to the surprise that would be elicited by the "abruptly halted object" referenced by Freyd and Finke. Surprise demonstrates the presence of the representational momentum in the listener's mind, for there would be no surprise were there no expectation that the phenomenon—either the ball flying through the air, or the melody sequencing—would continue.

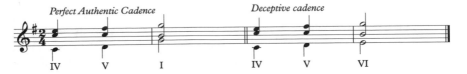

Fig. 3 Perfect authentic cadence; deceptive cadence

Representational momentum and the deceptive cadence The canonical finish to a musical phrase, the perfect authentic cadence, shown in the first two measures of Fig. 3, outlines the chordal sequence from the subdominant chord (IV), to the dominant (V), finally resolving to the tonic (I) chord. If completed, the listener expects a full stop to the musical phrase in progress. The music may go on, but one musical idea has stopped and another has begun.

The *deceptive cadence* (Fig. 3) subverts the listener's expectation of phrase completion. It begins like the perfect authentic cadence, but at the last chord, it "resolves" to the VI chord instead of the I. The triad on VI shares two of its three degrees with the tonic I triad, so the VI triad mimics the tonic enough so that the ear is not completely derailed by its substitution for the tonic. However, it is not the tonic, and until that moment, the listener expected resolution to the I chord, and is surprised when the VI is substituted, reengaging the listener's interest.

The deceptive cadence is the musical equivalent of "bait-and-switch", whereby what we are expecting is not what we get. Imagine you are a hunter in the woods and are about to bag a fat squirrel for dinner, but it slips away. This is the effect of the deceptive cadence on the ear. The listener is now more "hungry" for the proper cadence; the composer can now build up to a more charged climax.

In order to eat, the hunter must continue hunting after missing the squirrel; just so, after a deceptive cadence, the listener must continue to seek resolution. Composers use this to extend the duration of a musical phrase. Figure 4 shows a deceptive cadence and its continuation in the opening of the second movement of Mozart's *Piano Sonata in C*.

To the listener, the *meaning* of the deceptive cadence (using Meyer's definition) is that there is more to the current phrase that is still to come.

Fig. 4 Mozart *Piano Sonata in C*, K. 330, opening of 2nd movement

2   Information Theory

In 1928, Harry Nyquist proposed that a signal must be sampled at twice its highest frequency so as to have enough information to completely reconstruct the original signal from its sampled representation [7]. Therefore, a signalling system with bandwidth B has a maximum data rate $2B$. A transmission system having K distinct amplitude levels represented with binary encoded values has a maximum data rate D of:

$$D = 2B \log_2 K. \tag{1}$$

Shannon and Weaver [9] extended Nyquist to account for noise:

$$C = B \log_2(1 + S/N) \tag{2}$$

where: C = channel capacity (bits/second), B = hardware bandwidth, S = average signal power, N = average noise power, S/N is signal-to-noise ratio.

The channel capacity C required to send a signal depends upon its degree of regularity. If a signal is highly ordered or predictable, it has a high degree of *redundancy*, and a summary of the redundant components of the signal can be transmitted instead of the entire signal, requiring less channel capacity C. If a signal is highly unordered or unpredictable, it has a high degree of *entropy*. The higher the degree of entropy, the fewer of its components are redundant. Components that cannot be summarized must all be transmitted, requiring more channel capacity C.

Information theory borrowed the term entropy from chemistry, where entropy is the thermodynamic probability of a molecular system, that is, a measure of the ways in which the *energy* of a molecular system is distributed among the possible motions of its particles. In information theory, entropy is a measure of the ways in which the *information* of a signaling system is distributed among its possible communications [8].

Surprisal is a measure of the uncertainty in a communication. Surprisal is analogous to the experience of "surprise", and it relates to the probability of an expected outcome.

Probability ranges over the unsigned unit interval (0.0–1.0) where, for probability $p = 1.0$ corresponds to absolute certainty, $p = 0.0$ corresponds to absolute uncertainty. Classically, probability values are defined for all time—they do not change.

Surprisal is inversely related to probability. In the limit as the probability of an event goes from 1.0 to 0.0, surprisal goes from zero to infinity. That is, for surprisal s and probability $p = 1 \rightarrow s = 0$, $p = 0 \rightarrow s = \infty$.

If an event will occur no matter what ($p = 1$), then there is no surprisal. For example, a coin toss will be either heads or tails—no surprise there. On the other hand, if there is a vanishingly small probability that an event will occur, then the surprisal goes to infinity. For example, suppose you win the lottery—your surprise knows no bounds! Therefore, $p = \frac{1}{2^s}$. Solving for s:

$$s = \log_2 \frac{1}{p} = -\log_2 p = -\frac{\ln p}{\ln 2}. \tag{3}$$

Surprisal is the inverse log probability of a token appearing in a message. Surprisal s relates to the bandwidth required to communicate a particular message that has probability p.

Frequency and surprisal The frequency of probable events has an amplifying effect on expectation. Suppose you randomly find a dollar on the sidewalk one day: you are surprised. But if you randomly find a dollar on the sidewalk several days in a week, you are astonished! In information theory, frequency is how often a token appears in a message.

If there are N tokens in message X and the ith token occurs N_i times, then $\frac{N_i}{N}$ is its frequency.

Average surprisal The average surprisal of a message is the normalized sum of the expectancy of its tokens. In music, the surprisal of a melody is the normalized sum of the expectancy of its notes.

For example, let all the keys on a piano be independently played. Let each piano key be $k_i, i = 1, 2, 3, \ldots, M$, where M is the number of keys. If N notes can arise in a melody X, then its average surprisal H is:

$$H(X) = \frac{1}{N} \sum_{i=1}^{M} \frac{N_i}{N} s_i \tag{4}$$

We normalize the sum by the number of tokens in the message to facilitate comparing surprisal across messages of varying length.

Examples of surprisal Let us take the hypothesis that the keys near middle-C are most frequently played on the 88-key piano keyboard. The normal (Gaussian) probability distribution function with mean $\mu = 44$ (corresponding to the center key of the keyboard, which has the pitch E4, that is, the pitch E above middle-C) and standard deviation $\sigma = 1$ is shown in Fig. 5. The corresponding normalized average surprisal is shown in Fig. 6.

If the hypothesis is correct, then we should expect to hear the keys near the center of the keyboard played most frequently on the piano, and if our expectation is violated, we are surprised. Thus, by Eq. 4 we would be surprised by a melody played entirely by high and low keys, and little surprised by a melody played near the center of the keyboard.

Taking the average surprisal function shown in Fig. 6, we can calculate the surprisal of various melodies played on the piano, as follows. The melody of Antonio Carlos Jobim's *One Note Samba* is sung on a single note. ("Eis aqui este sambinha, feito numa nota só ...") If the melody is played on E4, then the average surprisal

Fig. 5 Probability density function

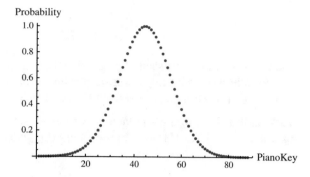

Fig. 6 Corresponding surprisal

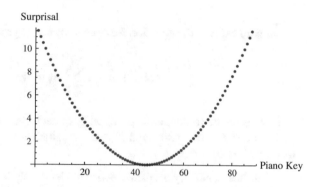

of the first 32 notes of this melody is 0. The average surprisal of a chromatic scale played in the middle of the piano keyboard would be very low, on the order of 0.006. The average surprisal of a chromatic scale far from the center of the keyboard would be higher, on the order of 0.8. The average surprisal of a random 12-note melody would be about 0.33.

Clearly, the meaningfulness of surprisal depends on the validity of the hypothesis. The relevance of information theory to music is its formalization of expectation and surprisal; but its ultimate usefulness to music theory depends upon the development of a corpus of theories that correctly capture the actual experience of listeners. It is not clear that this is possible to do in absolute terms. Given the evident variety of music around the world and through time, one assumes that the relevant musical schema depend upon a highly contextual field of cultural antecedents that are difficult to elicit, let alone classify.

Uncertainty As the total number of events in a message N increases to infinity, the event frequency $\frac{N_i}{N}$ tends to its static probability p_i. By combining Eq. 4 with the definition for surprisal s_i (Eq. 3) and substituting p_i for N_i/N, we have:

$$H(X) = -K \sum_{I=1}^{M} p_i \log_2 p_i \qquad (5)$$

where K is a positive constant of proportionality.

Uncertainty is the average surprisal per token for an infinite length sequence of symbols. (It is always the receiver that is uncertain.)

Information (Entropy) By suitable choice of K, we may choose any base for the logarithm in Eq. 5. Here is the definition of entropy given by Shannon and Weaver [9]:

$$H(X) = -K \sum_{i=1}^{M} p_i \ln p_i. \qquad (6)$$

Compare Eq. 6 to the equation for thermodynamic probability:

$$H(X) = -k \sum_{i=1}^{M} W_i \ln W_i, \qquad (7)$$

where W_i is the thermodynamic probability of each state, k is Boltzmann's constant, equal to 1.3807×10^{-23} J/K, and H is the resultant entropy. The similarity between Eqs. 6 and 7 is striking.

Only absolute certainty banishes entropy absolutely In the event that there is total pattern redundancy in a communication, there is zero entropy. "For a given n, H is a minimum when all the P_i are epsilon [vanishingly small] but one. This is intuitively the most certain situation" [9].

The most uncertain situation has the maximum entropy "For a given n, H is a maximum and equal to $\log n$ when all the P_i are equal (i.e., $1/n$). This is also intuitively the most uncertain situation" [9].

Redundancy is the complement of entropy $H(X)$ related to its theoretical maximum, $\log N$:

$$R(X) = 1 - \frac{H(X)}{\log N}. \qquad (8)$$

Redundancy $R(X)$ is what is left in a signal after subtracting its entropy. Information theory presents us with the somewhat counterintuitive outcome that the greatest amount of information is associated with the greatest degree of uncertainty. One way to view this is that entropy is the measure of the amount of information that is missing in the recipient prior to reception of the message.

While classical information theory is static, one-dimensional, and non-hierarchical, information theory offers crisp analogs to musical states of the listener: surprisal, expectation, and uncertainty. These theories help relate musical structure to the concomitant musical affect in the listener.

Conclusion I hope that these ideas can be used to help put music theory on an empirical basis. I believe that surprisal, expectation, and uncertainty are the universal underpinnings of music. I hope that this will encourage others to apply these ideas to the study of all forms of music.

References

1. Aristoxenus. C. 300 BCE: The Harmonic elements. In: Strunk, O. (ed.) Source Readings in Music History: Antiquity and the Middle Ages, pp. 27–31. W.W. Norton & Co., New York (1950)
2. Freyd, J.J.: Dynamic mental representations. Psychol. Rev. **94**(4), 427–438 (1987)
3. Freyd, J.J., Finke, R.A.: Representational momentum. J. Exp. Psychol. Learn. Mem. Cogn. **10**, 126–132 (1984)
4. Huron, D.: Sweet Anticipation: Music and the Psychology of Expectation. The MIT Press, Cambridge (2006)
5. Langer, S.: Philosophy in a New Key: A Study in the Symbolism of Reason, Rite, and Art. Harvard University Press, Cambridge (1957)
6. Meyer, L. B.: Emotion and Meaning in Music. University of Chicago Press, Chicago (1956)
7. Nyquist, H.: Certain topics in telegraph transmission theory. Trans. AIEE **47**, 617–644 (1928). Reprint as classic paper in: Proc. IEEE **90**(2), Feb 2002
8. Shannon, C.E.: A mathematical theory of communication. Bell Syst. Tech. J. **27**(379–423), 623–656 (1948)
9. Shannon, C.E., Weaver, W.: The Mathematical Theory of Communication. University of Illinois Press, Urbana (1949)

Gestural Dynamics in Modulation: (Towards) a Musical String Theory

Guerino Mazzola

Abstract We prove a modulation theorem for diatonic scales that is based on the theory of hypergestures and vector fields derived from inner symmetries of diatonic scales and Lie bracket fields. It yields the same modulation degrees as the classical model (Mazzola, Gruppen und Kategorien in der Musik, 1985, [1], Geometrie der Töne, 1990, [2], Mazzola et al., The Topos of Music-Geometric Logic of Concepts, Theory, and Performance, 2002, [3]), which confirmed Schoenberg's modulation theory (Schoenberg, Harmonielehre 1911, Universal Edition, Wien 1966, [4]). In this hypergestural model, integration of differential forms is considered. In this context, we can model and prove Stokes' theorem for hypergestures, generalizing the classical case. Stokes' theorem is a central result in differential geometry, relating the integral of the derivative of a form to the boundary of its domain of integration. It has important application in physics, such as in mechanics (integral invariants, see (Abraham, Foundations of Mechanics, 1967, [5])) or in electrodynamics (relating differential and integral forms of Maxwell's equations (Jackson, Classical Electrodynamics, 1998, [6])). The basic form of this theorem deals with integration on singular hypercubes. In (Mazzola, J Math Music 6(1):49–60, 2012, [7]) we have extended singular homology on hypercubes to singular homology on hypergestures. It was therefore straightforward to try to extend Stokes' theorem to hypergestures.

1 Introduction

In a recent publication [8], we have opened the discussion of a hypergestural restatement of mathematical counterpoint theory. The present paper aims at a discussion in the same vein of the classical mathematical modulation theory [1–3]. Following that

G. Mazzola (✉)
School of Music, University of Minnesota, 2106 Fourth Street South, Minneapolis, MN 55455, USA
e-mail: mazzola@umn.edu

G. Mazzola
Institut Für Informatik, Universität Zürich, Binzmühlestrasse 14, 8050 Zürich, Switzerland

© Springer International Publishing AG 2017
G. Pareyon et al. (eds.), *The Musical-Mathematical Mind*,
Computational Music Science, DOI 10.1007/978-3-319-47337-6_18

approach, it can be proved that tonal modulation as described by Arnold Schoenberg [4] can be modeled using symmetries S between scales underlying the involved tonalities. For example, to modulate from C-major to F-major, Schoenberg proposes the three modulation degrees II_F, IV_F, VII_F. These degrees also come out from the mathematical model, where the C scale is mapped to the F scale using the inversion symmetry $S = T^9. - 1 = U_{e/f}$ between e and f. The mathematical model yields exactly Schoenberg's modulation degrees in all cases where he describes a direct modulation, namely for fourth and fifth circle distances 1, 2, 3, 4.

The present approach is based on the idea that degrees in the start tonality are interpreted as being gestures that move to degrees (qua gestures) of the target tonality be means of hypergestures. This means that the symmetries relating tonalities in the classical setup are replaced by hypergestures that connect gesturally interpreted degrees.

The present hypergestural model solves the problem, but it opens more questions than it solves in the sense that the construction of hypergestures that replace the classical inversion symmetries is by no means unique. We are still in search for a theory that might generate natural "minimal action" hypergestures in the sense of Hamilton's variational principle in mechanics. In fact, the classical modulation model was driven by the idea of elementary fermion particles in physics, interacting via bosons that materialize interaction forces. The hypergestural restatement would view symmetry-corresponding degrees $X, S(X)$ as being musical fermions being connected via a boson hypergesture $h : X \to S(X)$. More precisely, the homological boundary $\partial h = (S(X) - X, -h_1^\square)$ has the first component $S(X) - X$ as difference of the involved fermions, whereas the second component $-h_1^\square$ is the boson deduced from the face operator $?^\square$ acting on the Escher-inverted h_1 of h, but see [7] for details. An intuitive illustration in Fig. 1 shows this situation, where X is given as a pitch class gesture $C \to B, B \to A, C \to A$, $S(X)$ is given by the gesture $C^* \to B^*, B^* \to A^*, C^* \to A^*$, and the hypergesture h deforms X to $S(X)$ along the lines from A to A^* etc., whereas the Escher-inverted perspective h_1 consists of the

Fig. 1 For a pitch class gesture X, with curves $C \to B, B \to A, C \to A$, the target gesture $S(X)$ is given by the gesture $C^* \to B^*, B^* \to A^*, C^* \to A^*$, and the hypergesture h deforms X to $S(X)$ along the lines from A to A^*, B to B^*, C to C^*

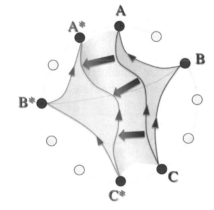

hypergesture deforming the line $C \to C^*$ to $A \to A^*$, the line $C \to C^*$ to $B \to B^*$, and the line $B \to B^*$ to $A \to A^*$.

The general procedure will be as follows: We first model gestures and hypergestures in the topological space $\mathbb{R}^2$, where the usual pitch class set $\mathbb{Z}_{12}$ is embedded on a circle. We then look at triadic degrees X of pitch class points, which are represented as gestures of lines connecting these points. Next, we construct vector fields on $\mathbb{R}^2$ whose integral curves give rise to hypergesture curves that deform the gestural degrees. Then we discuss cadences of such triadic degrees and their behavior under hypergestural deformation. We shall prove that for a specific choice of such vector fields, the inversion symmetries used in the classical model map pitch classes x into pitch classes living in the same integral curve as x. Next we consider the trajectories of the curves of the Escher-inverted perspective and calculate energy integrals of such curves. Under the condition of non-vanishing energy, we can then exhibit the admitted degrees. These integrals refer to Stokes' theorem, and we therefore need to think about Stokes' theorem for hypergestures. Its statement and proof are found in the concluding sections of this paper and might be of more general interest.

Short Recapitulation of the Classical Model's Structure

The classical model is described in [3, 27.1], we only give a short and not exhaustive recapitulation thereof here. For a modulation from major tonality X to major tonality Y, the triadic modulation degrees (in the sense of Schoenberg) in Y are calculated by means of a modulation quantum Q, which is a set of pitch classes. Its intersection $Y \cap Q$ is, by construction, the union of the modulation degrees. This modulation quantum is defined by a number of properties:

1. Q has an inner symmetry that transforms X to Y.
2. For a given cadential set J of Y, all degrees of J are subsets of Q.
3. The intersection $Q \cap Y$ is rigid, i.e. it has no nontrivial inner symmetries (in the group of transpositions and inversions).
4. The quantum Q is minimal with the first two properties.

The motivation of such a quantum is that, by (i), it "materializes" a symmetry qua "force" that transforms X to Y, that, by (ii), it is rich enough to determine Y by a cadence, that, by (iii), the symmetry of Q that transforms X to Y is uniquely determined by Q. And (iv) is a Hamilton principle: we want Q to be minimal with the first two properties. Observe that this setup does not guarantee the existence of modulation quanta. The modulation theorem [3, Theorem 30, 27.1] for 12-tempered tuning guarantees the existence of such quanta. This theorem is also valid for just tuning [3, 27.1.6], but in the present paper we focus on 12-tempered tonalities.

2 Hypergestures Between Triadic Degrees that are Parallel to Vector Fields

As said above, we embed the set $\mathbb{Z}_{12}$ of pitch classes in a circle with diameter π as a part of the real plane $\mathbb{R}^2$. A triadic degree, and more generally any pitch class set X is enriched by by a system of differentiable curves $l(x, y)$ from x to y (or vice versa), one for every *un*ordered pair x, y of different points in X, the selected direction is irrelevant (why will be seen later), the corresponding gesture is denoted by $\overrightarrow{X}$. An example is shown in Fig. 1 for the set $X = \{A, B, C\}$. If such a gesture has skeleton Σ, it is an element of $\Sigma \overrightarrow{@} \mathbb{R}^2$. We shall then consider hypergestures $h \in \uparrow \overrightarrow{@} \Sigma \overrightarrow{@} \mathbb{R}^2$ that connect two pitch class set gestures (of same skeleton) $\overrightarrow{X}, \overrightarrow{Y}$, i.e. $\overrightarrow{X} = h(0)$, $\overrightarrow{Y} = h(1)$. We shall now look at vector fields $\mathcal{X}$ on $\mathbb{R}^2$ that are smooth enough to have integral curves, fields that are locally Lipschitz, to be precise. For every point $x \in \mathbb{R}^2$, there is a uniquely defined maximal integral curve $\int_x \mathcal{X} : D(x) \to \mathbb{R}^2$, defined on the open domain $D(x) \subset \mathbb{R}$, starting at x, i.e. $\int_x \mathcal{X}(0) = x$, and $T(\int_x \mathcal{X})(\lambda) = \mathcal{X}(\int_x \mathcal{X}(\lambda))$ for all parameters $\lambda \in D(x)$ of $\int_x \mathcal{X}$.

Definition 1 Given a hypergesture $h \in \uparrow \overrightarrow{@} \Sigma \overrightarrow{@} \mathbb{R}^2$, connecting $\overrightarrow{X} = h(0)$ to $\overrightarrow{Y} = h(1)$, we say that it is *parallel to a vector field* $\mathcal{X}$ if for every point x in X, there is a function $f : I \to D(x)$ of the unit interval $I = [0, 1]$ into the domain $D(x)$ such that the $\uparrow$-gesture $h_1(x)$ of h_1 starting at x has values $h_1(x)(\lambda) = \int_x \mathcal{X}(f(\lambda))$ for all $\lambda \in I$.

The terminology is justified for a differentiable such function f since then, its tangent function Tf evaluates to vectors parallel to the vector field's vectors. The following lemma enables us to construct parallel hypergestures from curves on vertices of a pitch class set X.

Lemma 1 *Given a vector field* $\mathcal{X}$, *a pitch class set X with a gesture* $\overrightarrow{X}$ *in* $\Sigma \overrightarrow{@} \mathbb{R}^2$, *and a pitch class set Y such that for every point $x \in X$, there is a curve* $f_x : I \to D(x)$ *such that* $\int_x \mathcal{X}(f(1)) =: y(x)$ *defines a bijection* $X \xrightarrow{\sim} Y$, *then there is a hypergesture* $h \in \uparrow \overrightarrow{@} \Sigma \overrightarrow{@} \mathbb{R}^2$ *connecting* $\overrightarrow{X}$ *with a gesture* $\overrightarrow{Y}$ *that is parallel to* $\mathcal{X}$.

The critical point here is the question whether we can find vector fields that connect degrees X, Y that are symmetric images of each other, i.e. $Y = S(X)$ for a symmetry S connecting two tonalities, by parallel hypergestures.

3 Lie Brackets Generate Vector Fields that Connect Symmetry-Related Degrees

In this section we define vector fields associated with pairs of tonalities and which fulfill the conditions explained above. Although such vector fields can be defined for quite general situations of tonality pairings, we want to restrict our attention

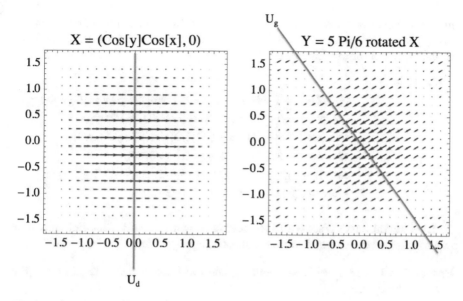

Fig. 2 Field X is $\mathcal{X}_C$ (*left*), field Y is $\mathcal{X}_F$ (*right*)

to the pairing of two tonalities that are one fourth apart from each other, and we may choose the concrete situation of C-major and F-major. For each such tonality T, which we identify with its scale for this special discussion, we define a vector field $\mathcal{X}_T$ that is motivated by the unique inner symmetry S_T of T. For $T = C$ this is the inversion $S_C = U_d$, for $T = F$, it is $S_F = U_g$. To have a simple representation of symmetries and fields, we choose a labeling of the pitch classes in $\mathbb{Z}_{12}$ such that $0 = d, 1 = d\#, 2 = e, 3 = f, 4 = f\#, 5 = g, 6 = g\#, 7 = a, 8 = a\#, 9 = b, 10 = c, 11 = c\#$. With this notation, and 0 being on top, and 3 to the right of the circular representation (like normal time visualisation), the symmetry S_C is the reflection at the vertical diameter through the pitch class circle. We now represent this reflection as a movement in horizontal direction from left to right, thinking of a $180°$-rotation in $\mathbb{R}^3$. This can be represented by a vector field $\mathcal{X}_C(x, y) = (\cos(y)\cos(x), 0)$. Similarly, for tonality F, we define its vector field $\mathcal{X}_F$ as being the clockwise rotation of $\mathcal{X}_C$ by $5\pi/6$. More generally, if R is a nonsingular linear transformation of $\mathbb{R}^2$, we construct a vector field X^R from X by $X^R(x) := R(X(R^{-1}(x))$. The we have $Y = X^R$ if R is the clockwise rotation by $5\pi/6$. Figure 2 shows these fields in a graphic generated by Mathematica® software.

The next step is mathematically well-defined, but we actually don't know why it works. To generate the field which will eventually guide the hypergestural lines, we consider the Lie bracket $[\mathcal{X}_C, \mathcal{X}_F]$ of the fields associated with the inner symmetries S_C, S_F. They are defined using the fact that vector fields are in one-to-one correspondence with derivations on functions, and then using the fact that the commutator of such derivations is again a derivation. Lie brackets are very important in mathematical physics, and in particular in Lagrangian and Hamiltonian mechanics. See [5] for the calculus of Lie brackets and its application to mechanics. Here is the explicit formula for this Lie bracket:

$$[\mathcal{X}_C, \mathcal{X}_F](x, y) =$$
$$(-(1/2)\sqrt{3}\cos((\sqrt{3}x)/2 + y/2)\cos(y)\cos(x/2 - (\sqrt{3}y)/2)\sin(x) -$$
$$1/2\cos(x)\cos((\sqrt{3}x)/2 + y/2)\cos(x/2 - (\sqrt{3}y)/2)\sin(y) +$$
$$\cos(x)\cos(y)(3/4\cos(x/2 - (\sqrt{3}y)/2)\sin((\sqrt{3}x)/2 + y/2) +$$
$$1/4\sqrt{3}\cos((\sqrt{3}x)/2 + y/2)\sin(x/2 - (\sqrt{3}y)/2)),$$
$$\cos(x)\cos(y)(1/4\sqrt{3}\cos(x/2 - (\sqrt{3}y)/2)\sin((\sqrt{3}x)/2 + y/2) +$$
$$1/4\cos((\sqrt{3}x)/2 + y/2)\sin(x/2 - (\sqrt{3}y)/2)))$$

The integral curve display of this field is shown in Fig. 3. This field has four remarkable properties which we want to list as a proposition:

Proposition 1 *With the above notations, the Lie bracket field $[\mathcal{X}_C, \mathcal{X}_F]$ has the following properties:*

1. *The twelve pitch class points are contained in three closed integral curves: C_b through $\{b, a\#\}$, C_c through $\{c, c\#, d, d\#, f\#, g, g\#, a\}$, and C_e through $\{e, f\}$.*
2. *The curves C_b, C_c, C_e are symmetrical with respect to the modulator symmetry $U_{e/f}$ that maps C to F in the sense that every pitch class p in its integral curve C_b, C_c or C_e is mapped to $U_{e/f}(p)$ that is contained in the* same *integral curve.*
3. *If R is the $180°$-rotation in $\mathbb{R}^2$, we have $[\mathcal{X}_C^R, \mathcal{X}_F^R] = [\mathcal{X}_C, \mathcal{X}_F]^R = [\mathcal{X}_C, \mathcal{X}_F]$.*
4. *If $R = U_{e/f}$ then $\mathcal{X}_F = -\mathcal{X}_C^R$, and we have $-[\mathcal{X}_C, \mathcal{X}_F]^R = [\mathcal{X}_C, \mathcal{X}_F]$. The latter is also true if R is the reflection orthogonal to $U_{e/f}$. These formulas mean intuitively that the two reflection axes that are visible in the left part of Fig. 3 transform the Lie bracket field into its negative.*

Recall that if $J(X)$ denotes the Jacobian of a vector field X on $\mathbb{R}^2$, then $[X, Y] = J(Y)X - J(X)Y$. Property (iii) is evident since for the $180°$-rotation R, $\mathcal{X}_C^R = -\mathcal{X}_C$ and $\mathcal{X}_F^R = -\mathcal{X}_F$, whence $[\mathcal{X}_C^R, \mathcal{X}_F^R] = [-\mathcal{X}_C, -\mathcal{X}_F] = [\mathcal{X}_C, \mathcal{X}_F]$. The equation $[\mathcal{X}_C, \mathcal{X}_F]^R = [\mathcal{X}_C, \mathcal{X}_F]$ follows immediately from the Jacobian formula. The last property in (iv) follows from (iii) and the first part of (iv). To prove this one, we need two easy auxiliary result about Lie brackets. The first result relates to the Jacobian of a vector field $XT(x) := X(T(x))$ deduced from a non-singular linear transformation T on $\mathbb{R}^2$. We have $J(XT)(x) = J(X)(T(x))T$. Using this result, if R is a linear involution ($R^2 = Id$), then we have $[X, -X^R] = -[X, -X^R]^R$. Property (iv) now follows from this last result since in our case, $\mathcal{X}_F = -\mathcal{X}_C^R$ for $R = U_{e/f}$. Property (ii) follows from property (i) and property (iv). Property (i) of the Lie bracket field not evident. We don't know why the twelve pitch classes are grouped in just three integral curves that are invariant under $U_{e/f}$. We have no mathematical proof of this proposition in the sense that we ere not able to calculate symbolically (with explicit formulas) those three symbolic integral curves C_b, C_c, C_e and to prove that the subsets of pitch classes are precisely contained in those curves. Also, Mathematica®

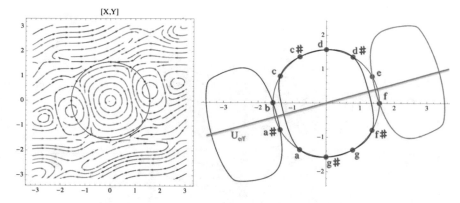

Fig. 3 The global display of the integral curves of the Lie bracket field $[\mathcal{X}_C, \mathcal{X}_F]$ (*left*), and (*right*) the three closed integral curves comprising all twelve pitch classes, and this in such a way that the modulator transformation $U_{e/f}$ from C to F maps pitch classes into pitch classes within the same integral curves

did not yield a solution using its DSolve function, our result is obtained using the numerical integration function NDSolve. (QED)

Using this proposition, we can now find hypergestures h, parallel to $[\mathcal{X}_C, \mathcal{X}_F]$, that map degrees of C-major or more general pitch class sets to symmetry-connected degrees or pitch class sets, respectively, in F-major. In fact, referring to the notations of Lemma 1, given a pitch class set X in C, we can find by Proposition 1 a curve $f_x :$ $I \to D(x)$ for every $x \in X$, such that $\int_x [\mathcal{X}_C, \mathcal{X}_F](f(1)) = S(x)$ defines a bijection with the symmetric pitch class set $Y = S(X)$. Therefore, by Lemma 1, there is a hypergesture h, parallel to $[\mathcal{X}_C, \mathcal{X}_F]$, that maps X to $S(X)$.

4 Selecting Parallel Hypergestures that are Admissible for Modulation

The next step consists of the selection of "good" hypergestures for the intended modulation. To this end, we look at the hypergestures $h_{x,y}$ obtained from the above parallel hypergestures h when restricting them to the single curves $l(x, y)$ in $\overrightarrow{X}$, being deformed under $h_{x,y}$ to curves $l(S(x), S(y))$ that define $\overrightarrow{S(X)}$. Such a deformation hypergesture consists of a (smooth) curve of curves $h_{x,y}(\lambda)$, $\lambda \in I$, whose endpoints x_λ, y_λ are all moving within one of the three integral curves C_b, C_c, C_e, each, and $h_{x,y}(0) = l(x, y), h_{x,y}(1) = l(S(x), S(y))$, see Fig. 4 for an example, starting at a curve from $l(c, e)$ and ending at curve $l(S(c) = a, S(e) = f)$, the intermediate curves $h_{x,y}(\lambda)$ all move along the integral curves C_c, C_e with their endpoints.

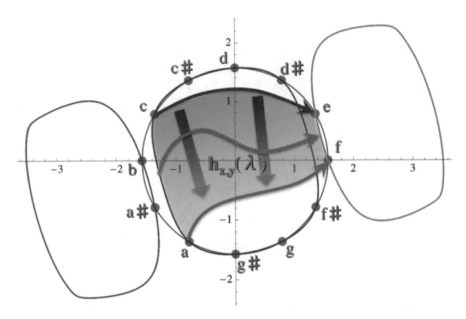

Fig. 4 Starting at a curve from $l(c, e)$ and ending at curve $l(S(c) = a, S(e) = f)$, the intermediate curves $h_{x,y}(\lambda)$ move along the integral curves C_c, C_e with their endpoints

Definition 2 With the above notation, such a hypergesture $h_{x,y}$ from curve $l(x, y)$ to curve $l(S(x), S(y))$ is called non-singular if for every parameter $\lambda \in I$, the gesture $h_{x,y}(\lambda)$ is not a loop.

Although this definition looks only geometric, it has an interpretation in terms of energy function. Suppose that $E(x, y)$ is a differentiable potential function on $\mathbb{R}^2$. Then we may consider the usual line integral $\int_{h_{x,y}(\lambda)} dE$, expressing the work to move from $h_{x,y}(\lambda)(0)$ to $h_{x,y}(\lambda)(1)$ under the given potential E. If we suppose that a Stokes theorem can be proved for hypergestures, we have $\int_{h_{x,y}(\lambda)} dE = \int_{\partial h_{x,y}(\lambda)} E = E(h_{x,y}(\lambda)(1)) - E(h_{x,y}(\lambda)(0))$. This latter vanishes if the curve $h_{x,y}(\lambda)$ is a loop. The converse is not true, but we can enforce the converse if we find enough potentials such that the vanishing of the integral for all these potentials implies that the curve is a loop. In fact, in our situation it is possible to find two simple potentials, $E_C(x, y) = x$ and its clock-wise rotation by $5\pi/6$, $E_F(x, y)$ (similar to the vector field construction). Evidently, $h_{x,y}(\lambda)$ is a loop if and only if $d(\lambda) := (\int_{h_{x,y}(\lambda)} dE_C)^2 + (\int_{h_{x,y}(\lambda)} dE_F)^2 = 0$. This will be our condition for an *admissible (parallel) hypergesture h from pitch class set X to $S(X)$, namely that all of its curve sub-hypergestures $h_{x,y}$, $x \neq y$, are non-singular.* The Stokes theorem can in fact be proved for hypergestures, we refer to the last part of this paper, starting from Sect. 6, for a thorough discussion of a hypergestural Stokes theorem.

In the classical modulation model, one looks at all minimal cadential sets of triadic degrees [3, 26.2.1]. Here they are:

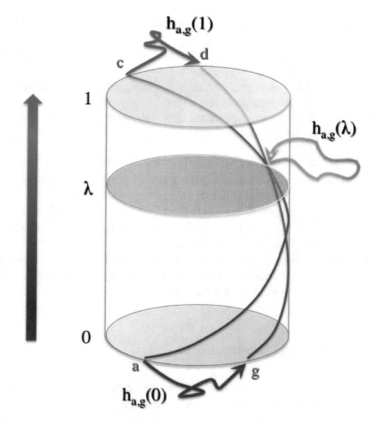

Fig. 5 The hypergesture from curve $h_{a,g}(0) = l(a, g)$ to curve $h_{a,g}(1) = l(c, d)$ enforces an intermediate singular loop position

$$J_1 = \{II, III\}, J_2 = \{II, V\}, J_3 = \{III, IV\}, J_4 = \{IV, V\}, J_5 = \{VII\}.$$

One then considers the S-transformed cadential sets. These involve all degrees II, III, IV, V, VII (in both scales C, F since S switches II_C to V_F, III_C to IV_F, IV_C to III_F, V_C to II_F, and VII_C to VII_F). We have this proposition:

Proposition 2 *For every triadic degree* $X_C = I_C, II_C, III_C, IV_C, V_C, VI_C, VII_C,$ *there is a non-singular parallel hypergesture* h_{X_C} *from* $\overrightarrow{X_C}$ *to* $\overrightarrow{S(X_C)}$ *for the Lie bracket field* $[\mathcal{X}_C, \mathcal{X}_F]$.

The proof of this proposition is an easy verification. Therefore each triadic degree can be connected hypergesturally to its symmetric counterpart. However, if we look at the cadential sets and the pitch class sets they define by union of their degrees, such as $\cup(J_1) := II_C \cup III_C$, such a connection is no more possible in general for corresponding gestures. Here are the obstructions, and Fig. 5 visualizes the singular situation for the hypergestural movement:

- For $J_1 = \{II_C, III_C\}$, the hypergesture h has a singular part for the curve $l(a, g)$ $(a \in II_C, g \in III_C)$ that maps to $l(c, d)$.
- For $J_2 = \{II_C, V_C\}$, the hypergesture h has a singular part for the curve $l(a, g)$ $(a \in II_C, g \in V_C)$ that maps to $l(c, d)$.
- For $J_3 = \{III_C, IV_C\}$, the hypergesture h has a singular part for the curve $l(a, g)$ $(a \in IV_C, g \in III_C)$ that maps to $l(c, d)$.
- For $J_4 = \{IV_C, V_C\}$, the hypergesture h has a singular part for the curve $l(a, g)$ $(a \in IV_C, g \in V_C)$ that maps to $l(c, d)$.

Therefore the only admissible hypergestural transformation is that from $\overrightarrow{VII_C}$ to $\overrightarrow{VII_F}$. This is the selection we find using the present hypergestural arguments. Then, going back to the construction of the modulation quantum in the classical model, we have to look at the intersection $F \cap (VII_C \cup VII_F) = F \cap \{b, d, f, e, g, a\#\} = \{d, f, e, g, a\#\} = II_F \cup IV_F \cup VII_F$, and the latter is exactly the set of modulation degrees described in the classical model and by Schoenberg.

This model also works for the fifth circle modulation from C to G, it is an easy exercise to go through all steps for this movement, and we get the classical modulation degrees III_G, V_G, VII_G as in the classical case.

5   The Other Direct Modulations

For other modulation types to more distant fourth circle tonalities, from C to $A_\flat$, say, we propose the following hypergestural construction. We factor the movement into fourth circle steps, e.g. C to F, then F to $B_\flat$, then $B_\flat$ to $E_\flat$, then $E_\flat$ to $A_\flat$. The corresponding integral curves through the twelve pitch classes are shown in Fig. 6.

But this is not factorizing the modulation steps, i.e. we only factor the hypergestural curves and then exhibit those hypergestures which have non-singular steps. Figure 7 shows such a factorization for the hypergesture moving e to d in a modulation $C \to B\flat$. The first part of the curve moves e to f on the closed integral curve C_e, the second part of the curve moves f to d on F_c. We shall realize this

$$C \longrightarrow F \qquad F \longrightarrow B_\flat \qquad B_\flat \longrightarrow E_\flat \qquad E_\flat \longrightarrow A_\flat$$

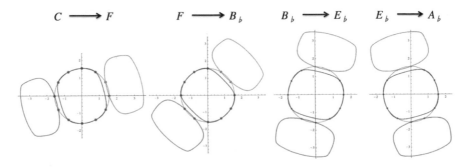

Fig. 6 The four closed integral curves for fourth circle modulations starting from $C, F, B_\flat, E_\flat$

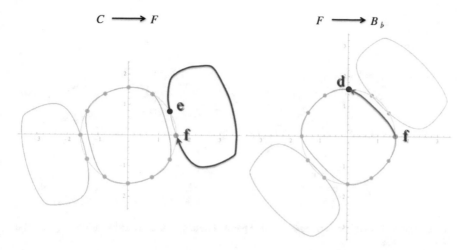

Fig. 7 The hypergestural curve from e to d factors through f on two integral curves, C_e, and F_c, while a direct movement is not possible for any of these closed integral curves

model for all fourth circle relations $C \to B_\flat, C \to E_\flat, C \to A_\flat$ (and of course for the corresponding fifth circle relations $C \to D, C \to A, C \to E$). The result will again yield the same modulation degrees as with the classical model.

The precise setup for modulations $C \to B_\flat, C \to E_\flat, C \to A_\flat$ is that we look for sequences of admissible parallel hypergestures. Denote by $S_C, S_F, S_{B_\flat}, S_{E_\flat}$ the four inversions mapping $C \to F, F \to B_\flat, B_\flat \to E_\flat, E_\flat \to A_\flat$. For example, for $C \to E_\flat$, and for a set X of pitch classes in C, we look for a sequence of admissible parallel hypergestures $h_C, h_F, h_{B_\flat}$ where h_C connects X to $S_C(X)$, h_F connects $S_C(X)$ to $S_F(S_C(X))$, and $h_{B_\flat}$ connects $S_F(S_C(X))$ to $S_{B_\flat}(S_F(S_C(X)))$, the latter being the target set in $E_\flat$. The concatenation $h = h_{B_\flat} \circ h_F \circ h_C$ of these three hypergestures is what we call an *admissible parallel hypergesture connecting a gesture* $\overrightarrow{X}$ to $\overrightarrow{S_{B_\flat}(S_F(S_C(X)))}$.

Let us make an example to understand the special character of such concatenations. We again look at the above concatenation $h = h_{B_\flat} \circ h_F \circ h_C$, and we start with a pitch class set $X = V_C$. We are looking for three admissible parallel hypergestures $h_{B_\flat}, h_F, h_C$ that connect V_C to $II_{E_\flat} = S_{B_\flat}(S_F(S_C(V_C)))$. Figure 8 shows that this is possible. The only non-trivial step is the first hypergesture, we have shown to the left the non-singularity of this hypergesture.

With this approach we now look at cadence sets $J_1, \ldots, J_5$ in C which (more precisely, as above: the unions of their members, e.g. $\cup(J_1) = II_C \cup III_C$ etc.) can be connected by admissible parallel hypergestures to corresponding cadence sets in the target tonality. If such hypergestures between cadence set J_k in C and cadence set J'_l in the target tonality exist, we proceed as before: We take the union $(\cup(J_k)) \cup (\cup(J'_l))$ and check whether their intersection $T \cap (\cup(J_k)) \cup (\cup(J'_l))$ with the target tonality T is rigid. The difference to the classical algorithm is that we

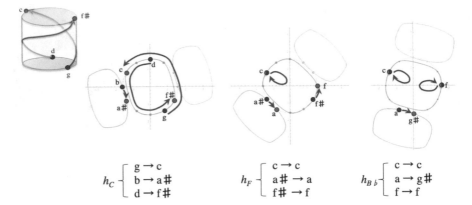

$$h_C \left\{ \begin{array}{l} g \rightarrow c \\ b \rightarrow a\# \\ d \rightarrow f\# \end{array} \right. \qquad h_F \left\{ \begin{array}{l} c \rightarrow c \\ a\# \rightarrow a \\ f\# \rightarrow f \end{array} \right. \qquad h_{B\flat} \left\{ \begin{array}{l} c \rightarrow c \\ a \rightarrow g\# \\ f \rightarrow f \end{array} \right.$$

Fig. 8 The concatenation of three admissible parallel hypergestures, connecting $V_C = \{g, b, d\}$ to $II_{E_\flat} = \{f, g\#, c\}$

don't check for minimality anymore. This condition has been taken care of by the distinguished hypergestural connection described by the integral curves of the Lie bracket vector fields. Minimality seems to be taken care of by the hypergestural transformation. The result is this:

Proposition 3 *With the above procedure, when applied to all fourth and fifth circle modulations for up to four circles, the resulting modulation steps coincide with the steps calculated in the classical model (coinciding with Schoenberg's steps).*

The proof (which we omit here) is lengthy, but easy, one has to go through all possible admissible parallel hypergestures and then to calculate the modulation steps as described above.

6 Stokes' Theorem for Hypergestures

Stokes' classical theorem states the formula

$$\int_C d\omega = \int_{\partial C} \omega,$$

where C is a compact oriented k-dimensional manifold with boundary and ω is a $k - 1$-form on C. The operator $d\omega$ is the exterior derivative of ω, and ∂C is the boundary of C. It is well known that this formula is valid for slightly more general situations, namely, where the boundary is not a manifold, but has singularities such as "corners" and the like, see [9, 11.4].

Stokes' theorem is of primordial importance in many fields of physics, e.g. in mechanics (integral invariants, see [5]) or in electrodynamics (relating differential and integral forms of Maxwell's equations [6]). The reason why we are interested in such a theorem for mathematical music theory is twofold: On the one hand, we have initiated a homological study of hypergestural structures [7] which has also provided us with applications to counterpoint theory [8]. As singular homology is strongly related to de Rham cohomology, in particular by Stokes' theorem, it is natural to ask for such a theorem in our context of hypergestures. But there is a deeper reason for such a project, namely the idea that music theory of hypergestures could provide us with models of energy exchange in gestural interaction. In such a (still hypothetical) theory, Stokes' theorem would play a crucial role regarding questions of energy conservation (integral invariants).

7 Almost Regular Manifolds, Differential Forms, and Integration for Hypergestures

We first need to specify the basic concepts that contribute to the Stokes statement. We are aware of the somewhat sloppy style in this quite standard part of the paper, the readers are kindly asked to fill out the standard technical details.

7.1 Locally Almost Regular Manifolds

Hypergestures in topological spaces were introduced in [10] and later generalized to hypergestures in topological categories [11]. In the present context we need hypergestures in manifolds since we are dealing with differentiable structures. We however need quite general manifolds in the sense of what are called "almost regular manifolds" in [9] or even more singular manifolds, where the boundaries have corners. To understand our requirement we look at typical manifolds in the context of hypergestures. In [10], we have introduced a standard topological space $|\Sigma|$ associated with a digraph Σ. It is the colimit of the digraph's arrow set A_Σ, the gluing operation being performed on the digraph vertices set V_Σ. This topological space specifies one line chart $|a| \xrightarrow{\sim} I = [0, 1]$ per arrow a and a point chart $|x|$ for each isolated vertex x. The specification of this atlas is mandatory since we don't want to glue two consecutive arrows $x \xrightarrow{a} y \xrightarrow{b} z$ to one line. The differentiability in the connecting vertex y is suspended. Or it may also happen that three or more arrows share a vertex, and then the differentiability in such a vertex would not make sense. Call *skeletal space* the manifold $|\Sigma|$ associated with skeleton Σ.

The best conceptual approach to this situation is to embed such a manifold in a differentiable manifold M as a subset whose charts are manifolds with boundary isomorphic to the unit interval I or to a zero-dimensional point manifold 0. We next need cartesian products of such manifolds when hypergestures are discussed. This means that we have to consider products of type $|\Sigma_1| \times |\Sigma_1| \times \ldots |\Sigma_n|$. These

Fig. 9 The cartesian product
$|\Sigma_1| \times |\Sigma_2|$ for
$\Sigma_1 = \bullet \to \bullet$,

manifolds are living in cartesian products of their carrier manifolds $M_1, M_2, \ldots, M_n$, and the typical boundary of a product $|\Sigma_1| \times |\Sigma_2|$ is $\partial(|\Sigma_1| \times |\Sigma_2|) = \partial|\Sigma_1| \times |\Sigma_2| \cup |\Sigma_1| \times \partial|\Sigma_2|$, see Fig. 9 for an example.

But observe that due to singular points in digraphs, such products can be inhomogeneous in their dimension. A product may be a disjoint union of submanifolds of different dimensions.

To get a reasonable category of such manifolds, we consider differentiable morphism $L \to M$ of the carrier manifolds L, M of $\mathcal{L}, \mathcal{M}$, respectively, that restrict to atlas-compatible maps $f : \mathcal{L}^I \to \mathcal{M}^J$, where I, J designate the atlases of $\mathcal{L}, \mathcal{M}$, respectively. Atlas-compatibility means that, like in mathematical music theory of global compositions, we are also given a map $g : I \to J$ such that f sends I-chart $\mathcal{L}_i$ to J-chart $\mathcal{M}_{g(i)}$. We denote this *category of locally almost regular manifolds* by $LARM$. Such a manifold need not have a determined dimension, but several dimensions according to connected components and charts. In what follows, we shall call dimension $dim(\mathcal{L})$ of an almost regular manifold $\mathcal{L}$ the maximum of dimension of such components. The submanifold of $\mathcal{L}$ of a determined dimension k will be denoted by $\mathcal{L}^k$.

The most important application of $LARM$ for the Stokes theory lies in a reinterpretation of hypergestures. Suppose we are given a hypergesture $c \in \Sigma_1 \Sigma_2 \ldots \Sigma_n \overrightarrow{@} \mathcal{L}$ over n skeleta $\Sigma_1, \Sigma_2, \ldots, \Sigma_n$ with values in a locally almost regular manifold $\mathcal{L}$. By the very definition of hypergestures, and by the adjointness property of the manifold $|\Sigma|$ associated with skeleton Σ [10, Proposition 5.1], as well as the adjointness of the cartesian product and repeated function spaces (also known as currying in computer science), $\Sigma_1 \Sigma_2 \ldots \Sigma_n \overrightarrow{@} \mathcal{L} \xrightarrow{\sim} |\Sigma_1| \times |\Sigma_2| \times \ldots |\Sigma_n| \textcircled{@} \mathcal{L}$, the set of continuous functions from the cartesian product of the skeletal manifolds to $\mathcal{L}$. Within this function set, we exhibit the differentiable morphisms and denote their set by $|\Sigma_1| \times |\Sigma_2| \times \ldots |\Sigma_n| \textcircled{d} \mathcal{L}$. The morphisms in the latter (more precisely: their corresponding hypergestures) are called *differentiable hypergestures*, the set of these hypergestures is also denoted by $\Sigma_1 \Sigma_2 \ldots \Sigma_n \textcircled{d} \mathcal{L}$. In the context of the Stokes theorem, we need differentiable singular n-cubes. Their generalization to hypergestures are differentiable gestural n-cubes, namely the elements of $\Sigma_1 \Sigma_2 \ldots \Sigma_n \textcircled{d} \mathcal{L}$. The free module $\mathbb{R} \Sigma_1 \Sigma_2 \ldots \Sigma_n \textcircled{d} \mathcal{L}$ of $\mathbb{R}$-linear combinations of differentiable gestural n-cubes (the module basis) defines the *(differentiable) n-chains over* $\Sigma_1, \Sigma_2, \ldots, \Sigma_n$ with values in $\mathcal{L}$.

7.2   Differential Forms

On a locally almost regular manifold $\mathcal{L}$ (we omit the atlas if ever possible to ease notation), differential forms can be considered in the sense that they are defined on each chart as usual. If such a chart $\mathcal{L}_i$ has dimension n, the differential forms of dimension $k \leq n$ define non-trivial real vector spaces $\bigwedge^k \mathcal{L}_{i,x}$ in each point x of $\mathcal{L}_i$. A differential k-form ω on $\mathcal{L}$ is a differentiable section in each chart $\bigwedge^k \mathcal{L}_i$. Since our manifolds are of different dimensions locally, we will have to deal with forms that don't have the same dimension everywhere, they are not homogeneous. We therefore consider the direct sum $\bigwedge^{\oplus k} \mathcal{L} = \bigoplus_{l \leq k} \bigwedge^l \mathcal{L}$. If we take a differential form $\omega \in \bigwedge^{\oplus k} \mathcal{L}$, its l-component will be denoted by ω_l. As in the classical case, for a morphism $f : \mathcal{L} \to \mathcal{M}$ of locally almost regular manifolds, one has the canonical inverse image $f^* \omega \in \bigwedge^k \mathcal{L}$ for $\omega \in \bigwedge^k \mathcal{M}$.

In the classical case, one has the exterior derivative operator $d : \bigwedge^k \mathcal{L} \to \bigwedge^{k+1} \mathcal{L}$ with $d^2 = 0$. For the non-homogeneous case mentioned above, we need a derivative operator $d^\oplus$ defined by $d^\oplus \omega = (\omega_0, d\omega_0, d\omega_1, d\omega_2, \ldots)$ for $\omega = (\omega_0, \omega_1, \omega_2, \ldots)$. For this operator, we have $d^{\oplus 2} \omega = (\omega_0, d\omega_0, 0, \ldots)$. And as in the classical case, the operators d and $d^\oplus$ commute with inverse images.

7.3   Integration

Modulo linear extensions to n-chains, we need to define $\int_c \omega$ for a gestural n-cube $c \in \Sigma_1 \Sigma_2 \ldots \Sigma_n \textcircled{\tiny D} \mathcal{L}$. As usual, the formula is defined to mean $\int_{|\Sigma_1| \times |\Sigma_2| \times \ldots |\Sigma_n|} c^* \omega$, which amounts to restrict to the special case $\mathcal{L} = |\Sigma_1| \times |\Sigma_2| \times \ldots |\Sigma_n|$. We shall define the integral by recursion on the hypergestural parameters and recalling the Fubini theorem for iterated integration [12, Theorem 3-1]. Let $(\lambda, t) \in T|\Sigma_1|_\lambda$, the tangent space at $\lambda \in |\Sigma_1|$. This argument defines a form $c^* \omega_{\lambda, t} \in \bigwedge^{\oplus (n-1)} |\Sigma_2| \times \ldots |\Sigma_n|$, and we may suppose by recursion that $I(\lambda, t) = \int_{|\Sigma_2| \times \ldots |\Sigma_n|} c^* \omega_{\lambda, t}$ is defined, which yields an element of $\bigwedge^{\oplus 1} |\Sigma_1|$. So we are left with the definition of the integral for $n = 0, 1$. If $n = 0$, $c \in \mathcal{L}$, and $\omega \in \mathcal{F}(\mathcal{L})$ is a function. Then we set $\int_c \omega = \omega(c)$. In dimension $n = 1$, there are three cases for Σ_1:

1. If $A_{\Sigma_1} = \varnothing$, then set $\int_c \omega = \sum_{i \in V_{\Sigma_1}} \omega_0(c(i)) = \sum_{i \in V_{\Sigma_1}} \int_{c(i)} \omega_0$.
2. Recall from [7, Sect. 3] that for an arrow a of Σ_1, a^- denotes the subskeleton of Σ_1 after taking away the tail $t(a)$ and all arrows connected to $t(a)$. And a^+ denotes the subskeleton of Σ_1 after taking away the head $h(a)$ and all arrows connected to $h(a)$. In this second case, we suppose that there is at least one arrow a, but both, A_{a^-} and A_{a^+} are empty. This means that, besides isolated vertices, there are either a number of loops on a single vertex or else there is a number of arrows between two distinct points. This is the classical one-dimensional situation for integration on the unit interval. So we define $\int_c \omega = \sum_{a \in A_{\Sigma_1}} \int_a \omega_1 + \int_{\text{isolated vertices}} \omega$, where $\int_a \omega_1$ is the evident classical integration.

3. In the third case, there is an arrow a such that $A_{a^-} \cup A_{a^+} \neq \emptyset$. We then set the recursive formula $\int_c \omega = \sum_{a \in A_{\Sigma_1}} (\int_{c|a^-} \omega - \int_{c|a^+} \omega)$, a formula that reminds us of the definition of the face operator $?^\square$ given in [7, Definition 3.1].

8 Stokes' Theorem

For the proof of Stokes' theorem for hypergestures, we need a technical lemma. It refers to the Escher theorem operation on chains $c \in \Sigma_1 \Sigma_2 \ldots \Sigma_n \textcircled{0} \mathcal{L}$ which generates a chain $c_j \in \Sigma_j \Sigma_1 \Sigma_2 \ldots \widehat{\Sigma_j} \ldots \Sigma_n \textcircled{0} \mathcal{L}$.

Lemma 2 *If $c \in \Sigma_1 \Sigma_2 \ldots \Sigma_n \textcircled{0} \mathcal{L}$ is a differentiable n-cube, $1 \leq j \leq n$, $a \in A_{\Sigma_j}$, and $\lambda \in |\Sigma_1|$, then we have*

$$(c_j | a^\pm)^\square(\lambda) = (c(\lambda)_j | a^\pm)^\square,$$

and therefore also

$$(c_j)^\square(\lambda) = (c(\lambda)_j)^\square.$$

The lemma follows from the observation that (1) the face operator yields the same linear combination on both sides since it acts on the same $\Sigma_j | a^\pm$, and (2) the evaluation at λ is taken on the same face operator result.

Theorem 1 (Stokes' Theorem for Hypergestures) *Let $c \in \mathbb{R}\Sigma_1 \Sigma_2 \ldots \Sigma_k \textcircled{0} \mathcal{L}$ be a k-chain in a k-dimensional locally almost regular manifold $\mathcal{L}$, and let $f \in \bigwedge^{k-1} \mathcal{L}$. Then*

$$\int_c d^\oplus f = \int_{\partial c} f.$$

Proof We can of course restrict to gestural k-cubes. For $k = 1$, f is a function on $\mathcal{L}$ and $c \in \Sigma \textcircled{0} \mathcal{L}$. Let first $A_\Sigma = \emptyset$. Then $\int_{\partial c} f = \sum_{i \in V_\Sigma} f(c(i))$, whereas $\int_c d^\oplus f = \sum_{i \in V_\Sigma} (d^\oplus f)_0(c(i)) = \sum_{i \in V_\Sigma} f(c(i))$ yields the same. For the second case, $A_{a^-} \cup A_{a^+} = \emptyset$, but arrows exist, we may focus on the subskeleton bearing those arrows, the discrete part having been already dealt with. Here,

$$\int_c d^\oplus f = \sum_{a \in A_\Sigma} \int_a df$$

$$= \sum_{a \in A_\Sigma} \int_{\partial a} f$$

$$= \sum_{a \in A_\Sigma} f(c(h(a))) - f(c(t(a)))$$

$$= \int_{\partial c} f,$$

this is the classical case. For the third case, $A_{a^-} \cup A_{a^+} \neq \emptyset$, we have

$$\int_c d^\oplus f = \sum_{a \in A_\Sigma} \int_{c|a^-} d^\oplus f - \int_{c|a^+} d^\oplus f$$

$$= \sum_{a \in A_\Sigma} \int_{\partial(c|a^-)} f - \int_{\partial(c|a^+)} f$$

$$= \sum_{a \in A_\Sigma} \int_{(c|a^-)^\square} f - \int_{(c|a^+)^\square} f$$

$$= \int_{\partial c} f$$

by recursion and since ∂ and $?^\square$ coincide in dimension one.

The case of higher dimensions runs as follows:

$$\int_c d^\oplus f = \int_{\lambda \in \Sigma_1} \int_{c(\lambda)} d^\oplus f$$

$$= \int_{\lambda \in \Sigma_1} \int_{\partial c(\lambda)} f \quad \text{(recursion)}$$

$$= \int_{\lambda \in \Sigma_1} \sum_j (-1)^j \int_{(c(\lambda)_j)^\square} f$$

$$= \sum_j (-1)^j \int_{\lambda \in \Sigma_1} \int_{(c(\lambda)_j)^\square} f$$

$$= \sum_j (-1)^j \int_{\lambda \in \Sigma_1} \int_{(c_j)^\square(\lambda)} f \quad \text{(Lemma 2)}$$

$$= \sum_j (-1)^j \int_{(c_j)^\square} f$$

$$= \int_{\partial c} f.$$

This terminates the proof of Stokes' theorem.

References

1. Mazzola, G.: Gruppen und Kategorien in der Musik. Heldermann, Berlin (1985)
2. Mazzola, G.: Geometrie der Töne. Birkhäuser, Basel (1990)
3. Mazzola, G., et al.: The Topos of Music-Geometric Logic of Concepts, Theory, and Performance. Birkhäuser, Basel (2002)
4. Schoenberg, A.: Harmonielehre (1911). Universal Edition, Wien (1966)
5. Abraham, R.: Foundations of Mechanics. Benjamin, New York (1967)

6. Jackson, D.: Classical Electrodynamics. Wiley, New York (1998)
7. Mazzola, G.: Singular homology on hypergestures. J. Math. Music **6**(1), 49–60 (2012)
8. Mazzola, G.: Hypergesture homology for counterpoint. In: Agustín-Aquino O., Junod, J., Mazzola, G. (eds.) Computational Counterpoint Worlds. Springer, Heidelberg (2014)
9. Loomis, L.H., Sternberg, S.: Advanced Calculus. Addison-Wesley, Reading (1968)
10. Mazzola, G., Andreatta, M.: Diagrams, gestures and formulae in music. J. Math. Music **1**(1), 23–46 (2007)
11. Mazzola, G.: Categorical gestures, the diamond conjecture, Lewin's question, and the Hammerklavier sonata. J. Math. Music **3**(1), 31–58 (2009)
12. Spivak, M.: Calculus on Manifolds. Benjamin, New York (1968)

Manuel M. Ponce's Piano Sonata No. 2 (1916): An Analysis Using Signature Transformations and Spelled Heptachords

Mariana Montiel

Abstract In the present work an analysis is made of several passages from Manuel M. Ponce's Sonata No. 2 for piano (Ponce, Sonata No. 2 for Piano, 1916/1968, [1]), employing Julian Hook's theoretical development of *signature transformations* and *proper spelled heptachords*. A signature transformation reinterprets a diatonic object in the context of a different key signature. The signature transformations form a cyclic group of order 84; indeed, the chromatic transpositions (Tn) and the diatonic transpositions (t_n) form subgroups of this cyclic group, hence contributing with yet another way of unifying diatonic and chromatic structures. After giving an introduction to the theory behind the signature transformations, we proceed to an analysis of illustrative passages of the Sonata, using units of varying size called *diatonic fragments*. During this analysis we realized that the classes of proper spelled heptachords, a generalization of the signature transformations, could explain the constant transition between 7-note nearly diatonic scales. These classes also have a clear mathematical structure, with a transposition operator τ (they are also called τ-classes), and possess some of the symmetries as well as the seven modes of the diatonic class. This analysis made us look for both intra-class transformations, similar to the ones we find in the diatonic class, and inter-class transformations that can explain the fluid movement between classes found not only in this sonata, but in other pieces that are classified as "chromatic" without more detail.

1 Introduction

Ponce's Sonata no. 2, without a doubt, has a nationalist character. The two themes of the first movement are borrowed from two folksongs, *El sombrero ancho* and *Las mañanitas* and the first theme of the second movement is based on *Pica, pica, perico*. The date of this composition, 1916, falls in what is still considered Ponce's "romantic period" as opposed to his "modern style" of later years [2]. Nevertheless, when studying this piece one finds a style that is far from the formal characterization

M. Montiel (✉)
Georgia State University, Atlanta, GA 30302, USA
e-mail: mmontiel@gsu.edu

© Springer International Publishing AG 2017
G. Pareyon et al. (eds.), *The Musical-Mathematical Mind*,
Computational Music Science, DOI 10.1007/978-3-319-47337-6_19

of the period we know as Romanticism; the first movement of the Sonata is full of non-traditional chord progressions, of dissonance, and the influence of the impressionism of Debussy, so admired by Ponce.

Within the neo-Riemannian focus there have arisen several forms of carrying out theoretical analysis of a score by means of mathematical transformation groups. There is an undeniable coincidence among these forms but, at the same time, each one offers unique aspects that privilege the specificities of the piece itself and the needs of the analyst. In this work we will make use of *signature transformations*, fruit of the theoretical development of [3], a tool we thought could serve to analyze and comprehend many of the melodic and harmonic transformations that Ponce carries out during the development of the sonata. Hook's signature transformations, that capture tonality in the seven diatonic modes, offer the possibility of tracing the *diatonic organization* [4]. We thought it would be interesting to experiment with this theoretical development in a piece like Ponce's Sonata no. 2, which definitely possesses the characteristics of twentieth century musical modernity, in spite of its classification within the Romantic period of this great composer. However, in this process, we realized that Hook's classes of *proper spelled heptachords* could explain the constant transition between 7-note nearly diatonic scales. These classes also have a clear mathematical structure, with a transposition operator τ (they are also called τ-*classes*), and possess some of the symmetries as well as the seven modes of the diatonic class. This analysis made us look for both intra-class transformations, similar to ones we find in the diatonic class, and inter-class transformations that can explain the fluid movement between these exotic scales (classes) found not only in this Sonata, but in other pieces that are classified as "chromatic", without more detail.

2   Signature Transformations

Signature transformations, as created and defined by [3], offer a novel way of tracing the transformations in a musical work from a diatonic perspective. In a study concerned with the diatonic organization in Vaughan Williams, [4] uses three representations to analyze what he calls *fixed-domain diatonic relations*. The three types of fixed-domain tonal relations are key signature, scale type, and tonic. Signature transformations realize the third type, that is, they trace changes in the tonalities of the seven diatonic modes that share a tonic. The fact that they share a tonic forces the change —*transformation*— of the key signatures. In Bates' study, the three forms were combined; in the present work we will concentrate exclusively on the signature transformations.

Signature transformations act on the set of *fixed diatonic forms* [see 1, 140–142]. Fixed diatonic forms are equivalence classes of fragments of diatonic music, with a key signature and a clef. These fragments are in the same equivalence class if their pitch-class content is the same (modulo 12), and if they determine the same diatonic collection up to enharmonic equivalence.

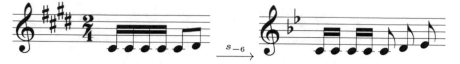

Fig. 1 Signature transformation

We will use the notation s_n, $n \in \mathbb{N}$, for the number n of sharps that are added and s_{-n} for the number $-n$ of flats that are added. The operation of adding sharps (or subtracting flats) is "positive", and the operation of subtracting sharps (or adding flats) is "negative". For example, in Fig. 1, s_{-6} reduces the key signature by 4 sharps and then we continue to count negatively by adding flats:

The content of the following paragraphs is related to Hook's theorem [3, 142–144]. This theorem provides the theoretical basis that establishes the resulting method. The signature transformations form a cyclic group of 84 elements (they pass through the twelve pitches of the chromatic scale and the seven diatonic modes) generated by s_1, although it is not expected that 84 sharps would be added to a key signature! Indeed, even though the signature transformations form a cyclic group, the s_n and s_{-n} can be reached through compositions with T_n and t_n, the chromatic and diatonic transposition operators respectively.

If we add seven sharps to a key signature we will transpose the diatonic collection a semitone (for example, from C major to C$^\sharp$ major). Therefore, s_7 operates in the same way as T_1 and, analogously, s_{-7} acts as T_{11}. It should be mentioned that, while the chromatic transposition operator implicitly changes the key signature as well as the actual notes, the diatonic transposition operator does not change the key signature; that is, the diatonic transposition operator transposes within its diatonic scale. Now, if we apply t_1 to a *diatonic fragment* —or *diatonic form*—, without changing the key signature, we have the same pattern in pitches but transposed up a scale step. However, if we apply s_{12} we also transpose a scale step (see Fig. 2 taken from [3, p. 143], with a diatonic fragment of four notes, where a key signature with six flats arrives to a key signature of twelve sharps by applying s_1 twelve times). Thus, every transition operator, whether chromatic or diatonic, can be written as an s_n for some n. Similarly, as a weak converse, any s_n can be written as a composition of some T_n and t_n as we can obtain the generator, s_1, in the following way:

$$t_3 T_7 = (t_1)^3 (T_1)^7 = (s_{12})^3 (s_7)^7 = s_{36} s_{49} = s_{85} = s_{84} s_1 = s_1 \text{(we are calculating modulo 84)}.$$

Signature transformations can explain transformational aspects of music that translates its content between different diatonic forms. This means that the transformations always occur within a diatonic context that must be identified, something that is not a requisite for other neo-Riemannian type transformations, such as P, L, and R.

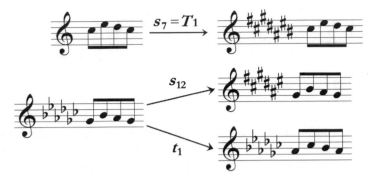

Fig. 2 Transition operators as signature transformations in [3]

E♭♭	B♭♭	F♭	C♭	G♭	D♭	A♭	E♭	B♭	F	C	G
-12	-11	-10	-9	-8	-7	-6	-5	-4	-3	-2	-1
D	A	E	B	F#	C#	G#	D#	A#	E#	B#	F##
0	1	2	3	4	5	6	7	8	9	10	11
C##											
12											

Fig. 3 The *line of fifths*

3 Proper Spelled Heptachords

Spelled heptachords are sets of seven pitch classes in which each letter name only appears one time. Any diatonic scale is a spelled heptachord. Many "almost diatonic" scales are spelled heptachords which are proper: free of enharmonic doublings or voice crossings (Fig. 3).

Let τ_k represent the transposition by fifths according to the following table:

Here the symmetry is around D because of its symmetric position in the line of fifths. Then we obtain 66 τ-classes (fields) of proper spelled heptachords, in which τ_k is the transposition operator within each of the classes. That is, each of the proper spelled heptachords can be expressed in any of the 12 keys; it is also important to mention that every spelled heptachord has seven modes [3, 5].

4 Ponce's Sonata No. 2

Ponce's Sonata no. 2 has been described as "modal" in certain parts [2]. However, we did not find this to be evident, as the signature transformations would have shown this characteristic, and their presence in the piece was virtually inexistent as direct transformations. As will be seen below, we could use the signature transformation

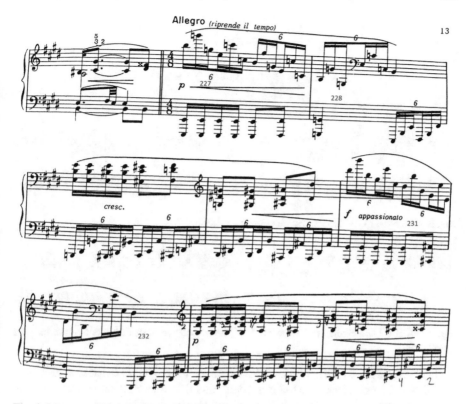

Fig. 4 Measures 226–242 of Ponce's Sonata no.2, first movement

perspective to follow certain transpositions in a gestural way [6], but that implies passages are not actually found in the piece.

This can be seen in the following analysis with a passage and its diatonic fragment that correspond to measures 227 and 228, and to measures 231 and 232, of the first movement of Ponce's Sonata no. 2. Measures 227 and 228 are in E Aeolian, which only has one sharp. This is reflected in Fig. 5, but in the original piece the key signature has four sharps, as can be seen in Fig. 4. Measures 231 and 232 are in $G^\sharp$ Aeolian, which has five sharps. To travel from E Aeolian to $G^\sharp$ Aeolian we must add four sharps by the application s_4 (which places us in E Lydian) and then transpose diatonically by two tones. We emphasize that, as the composition is commutative, it could have been carried out in the inverted order (although there are examples in which it is not possible musically to carry out some s_n in particular, due to the diatonic context). Hence, the signature transformation is $t_2 s_4 = (t_1)^2 s_4 = (s_{12})^2 s_4 = s_{28} = s_4 t_2$. Of course, we can look at this transformation as simply $T_4 = (T_1)^4 = (s_7)^4 = s_{28}$. As was mentioned above, the inclusion of E Lydian is gestural, given that it does not appear in the piece but, according to the signature transformation perspective, is implied.

Fig. 5 E Aeolian to G♯ Aeolian, passing through E Lydian

Fig. 6 From $C^\sharp$ Mixolydian to Dorian mode of the acoustic scale or fifth mode of the melodic minor scale

In measures 89–91 we have the $C^\sharp$ Mixolydian mode. When one looks at measures 354–356 it seems that we have found an ideal candidate for a signature transformation that would change the mode while leaving the tonic fixed. This signature transformation should be s_{-1}, given that one sharp is eliminated. However, the eliminated sharp is $A^\sharp$ and, diatonically, it is not possible to go from six to five sharps by removing $A^\sharp$, it would have to be $E^\sharp$ that would disappear. What occurs here is that in measures 354–356, with $C^\sharp$ as tonic we do not have any of the seven diatonic modes; what we have is the Hindu scale whose pattern is 2212122 (also known as the Dorian mode of the acoustic scale, or the fifth mode of the melodic minor scale) (Fig. 6).

Hook's work on spelled hexachords [7] does address non-diatonic collections and actually classifies a rotation of the pattern of the scale identified in measures 354, 355, and 356 under the name of MMIN (for melodic minor). Hence, although in his generalization of the signature transformations there is not, until now, a mathematical function that represents the change from a diatonic context to a non-diatonic one, it can be categorized within the theory developed in this article on spelled hexachords. According to Hook,

> In the X/H notation from my paper, the mod-7 musical material X (the "dpc structure") does not change at all; only the heptachord H changes, in this case from DIA(+6) to MIN(+5).[1] This *field change* is similar to the *field transposition* in Fig. 1 [7, 91], but it cannot literally be a transposition since it's not the same type of field. (Personal communication, Sept. 2014).

---

[1] DIA(+6) means diatonic with 6 sharps, and MMIN(+5) means melodic minor (or *acoustic*) with 5 sharps.

2029-33

Fig. 7 Ponce's Sonata no. 2, measures 55–70

However, if we look at the following passages from the Sonata, we find constant transformations between heptachords, both intra- and inter-classes.

In Figs. 7 and 8 we can see a passage from measures 57–91, skipping measures 71–75 for reasons of space and the fact that they are not relevant to what we are showing. We see the Locrian mode of the ascending minor scale starting on $G^\sharp$ (or Lydian mode of the acoustic scale) in measures 57–59. We find $\tau_3(G^\sharp)$, that is, a heptachord in the same τ-class starting on $E^\sharp$, in measures 79–81. However, before this τ transformation, we find the Hungarian Gypsy scale in $C^\sharp$ in measures 61–63 (which will be transformed by τ_3 to $A^\sharp$ in measures 83–85) as well as the Mela-Dhenuka scale in $F^\sharp$ in measures 67–69 and the Mixolydian mode of the diatonic scale starting on $D^\sharp$ in measures 89–91. This is only one example of this constant change of heptachord classes, as well as within the heptachord classes by the τ transformation, in several passages of this Sonata.

Fig. 8 Ponce's Sonata no. 2, measures 71–91

The 66 τ-classes of proper spelled heptachords, plus the 462 spelled pitch class structures that are generated by *complete diatonic structures*, provides a formal, mathematical and, above all, detailed, way to analyze music that has often been labeled as "chromatic", without any further classification. Reference [5] show (visually and audibly) the 462 modes of the diatonic bell, that is, 66 representatives of Hook's τ-classes and their 7 rotations (modes). However, is it possible to find algebraic mathematical functions that represent the changes between different classes of spelled heptachords (scales)? We know that:

a. The signature transformations are restricted to DIA;
b. The τ-classes are classified in terms of the fifth transpositions;
c. Every proper spelled heptachord has seven modes.

5   One Approach to these Transitions

In this final part we present an idea that arose when trying to answer the previous questions. For reasons of space and the nature of this report, this approach will be developed in a later work.

Let C_i be a category that corresponds to any of the 66 τ-class heptachords. The objects are the heptachords in the particular class, that is, the seven "modes" of the representative of the class and the 12 keys on which each of the modes can begin, 84 "objects". The morphisms are $\rho_0, \rho_1, \ldots \rho_6$, the seven rotations that produce the seven modes, and $\tau_0, \tau_1, \ldots, \tau_6$ which are the transpositions by fifths explained above, as well as the compositions of the ρ_k and the τ_l. The functor that takes C_1 to C_2, $F : C_1 \rightarrow C_2$ is contrived in the following way. If H_i is a heptachord in a particular key and mode in C_1, then $F(H_i)$ is a heptachord in the same key and mode in C_2. F complies with the conditions of a covariant functor, given that $F(\rho_k \tau_l) = F(\rho_k)F(\tau_l) = \rho_k \tau_l$ and the same is true for any of the possible compositions and their orders. Similarly, $F(I_{C_1}) = I_{C_2}$. Hence this approach permits the transitions between the different τ-classes of heptachords seen as categories, and gives a simple and well known frame, that of rotations and translations (transpositions) to carry it out.

References

1. Ponce, M.M.: Sonata No. 2 for Piano. Peer International Corporation, New York (1916/1968)
2. Guerra, D.M.: Manuel M. Ponce: a study of his solo piano works and his relationship to Mexican musical nationalism. Dissertation. UMI Microform 9722750 Copyright 1997 by UMI Company, Ann Arbor, Michigan (2012)
3. Hook, J.: Signature transformations. In: Douthett, J., Hyde, M., Smith, C. (eds.) Music Theory and Mathematics: Chords, Collections, and Transformations, pp. 137–160. University of Rochester Press, Rochester (2008)
4. Bates, I.: Vaughan Williams's five variants of Dives and Lazarus: a study of the composer's approach to diatonic organization. Music Theory Spectr. **34**(1), 3450 (2012)
5. Audetat, P., Junod, J.: The Diatonic Bell. https://www.cloche-diatonique.ch/atlas/preview.php?lang=en&c=12&d=7 (2006). Accessed 4 Dec 2014
6. Mazzola, G.: Mathematical music theory: status quo 2010. In: Agustin-Aquino, O.A., Lluis-Puebla, E. (eds.), Memoirs of the Fourth International Seminar on Mathematical Music Theory. https://www.sociedadmatematicamexicana.org.mx/SEPA/ECMS/resumen/PME1_1.pdf (2011). Accessed 11 Oct 2011
7. Hook, J.: Spelled heptachords. In: Agon, C., Andreatta, M., Assayag, G., Amiot, E., Bresson, J., Mandereau, J. (eds.) Mathematics and Computation in Music MCM Proceedings, pp. 84–97. Springer, Heidelberg (2011)

Textural Contour: A Proposal for Textural Hierarchy Through the Ranking of Partitions *lexset*

Daniel Moreira de Sousa

Abstract This paper proposes an organisation of textural progressions through the ranking of partitions *lexical set* (*lexset*). Departing from the Musical Contour Theory, mainly developed by Michael Friedmann and Robert Morris, and extending its principles to the music textural domain, it is possible to generate the Textural Contour. The Partitional Analysis, which emerges from the approximation between Wallace Berry's approach and the Theory of Integer Partitions, is applied as a methodological approach for the textural parameter. The Textural Contour provides some tools for textural analysis based on the relative variation of textural complexity. The paper concludes with an example of methodological application of the Textural Contour and ranking of partitions.

1 Introduction

Textural Contour emerges from the junction of the Musical Contour Theory [4, 13, 14] and some concepts originated from the Partitional Analysis [5]. Such proposal has been developed and tested in preliminary studies [9, 11, 12] realized by the MusMat Research Group at Federal University of Rio de Janeiro (UFRJ). As a result of this research two computational tools were developed, intending to facilitate the implementation of the Textural Contour: *Partitional Operators* [6] and *Contour Analyzer* [10].

2 Musical Contour Theory

The term "contour", in a general sense, refers to a configuration that express the relation between two or more parameters or dimensions. A contour is defined by the ordering relation between the involved parameters, as in a meteorological

D.M. de Sousa (✉)
Federal University of Rio de Janeiro, Rua do Passeio, 98, Centro, Rio de Janeiro, Brazil
e-mail: danielspro@hotmail.com

© Springer International Publishing AG 2017
G. Pareyon et al. (eds.), *The Musical-Mathematical Mind*,
Computational Music Science, DOI 10.1007/978-3-319-47337-6_20

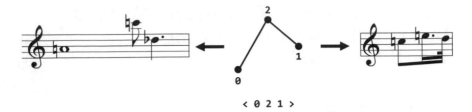

Fig. 1 Two different motives noted with the melodic contour < **021**>

cartography, for example, where the humidity is related to a geographic position. In music, the parameters most commonly associated are pitches and time, resulting in what is named melodic contour.

Musical Contour Theory (MCT) emerges as a formalization of the relations in a melodic contour, thus constituting a series of concepts and methodological tools with application in analytical and creative domains. MCT departs from an abstraction of the absolute pitch levels in a melodic structure, using a numerical representation, ordered from the lowest level (noted as 0) up to highest (noted as $n - 1$, where n is the number of different levels in the structure). This representation describes the relative position of the pitches among all levels according to some criteria. For example, a melodic contour noted < **021** > indicates a sequence that begins at the lowest pitch ascends to the highest one and ends at the intermediate one, disregarding the absolute pitches involved (Fig. 1).

This abstraction was intended to approximate MCT to Allen Forte's Musical Set Theory, allowing the establishment of relations among distinct melodic sequences based on their identity as well as the understanding of the transformational processes resulted from application of canonical operations (like inversion, retrograde, and retrograded inversion). This abstraction also enables the generation of derived contours using mathematical processes and describes information about a given contour's structure.

In spite of conventional focus of MCT on the pitch, there are several studies that deal with other structural parameters using the same principle of abstraction. Elizabeth Marvin [8] applied the principle of contour to the rhythm, proposing a contour based on the duration of the figures. Morris [14] and Marcos Sampaio [15] used chord density as a parameter for the establishment of a contour. They also applied the abstraction to the organizing the dynamic notation. Robert Clifford [3] proposed a textural contour, but in spite of using the term "textural", he relates it to movements of pitched events, mainly from the point of view of the registry.

The textural parameter is made measurable and comparable through Berry's [2] definition:

> The texture of music consists of its sounding components; it is conditioned in part by the number of those components sounding in simultaneity or concurrence, its qualities determined by the interactions, interrelations, and relative projections and substances of component lines or other component sounding factors. A set of interactions and interrelations between sounding components [2].

For the creation of a Textural Contour it is necessary to establish a hierarchy for textural configurations. For this purpose the concepts of the Partitional Analysis [5] are adopted in this study.

3 Partitional Analysis

Partitional Analysis (PA) is an original analytical and compositional tool resulted from the approximation between Berry's textural analysis and the Theory of Integer Partitions [1]. PA provides a formalization of the textural organization by numerical and graphic representations of concurrent musical ideas.

Berry proposes a formal methodology for analyzing textural configurations constructed from the comparison of basic features of the different sounding components, like rhythmic profiles and melodic contours. These configurations are read and processed as integer partitions in PA. For example, the partitions of number 3 can be used to organize the instrumental combination by the rhythmic criterion, forming: a three-part block (3); two-part block and a solo voice (2+1); or three-part polyphony (1+1+1) (Fig. 2).

From the observation of binary relations within the voicing configurations, Gentil-Nunes elaborated a pair of indices that express the relations of these configurations: the *agglomeration* index (*a*), which refers to the thickening of the internal elements, defined by its mutual collaboration (sound blocks); and the *dispersion* index (*d*), which represents the internal diversity, defined by the contraposition of its elements (polyphony). Gentil-Nunes also proposes the *partitional operators*, which express the process of internal transformation involved in the progression of one partition to the next one. The partitional operators are classified as positive or negative, according to the progressive or recessive characteristic of the corresponding transformations, and are subdivided into three groups: (1) *simple*, (2) *compound* and (3) *relational*. This paper is specifically concerned to the simple operators, namely, *resizing*, *revariance* and *simple transfer*.

Fig. 2 Textural progression with the partitions of number "3"

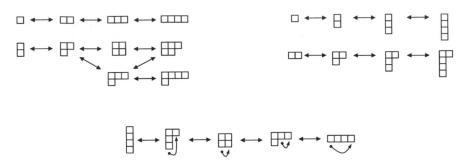

Fig. 3 Paths formed by *simple operators* using the Young's Diagram

Resizing (m) is related to the change of the thickness of one of the elements of the partition. It is derived from the inclusion relation, where the antecedent partition is contained in the consequent one. *Revariance (v)* is also derived from the inclusion relation and its occurrence concerns the addition or subtraction of a new component of density-number[1] equal to 1, changing the degree of polyphony.

Simple transfer (t) derives from the compound use of resizing and revariance operations, in compensatory movements (resulting in a constant number of sounding factors), caused by the internal reorganization of components in both thickness and number of parts. Since the number of factors in simple transfer is constant, its positive or negative direction is based on the common practice, where polyphonic partitions are considered as more complex than massive ones.

Each one of the simple operators represent specific connections between adjacent configurations of partitions, forming a path from the succession of positive movements (*m+*, *v+* and *t+*). According to the selected operator and partition, more than one adjacent connection will be possible (Fig. 3).

From these paths, a hierarchy for each individual process is established, resulting in a taxonomy of textural configurations based on the *Partitional Young Lattice* (PYL). PYL is an abstraction that encompasses all partitions (with their correspondent indexes a, d) from 1 to a given number, with the classification of their connections. The global list of partitions in this structure (partition *lexset*)[2] provides a vocabulary of available textural configurations for the composer and the analyst.

4 Ranking Partitions for the Textural Contour

A given *Textural Contour* is established in two stages: (1) by ordering its partitions according to relative textural complexity; (2) by applying the MCT abstraction to the resulted ranking. In this way, it is possible to compare two apparently different textural progressions, relating them to the same Textural Contour.

---

[1] Absolute number of simultaneous voices or lines present in a given musical segment [2].

[2] Collection of lexical set formed by all partitions from 1 to the number of involved sounding factors.

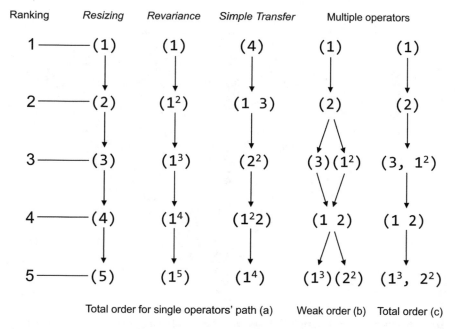

Fig. 4 Ranking orders of partitions in a single and multiple operators

The ranking of partitions of a textural progression is not a trivial problem, since the net of partitions operators forms a partially ordered set. The comparison of two partitions is based on the quality of involved operators and on the number of steps. The ranking is not linear and some partitions are, in fact, incomparable.

Ryszard Janicki [7] propose the pairwise comparisons as a method for ranking comparison of complex structures. This method consists on reducing an overall organization to the comparison of a pair of objects and then, creating a global ranking. According to Janicki's concepts, partitions form either a weak or a stratified order, i.e., the incomparable partitions are at the same ranking level, however it can be organized in a type of total order. Figure 4 shows a simple total order in a single operator path (a) and a weak order in a multiple operator path (b) organized into a total order (c).

The ranking of partitions *lexset* in a total order depends on a three-part comparison, instead of the two proposed by Janicki: the *ur-mesh*,[3] formed by (1), (2) and (1^2). The ur-mesh has all simple operators with a well-defined linear order. Such structure can be taken as basis to assess all partitions, demonstrating the levels of complexity inside PYL. From such an evaluation, it is elaborated a preliminary ranking of a specific partitions *lexset* (Fig. 5). The *h-related* partitions are those ones that share the same index pair.

---

[3] Partitions are noted in a compacted form with an index that expresses the multiplicity.

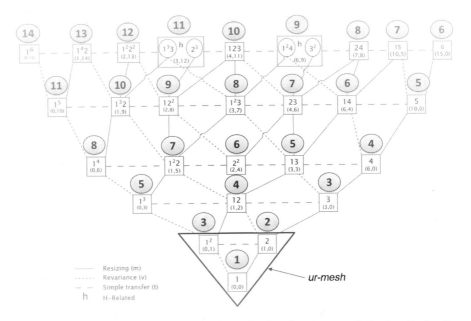

Fig. 5 Partitional Young Lattice (PYL) ranking orders based on the *ur-mesh* structure for density-number = 6

The difference between partitions that share the same level of complexity is expressed by including a sub-level that express the amount of real components. A sub-level notation is also intended to represent: (1) the relative difference of complexity between incomparable partitions in a refined form; (2) the structural relation between the amount of real components of each incomparable partition and their complexity levels, showing a possible attempt to balance the textural complexity through internal changes of configuration in different partitions.

Even with the use of sub-levels, some partitions remain in an equivalent level, i.e., the incomparable ones has the same number of real component. For example, the partitions (1 4) e (2^2) share the same level and both have two real components. The use of a sub-level with more digits, among other criteria, would not prevent this situation, as it is an intrinsic characteristic of the partitions set.

In the analytical methodology, the sub-levels are used when a group of incomparable partitions is presented. For example, the Introduction of the Fourth movement of Beethoven's String Quartet Op. 95 has nine different partitions, with only three groups of incomparable partitions, each one presenting two partitions. Partitions without their incomparable pair receive sub-level 0 (a). The contour formed is: < **4324324254303454345454141** > and the levels refined by sub-levels are 3, 4 e 5 (Fig. 6b).

(a)

Level	Sub-Level	Partitions
0	0	2
1	0	3
2	0	4
3	2	1 3
3	3	1^3
4	2	2 3
4	3	1^2 2
5	3	1^2 3
5	4	1^4

(b)

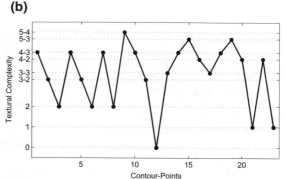

Fig. 6 Table with level, sub-level and partitions (**a**) and the graphic of the Textural Contour (**b**) of Introduction of the Beethoven's String Quartet, opus 95, Fourth movement

The Textural Contour shows that the lower level (0) is located at the middle, dividing the partitions in two groups, with different partitions, just intersected by level 4–3. The textural movement, from the point of view of complexity, is basically an alternation between levels (like neighbour tones), inside the global range of levels, sometimes presenting intermediate ones.

5 Conclusions

The present proposal for a Textural Contour is a new way to express the textural progression of a piece, creating tools for analysis and compositional manipulation. Ranking partitions makes possible the establishment of a textural hierarchy enabling the use of some methodological tools derived from MCT. The Textural Contour can be compared with other types of contours within a given piece in the search of mutual relations.

Further studies intend to create specific concepts for textural domain by analyzing a group of works from different periods and styles. The development of a new computational application for ranking partitions and plot the Textural Contour is the next objective of the present research.

References

1. Andrews, G.: The Theory of Partitions. Cambridge University, Cambridge (1984)
2. Berry, W.: Structural Functions in Music. Dover Publications, New York (1976)
3. Clifford, R.J.: Contour as a structural element in selected pre-serial works by Anton Webern. Thesis (Ph.D in Music). University of Wisconsin-Madison (1995)

4. Friedmann, M.: A methodology of the discussion of contour: its application to Schoenberg's music. J. Music. Theory **29**(2), 223–248 (1985)
5. Gentil-Nunes, P.: Partitional analysis: mediation between musical composition and the theory of integer partitions. Thesis (Ph.D. in Music). Rio de Janeiro: Federal University of Rio de Janeiro State (UNIRIO) (2009)
6. Gentil-Nunes, P., Moreira, D.: Partitional operators. Rio de Janeiro: Musmat. Ver. 1.32. http://www.musmat.org (2013)
7. Janicki, R.: Ranking with partial orders and pairwise comparisons. Paper presented at the meeting of the RSKT (2008)
8. Marvin, E.: A generalized theory of musical contour: its application to melodic and rhythmic analysis of non-tonal music and its perceptual and pedagogical implications. Thesis (Ph.D. in Music) University of Rochester (1988)
9. Moreira, D.: Contornos particionais: aplicações metodológicas na Introdução da Sagração da Primavera de Igor Stravinsky. Anais do 12 Colóquio of PPGM of School of Music, Federal University of Rio de Janeiro. Rio de Janeiro: UFRJ (2013)
10. Moreira, D.: Contour analyzer. Rio de Janeiro: Musmat. Ver. 1.5. http://www.musmat.org (2014)
11. Moreira, D., Gentil-Nunes, P.: Contornos musicais e os operadores particionais: uma ferramenta computacional para o planejamento textural. In: Congresso da Associação Nacional de Pesquisa e Pós-graduação em Música (ANPPOM), XXIV. UNESP, São Paulo (2014)
12. Moreira, D., Gentil-Nunes, P., Almada, C.: Contornos musicais: aplicações no indexograma e na curva derivativa. In: Congresso da Associação Nacional de Pesquisa e Pós-graduação em Música (ANPPOM), XXIII, UFRN: Natal, Rio Grande do Norte (2013)
13. Morris, R.: Composition with Pitch-classes: A Theory of Compositional Design. Yale University Press, New Haven (1987)
14. Morris, R.: New directions in the theory and analysis of musical contour. Music Theory Spectr. **15**, 205–228 (1993)
15. Sampaio, M.: A Teoria de Relações de Contornos Musicais: Inconsistências, Soluções e Ferramentas. Thesis (Ph.D in Music). School of Music, Federal University of Bahia, Salvador (2012)

The Sense of Subdominant: A Fregean Perspective on Music-Theoretical Conceptualization

Thomas Noll

Abstract Rameau [15] redundantly defines the *subdominant* (1) as the fifth under the tonic and (2) as the scale degree immediately below the dominant. In the context of scale theory this motivates the interpretation of this definition as an equation. It states that the *diazeuxis* (the difference between the generator and its octave complement) is a step interval of the scale. The appropriate scale-theoretic concept for the formulation of this equation is that of a *Carey–Clampitt Scale* (a non-degenerate well-formed scale). The Rameau equation then imposes a constraint on the associated *Regener transformation* which converts note intervals from generator/co-generator coordinates into step/co-step coordinates. The equation takes two forms depending on the sign of the diazeuxis (positive or negative). The solutions then come either with two (flatward directed) co-steps or two (sharpward directed) steps, accordingly. In addition to this generic characterization the paper closes with a corollary on the specific scale properties of reduced Clough–Myerson scales. These scales solve Rameau's equation if and only if they are *Agmon scales*.

1 Competing Motivations for the Term 'Subdominant'

In a thread called *subdominant versus predominant* of the Email-forum *smt-talk* Nicolas Meeùs[1] reminds the debaters about an enduring co-existence of two traditions in the interpretation of the term 'subdominant' which can be traced back to the first half of the 18th century. Two alternative motivations for the composition of the term 'dominant' with the prefix 'sub' correspond to two alternative theoretical prioritisations in the face of a conceptual ambivalence between two types of tone relations, both of which are thought to be constitutive for the diatonic scale. On the one hand 'subdominant' means a kind of *dominant under* (the tonic) and on the

---

[1]I wish to acknowledge Nicolas Meeùs' postings to smt-talk from february 24 (23:35:46 CET) and march 11 (14:37:58 CET) 2012 as particularly inspirational for the present paper.

T. Noll (✉)
Escola Superior de Música de Catalunya, Departament de Teoria,
Composició i Direcció, C. Padilla, 155 - Edifici L'Auditori, 08013 Barcelona, Spain
e-mail: thomas.mamuth@gmail.com

© Springer International Publishing AG 2017
G. Pareyon et al. (eds.), *The Musical-Mathematical Mind*,
Computational Music Science, DOI 10.1007/978-3-319-47337-6_21

207

other hand it means the scalar neighbor *below the dominant*. In the alphabetical table of terms within his book *Génération harmonique* [15] Jean Phillip Rameau (1737) combines both motivations of the term:

> Soudominante. C'est la quinte au-dessous, et par Renversement la Quarte du Son principal, dit Note-Tonique, et qui se trouve immédiatement au-dessous de la dominante dans l'ordre Diatonique. [Subdominant. It is the fifth under, and by inversion the fourth above the principal sound, said Note-Tonique, and which is found immediately under the dominant in the diatonic order.]

The first motivation renews Rameau's definition from his earlier book [16]. It pinpoints the subdominant on the left side of the triple proportion 1:3:9, i.e. on the flat side of a short chain of fifths, such as F - C - G. The second motivation has the virtue to be in coherence with the motivation of two other terms in that alphabetical table: 'super-tonic', 'super-dominant', where the prefix 'super' refers to the upper scalar neighbors of tonic and dominant degrees, respectively. Furthermore this is in accord with Jean-François Dandrieu's earlier motivation for 'subdominant' in the introduction to his *Principes de l'accompagnement* [10].[2]

A review of the irritations and debates in the wake of Rameau's double motivation of the term deserves a separate study. This is all the more so for a discussion of the theoretical tensions in the conceptualization of diatonic tone relations. Therefore, as a modest beginning and preparatory step for a thorough theoretical discussion the present paper investigates the logical conditions which make this ambivalence possible.

Both expressions 'dominant under the tonic' and 'scale degree below the dominant' have the same *reference*. They both refer to the same scale degree of the diatonic scale. But according to Gottlob Frege's distinction between reference and sense, we can observe that these expressions have different *senses*, as they provide different modes of access to that same scale degree. Following Frege there is a certain epistemic value to the equation between both expressions. If we apply this equation to the characterization of the interval between the dominant and subdominant scale degrees, we obtain the following condition: The diazeuxis, (i.e. the difference between fifth and fourth) is a step interval. This condition can be suitably studied in the context of mathematical scale theory.

2 Agmon's Diatonic Property: Gradus Ad Parnassum

This section provides a brief summary of basic scale theory with the aim to revisit Eytan Agmon's [1] concept of a *diatonic system* in the context of several other scale concepts. Starting point is the concept of *scale*, which connects the set $\mathbb{Z}_n$ of *scale degrees* with a set S of *pitch classes*. The cyclic group $\mathbb{Z}_n$ plays two roles, namely (1) that of the interval group and (2) that of the carrier set of the *Canonical Generalized Interval System* associated with this group ([12]). Pitch class scales are abstractions

---

[2]Joel Lester [13] cites Dandrieu as the earliest known source, where the term 'soudominant' occurs.

of pitch scales and embody the idea of a cyclic octave-periodic pitch scale. Therefore the set S is supposed to be a subset of the circle group $\mathbb{R}/\mathbb{Z}$, which also serves as the carrier set and interval group of the Canonical Generalized Interval System associated with $\mathbb{R}/\mathbb{Z}$.

Definition 1 (*Scale*) Consider a subset $S \subset \mathbb{R}/\mathbb{Z}$ of cardinality $n \in \mathbb{N}$ represented by real numbers in $[0, 1)$. Further consider a bijection $s : \mathbb{Z}_n \tilde{\to} S$, such that these representatives satisfy the order relation $0 = s(0) < s(1) < \cdots < s(n-1) < 1$. This bijection $s : \mathbb{Z}_n \tilde{\to} S$ is then called a (periodic) *scale* mod 1.

In the following definition intervals come into play. Intervals between scale degrees are called *generic*, while intervals between pitch classes are called *specific*. The map $spec_s : \mathbb{Z}_n \to \mathbb{R}/\mathbb{Z}$ connects the intervals of the two generalized interval systems:

Definition 2 (*Spectrum, Myhill's property*) Consider a scale $s : \mathbb{Z}_n \tilde{\to} S$ of cardinality n in the sense of Definition 1. For every $k = 0, \ldots, n-1$ consider the set $spec_s(k) := \{s(l+k) - s(l) \, mod \, 1 \mid l = 0, \ldots, n-1\}$, which is called the *spectrum* of the generic interval k. The scale $s : \mathbb{Z}_n \tilde{\to} S$ is said to have *Myhill's property*, iff each spectrum $spec_s(k)$ has precisely two elements for $k = 1, \ldots, n-1$.

Definition 3 (*Generated Scale*) Consider a real number $g \in \mathbb{R}, 0 < g < 1$ and the arithmetic sequence with period g of length $n \in \mathbb{N}$ along the circle $\mathbb{R}/\mathbb{Z}$. The points of this sequence can be represented by the set

$$S = \{x, x + g, x + 2g \, mod \, 1, \ldots, x + (n-1)g \, mod \, 1)\} \subset [0, 1).$$

Further consider two bijective maps $q, s : \mathbb{Z}_n \tilde{\to} S$ satisfying $q(k) = k \cdot g \, mod \, 1$ and $0 \le s(0) < s(1) < \cdots < s(n-1) < 1$. These maps are called the *generation order* and the *scalar order* of the g-generated scale S, respectively. The concatenation $s^{-1} \circ q : \mathbb{Z}_n \to \mathbb{Z}_n$ is called the *generation-order-to-scale-step-order-conversion*.

Definition 4 (*Well-formed Scale*) A g-generated scale of length n in the sense of Definition 3 is called *well-formed* iff its associated generation-order-to-scale-step-order-conversion $s^{-1} \circ q : \mathbb{Z}_n \to \mathbb{Z}_n$ is an affine map, i.e. iff there exist two elements $m, t \in \mathbb{N}$, such that $s^{-1} \circ q(k) = m \cdot k + t \, mod \, n$ for all $k = 0, \ldots, n-1$. A well-formed scale is called *degenerate*, iff $spec_s(1) = \{\frac{1}{n}\}$, i.e. iff it is entirely regular. Otherwise it is called *non-degenerate*.

The affine generation-order-to-scale-step-order-conversion on $\mathbb{Z}_n$ pertains the carrier set of the associated GIS. It induces a linear map on the interval group, namely the multiplication by $m : \mathbb{Z}_n \to \mathbb{Z}_n$.

The following proposition summarizes fundamental results by Norman Carey and David Clampitt (c.f. [2]):

Proposition 1 *Consider a g-generated scale s of length n in the sense of Definition 3 with the scale order and generation order maps $q, s : \mathbb{Z}_n \tilde{\rightarrow} S$. Let the natural number $m < n$ represent the residue class $s^{-1}(g) \in \mathbb{Z}_n$. Then the following three conditions are equivalent:*

1. *The scale has Myhill's property.*
2. *The scale is non-degenerate well-formed with $s^{-1}(q(k)) = m \cdot k + t \bmod n$.*
3. *The ratio $\dfrac{m}{n}$ is a semiconvergent of the generator g with $\dfrac{m}{n} \neq g$.*

For completeness we mention that in [3] Carey and Clampitt obtained the following stronger result:

Proposition 2 *Consider a scale of length n in the sense of Definition 1. Then the following two conditions are equivalent:*

1. *The scale has Myhill's property.*
2. *The scale is non-degenerate well-formed.*

In the light of this finding it is convenient to introduce a nickname for these scales: *Carey–Clampitt Scales.*

Definition 5 (*Maximal Even Scales, Clough–Myerson Scales*) A scale $s : \mathbb{Z}_d \tilde{\rightarrow} S$ of cardinality d in the sense of Definition 1 is called *maximally even*, iff there exists a natural number c, such that the spectrum of any generic interval k is either a singleton set $spec_s(k) := \left\{ \dfrac{m}{c} \right\}$ or a 2-element set of the form $spec_s(k) := \left\{ \dfrac{m}{c}, \dfrac{m+1}{c} \right\}$. If for all $k = 1, \ldots, d - 1$ the spectrum is 2-elemented, the scale s is called a *Clough–Myerson Scale.*[3]

Proposition 3 *If the step interval spectrum of a non-degenerate well-formed scale s is of the form $spec_s(1) := \left\{ \dfrac{1}{c}, \dfrac{2}{c} \right\}$, then s is a Clough–Myerson Scale. Such scales are called* reduced.

Proof Let m denote the span (=generic step size) of the generator and $m' = m^{-1} \bmod d$ the 'generator-span' of the generic step (i.e. the length of the chain of generators that represents the generic step interval). We inspect the spectrum of the generator $spec_s(m) = \left\{ \dfrac{q'}{c}, \dfrac{q}{c} \right\}$. Let us suppose that the specific genera-tor $g = \dfrac{q}{c}$ (of multiplicity $n - 1$) is larger than the single instance of $\dfrac{q'}{c}$. With respect to the specific step sizes this implies $2 = m'q \bmod c$ and $1 = m'(q - 1) + q' \bmod c$. The difference between both equations is $1 = q - q' \bmod c$ and hence $q = q' + 1$. Now consider the 'generator-span' $k' = k^{-1} \bmod d$ of any generic inter-val k with spectrum $spec_s(k) = \left\{ \dfrac{p'}{c}, \dfrac{p}{c} \right\}$ and suppose that $\dfrac{p'}{c} < \dfrac{p}{c}$. We obtain the

---

[3]Proposed term by the author.

two equations $p = k'q \bmod c$ and $p' = k'(q-1) + q' \bmod c$. Their difference is $p - p' = q - q' \bmod c$ and hence $p = p' + 1$. The case $q' > q$ works completely analogous.

$\blacklozenge$

Definition 6 (*Contradiction and Ambiguity*) Consider a scale $s : \mathbb{Z}_n \tilde{\rightarrow} S$ of cardinality n. Consider two generic intervals $k < l$ together with their spectra $spec_s(k)$ and $spec_s(l)$. Two specific intervals $i \in spec_s(k)$ and $j \in spec_s(l)$ are said form a *contradiction* (against the generic order) iff $j < i$. They are said form an *ambiguity* iff $j = i$. Scales with no contradictions are called *consistent*; scales with no contradictions and no ambiguities are called *coherent*; scales with no contradictions and one single instance of ambiguity are called *quasi-coherent*.

In [9] John Clough and Jack Douthett show (in their Lemma 2.1):

Proposition 4 *Maximally even scales have no contradictions.*

In [5, 6] Norman Carey examines the conditions for the occurrence of contradictions and ambiguities. Among other results he discloses the following facts about the case of well-formed scales. The following proposition paraphrases some details from Theorem 6 in [6] and a corollary from Propositions 3 and 4 above.

Proposition 5 (Contradiction and Ambiguity for Well-formed Scales) *Consider a non-degenerate well-formed scale of length n together with its specific step intervals* $spec_s(1) = \{\alpha, \beta\}$, *and suppose that* $\beta < \alpha$. *Let* $\rho = \dfrac{\alpha}{\beta} > 1$ *denote the ratio between the larger and the smaller step sizes, and let* n_α *and* n_β *denote their multiplicities in s accordingly, i.e.* $n_\alpha \alpha + n_\beta \beta = 1$, *and* $n_\alpha + n_\beta = n$.

1. *If* $n_\beta = 1$ *the scale s has no contradictions and no ambiguities.*
2. *If* $\rho < 2$ *the scale s has no contradictions and no ambiguities.*
3. *If the scale s has ambiguities then* $2 \leq \rho < \dfrac{n-1}{n_\alpha} + 1$.
4. *If* $\rho > 2$ *and* $n_\beta > 1$, *the scale s has contradictions.*
5. *If* $\rho = 2$ *the scale s has no contradictions.*

Facts (1)–(4) are contained in Theorem 6 in [6]. With respect to (5) we conclude from $\rho = 2$ that s fulfills the presuppositions of Proposition 3 and that therefore it is maximally even. Proposition 4 then states that s has no contradictions.

Thus, in order to avoid interval contradictions the ratio ρ between the larger and the smaller step sizes should be less or equal to 2. Eytan Agmon (1989) [1] proposes to counterbalance this condition by an *efficiency* criterium for the embedding of the scale into a chromatic ambient universe.

Definition 7 (*Efficiency*) A scale s of length n is called *efficient*, if the set $Int(s) = \{s(k) - s(l) \bmod 1 \mid k, l = 0, \ldots, n-1\}$ of its specific intervals forms a group.

Agmon's efficiency criterium demands basically that there are no redundant chromatic intervals, which are not already exemplified by specific intervals of the scale. The above definition might seem to differ from Agmon's at first glance, but it covers the same concept. If the finite set $Int(s)$ coincides with the generated subgroup $\langle Int(s) \rangle$ within the circle group $\mathbb{R}/\mathbb{Z}$, it has to be of the form $Int(s) = \frac{1}{c}\mathbb{Z}/\mathbb{Z}$, which means that the scale is efficient in Agmon's sense with respect to the c-note chromatic universe. Clough and Douthett ([9], Proposition 1.12) characterise efficient scales among maximally even scales in terms of their cardinalities and their specific step intervals. In the following proposition we stick to the special case of Clough–Myerson scales.

Proposition 6 *For any Clough Myerson scale* $s : \mathbb{Z}_d \tilde{\to} S \subset \frac{1}{c}\mathbb{Z}/\mathbb{Z}$ *the following three conditions are equivalent:*

1. *s is efficient.*
2. $\frac{c}{2} < d < c.$
3. *s is reduced, i.e.* $spec_s(1) := \left\{ \frac{1}{c}, \frac{2}{c} \right\}.$

Corollary 1 (Consistency and Efficiency) *A non-degenerate well-formed scale s of length n is consistent and efficient iff it is a rounded Clough Myerson scale.*

Corollary 2 (Coherence and Efficiency) *A non-degenerate well-formed scale s of length n is coherent and efficient iff it is a rounded Clough Myerson scale with a singleton minor step.*

Definition 8 *(Agmon's diatonic property)* A non-degenerate well-formed scale *s* of length *n* fulfills Agmon's *diatonic property* if it is efficient and quasi-coherent. A scale satisfying this property shall be called *Agmon Scale.*[4]

Proposition 7 *A non-degenerate well-formed scale s of length n fulfills Agmon's diatonic property iff it is a rounded Clough Myerson scale with two minor steps.*

Proof For the case (3) in Proposition 5 Carey provides an explicit formula for the number of instances of ambiguity. For the special value $\rho = 2$ this formula simplifies to $\frac{(n_\beta + 1)(n_\beta)(n_\beta - 1)}{6}$, where n_β denotes the multiplicity of the minor step. This expression yields the value 1 iff $n_\beta = 2$.   ◆

3 Regener Transformations and Rameau's Equation

Our goal is now to understand the condition that the interval of the diazeuxis coincides with a step interval of the scale. A suitable formalization of this condition shall be coined *Rameau's Equation*. The concept of Carey–Clampitt scale, (i.e. of a non-degenerate well-formed scale) provides an appropriate scope for the interpretation

---

[4]Proposed term by the author. Clough and Douthett also coined the term *hyper-diatonic scale.*

of this condition. First of all it is clear, that the double-sense of the subdominant involves a comparison of interval-measurement by generator and co-generator intervals (traditionally, by fifths and fourths), on the one hand, and by the step intervals, on the other. As the well-formedness property embodies the compatibility condition between both types of measurement, it provides an appropriate scope for the study of Rameau's equation. The degenerate case is uninteresting if the step itself is regarded as the generator. For other generators we obtain limiting cases of the more interesting non-degenerate situation: The scale offers the possibility to lift the linear automorphism $m : \mathbb{Z}_n \to \mathbb{Z}_n$ from generic scale degree intervals in $\mathbb{Z}_n$ to a free $\mathbb{Z}$-module of rank 2 of *note intervals*.

Lemma 1 *Consider a Carey–Clampitt scale s with generator x and co-generator $y = 1 - x$. The two step intervals can be expressed as linear combinations $a = lx - iy$ and $b = -jx + ky$, with $i, j, k, l \in \mathbb{N}$ and $k, l > 0$.*

The Lemma is a consequence of Proposition 1. If the semi-convergent $\dfrac{m}{n}$ is represented by the $SL(2, \mathbb{N})$ element $\begin{pmatrix} m_1 & m_2 \\ n_1 & n_2 \end{pmatrix}$ with $m = m_1 + m_2$ and $n = n_1 + n_2$, then one obtains $i = m_1, j = n_1 - m_1, k = m_2, l = n_2 - m_2$.

The positions of the minus-signs in the representations of a and b allow us to distinguish their roles as the (sharpward directed) *step* and the (flatward directed) *co-step*, respectively.

Definition 9 The $\mathbb{Z}$-modules $\mathbb{Z}[x, y]$ and $\mathbb{Z}[a, b]$ are different coordinate spaces for the *note interval system*, associated with the Carey–Clampitt scale s. The linear transformation $M_s = \begin{pmatrix} i & k \\ j & l \end{pmatrix} : \mathbb{Z}[x, y] \to \mathbb{Z}[a, b]$ converting generator/co-generator coordinates into step/co-step coordinates is called the *Regener transformation*.[5]

The right arrow in the diagram below shows the linear map $m : \mathbb{Z}_n \to \mathbb{Z}_n$ on generic scale degree intervals. The middle arrow shows the Regener-Transformation. The arrow to the left shows its restriction to the submodule, generated by the octave $x + y$ and the augmented prime $(l + j)x - (i + k)y$. These intervals are factored out in $\mathbb{Z}_n$.

$$0 \to \mathbb{Z}[x + y, (l + j)x - (i + k)y] \hookrightarrow \mathbb{Z}[x, y] \longrightarrow \mathbb{Z}_n \to 0$$

$$\downarrow \qquad\qquad \downarrow \begin{pmatrix} i & j \\ k & l \end{pmatrix} \qquad \downarrow \cdot m$$

$$0 \to \mathbb{Z}[(i + j)a + (k + l)b, a - b] \hookrightarrow \mathbb{Z}[a, b] \longrightarrow \mathbb{Z}_n \to 0$$

Hence, the Regener transformation lifts the linear map $m : \mathbb{Z}_n \to \mathbb{Z}_n$ on generic scale degree intervals to the entire free commutative group of note intervals.

---

[5] For the motivation of this term see [17].

Definition 10 (*Rameau's equation*) Consider a Carey–Clampitt scale s with generator x, co-generator y, step a and co-step b and the associated Regener transformation $M_s : \mathbb{Z}[x, y] \rightarrow \mathbb{Z}[a, b]$. Depending whether the sharpward directed diazeuxis $x - y$ is positive or negative, Rameau's equation takes the form:

$$\begin{pmatrix} i & j \\ k & l \end{pmatrix} \begin{pmatrix} 1 \\ -1 \end{pmatrix} = \begin{pmatrix} 1 \\ 0 \end{pmatrix} \text{ or } \begin{pmatrix} i & j \\ k & l \end{pmatrix} \begin{pmatrix} -1 \\ 1 \end{pmatrix} = \begin{pmatrix} 0 \\ 1 \end{pmatrix}.$$

By comparing the coordinates on both sides we obtain the following:

Proposition 8 $M_s = \begin{pmatrix} j+1 & j \\ 1 & 1 \end{pmatrix}$ *(for $x - y$ positive) and* $M_s = \begin{pmatrix} 1 & 1 \\ l & l+1 \end{pmatrix}$ *(for $x - y$ negative) are the unique solutions for Rameau's equation.*

This implies the following fact about the co-step-multiplicity n_b (or the step-multiplicity n_a) of the given scale:

Corollary 3 *The Regener transformation M_s associated with a Carey–Clampitt scale s with generator x solves Rameau's equation iff either $x > \dfrac{1}{2}$ and $n_b = 2$ or $x < \dfrac{1}{2}$ and $n_a = 2$.*

This characterization is entirely generic. But in the light of Propositions 6 and 7 we also obtain an implication between efficiency and the quasi-coherence:

Corollary 4 *Consider a reduced Clough–Myerson scale (i.e. $\mathrm{spec}_s(1) := \left\{ \dfrac{1}{c}, \dfrac{2}{c} \right\}$), whose generator satisfies $g > \dfrac{1}{2}$ and whose secondary step interval b is the smaller step interval β. The Regener transformation on the note interval group associated to this scale solves Rameau's equation iff the scale is an Agmon scale.*

A non-commutative refinement of this scale-theoretic investigation lifts the Regener transformation to Sturmian morphisms. The modes corresponding to the solutions of Rameau's equation are called *diazeuctic modes* (see [14]).

References

1. Agmon, E.: A mathematical model of the diatonic system. J. Music Theory **33**(1), 1–25 (1989)
2. Carey, N., Clampitt, D.: Aspects of well-formed scales. Music Theory Spectr. **11**(2), 187–206 (1989)
3. Carey, N., Clampitt, D.: Self-similar pitch structures, their duals, and rhythmic analogues. Perspect. New Music **34**(2), 62–87 (1996)
4. Carey, N.: Distribution Modulo 1 and Musical Scales, Ph.D. Dissertation, University of Rochester (1998)
5. Carey, N.: On coherence and sameness and the evaluation of scale candidacy claims. J. Music Theory **46**, 1–56 (2002)

6. Carey, N.: Coherence and sameness in well-formed and pairwize well-formed scales. J. Math. Music **2**(1), 79–98 (2007)
7. Clampitt, D., Noll, T.: modes, the height-width duality, and Handschin's tone character. Music Theory Online **17**(1) (2011)
8. Clough, J., Myerson, G.: Variety and multiplicity in diatonic systems. J. Music Theory **29**(2), 249–270 (1985)
9. Clough, J., Douthett, J.: Maximally even sets. J. Music Theory **35**, 93–173 (1991)
10. Dandrieu, J.-F.: 1719. *Principes de l'accompagnement*, Paris
11. Frege, G.: Über Sinn und Bedeutung, Zeitschrift für Philosophie und philosophische Kritik, NF 100, 1892, S. 25-50. English Translation On Sense and Reference by Max Black. https://en.en.wikisource.org/wiki/On_Sense_and_Reference. Accessed 3 Jan 2014
12. Kolman, O.: Transfer principles for generalized interval systems. Perspec. New Music **42**(1), 150–190 (2004)
13. Lester, J.: Compositional Theory in the Eighteenth Century. Harvard University Press, Massachusetts (1994)
14. Noll, Th.: In: Collins, T. et al. (ed.) Triads as Modes within Scales as Modes. MCM LNAI 9110, pp. 373–384 (2015)
15. Rameau, J.-P.: 1726. *Nouveau systême de musique théorique*. Paris
16. Rameau, J.-P.: 1737. *Génération harmonique*. Paris: Prault. 1737. Engl: Deborah Hayes: Harmonic Generation, Standford University (1968)
17. Regener, E.: Pitch Notation and Equal Temperament: A Formal Study. University of California Press, Berkeley (1973)

How Learned Patterns Allow Artist-Level Improvisers to Focus on Planning and Interaction During Improvisation

Martin Norgaard

Abstract In this paper, I argue that stored auditory and motor patterns are inserted into ongoing musical improvisations. This position aligns with the theoretical framework suggested by [11]. In support of this theory, I cite research in which artist-level improvisers describe their own thinking and mention learned patterns as one of the central mechanism underlying improvisation. I further outline how known solos by tonal jazz artists contain a large number of repeated patterns and that a computer algorithm using patterns is capable of producing new solos in similar style. Finally, an experimental study shows that improvisers use more patterns when their attention is diverted during improvisation. According to interviews with advanced improvisers in solo settings, their attention is focused on larger architectural structures during improvisation. Other research with advanced improvisers in group settings point out that interaction with other ensemble members also may be at the front of the improvisers mind in that setting [1, 5]. Therefore, I conclude that it may be the partially automatic process of inserting learned patterns into ongoing improvisations that allows the artist-level improviser to focus on planning and interaction.

1 Patterns in Music Improvisation

Improvisation is a component of musical practice across idioms and cultures, however, the cognitive mechanisms underlying extemporaneous musical performance are not well understood. Specifically, the function of learned auditory and motor patterns is hotly debated [4, 7, 11]. For example, when looking at tonal jazz it is well known that transcriptions of improvised solos include repeated melodic patterns [2, 3, 10]. In one view, these patterns appear by accident because the improviser follows tonal rules [4]. In an opposing view, musicians have a library of auditory and motor patterns that are inserted into ongoing improvisations [11]. In this paper I argue for the latter position and posit that the use of patterns allows improvisers to focus on higher level processes related to planning and interaction.

M. Norgaard (✉)
Georgia State University, Atlanta, GA 30302, USA
e-mail: mnorgaard@gsu.edu

© Springer International Publishing AG 2017
G. Pareyon et al. (eds.), *The Musical-Mathematical Mind*,
Computational Music Science, DOI 10.1007/978-3-319-47337-6_22

Interviews with artist-level jazz improvisers in solo settings reveal a conscious focus on planning processes [6]. In this study I interviewed advanced musicians about a solo just performed while they looked at approximate notation and listened to audio from the solo. They commented that their focus was often on planning larger architectural features of upcoming passages such as note density, register, or intervallic content. Concerning individual note decisions they described this process as being automatic and not under their conscious control. Specifically, they pointed out instances where the automatic process resulted in "unplanned" note choices. This is a further indication that the note level decisions are guided by subconscious processes outside of conscious control.

Explaining this subconscious component, the artist-level improvisers pointed to two main processes [6]. One process was described as inserting learned material from a bank of ideas. One participant used the analogy of concatenating "Lego blocks." The other process was described as connecting chord tones following tonal rules. Interestingly, these two processes align with the two views described above that emphasize either rules or patterns as the main guiding principle behind musical improvisation.

To further investigate the role of patterns, I conducted a corpus analysis study of solos by jazz great, Charlie Parker [7]. Previous analysis of patterns in Parker's solos was done by hand [10]. To get a more accurate measure of pattern use, I designed a computer algorithm that investigated tonal and rhythm patterns starting on each note. This eliminated the need for segmenting solos into discreet patterns which necessarily involves subjective judgments about pattern boundaries [10]. I found that 82% of all notes in Parker's solos begin a five note interval pattern. I also found patterns up to 50 intervals in length and showed that longer patterns appear in solos recorded in different sessions. These results support the idea that patterns are stored in an "idea bank" for use in different solos [11].

2   An Algorithmic Research on Improvisation

In a subsequent study, we built a simple computer algorithm capable of improvising using patterns [9]. This algorithm uses a Markov-chain based mechanism to create new improvisations by reusing material from a given corpus. Using the Parker solos analyzed in the previous study [7], the algorithm outputs improvisations that itself contains pattern structures similar to the original corpus [9]. Importantly, the output clearly reflects the style of the corpus on which the improvisation is based. We believe that this algorithm may use processes similar to human improvisers again supporting the idea that patterns are central to musical improvisation.

According to the descriptions of improvisational thinking by advanced improvisers, their focus of attention is centered around planning of larger architectural features [6]. If this is true, could it be that the use of patterns allows improvisers to focus on these larger musical structures? To test this idea, my team conducted a study in which advanced jazz pianists improvised while their attention was diverted to a secondary task [8]. We hypothesized that improvisations created while the musician's attention was focused elsewhere would contain more patterns. During the study, improvisers were asked to count taps on their shoulders while improvising on the 12-measure blues form. They completed 8 trials each containing 60 measures of improvised material. In half of the trials they were asked to count taps, in the other half they improvised normally. Conditions were counter-balanced across participants. Confirming our hypothesis, we indeed found that the solos performed while also counting shoulder taps included more patterns. Again this aligns with the idea that patterns are a central mechanism underlying musical improvisation. Furthermore the study showed that improvisers can focus elsewhere without interrupting an ongoing improvisation and that this may be possible in part due to a learned process of inserting patterns.

References

1. Berliner, P.F.: Thinking in Jazz. University Of Chicago Press, Chicago (1994)
2. Finkelman, J.: Charlie Christian and the role of formulas in jazz improvisation. Jazzforschung/Jazz Res. **29**, 159–188 (1997)
3. Gushee, L.: Lester young's shoe shine boy. In: Porter, L. (ed.) A Lester Young Reader, pp. 224–254. Smithsonian Institution Press, Washington (1991)
4. Johnson-Laird, P.N.: How jazz musicians improvise. Music Percept. **19**, 415–442 (2002)
5. Monson, I.: Saying Something: Jazz Improvisation and Interaction. The University of Chicago Press, Chicago (1996)
6. Norgaard, M.: Descriptions of improvisational thinking by artist-level jazz musicians. J. Res. Music Educ. **59**, 109–127 (2011)
7. Norgaard, M.: How jazz musicians improvise: the central role of auditory and motor patterns. Music Percept. **31**, 271–287 (2014)
8. Norgaard, M., Emerson, S.N., Braunsroth, K.D., Fidlon, J.: Creating under Pressure: Effects of divided attention on the improvised output of skilled jazz musicians. Music Percept. **33**, 561–570 (2016)
9. Norgaard, M., Spencer, J., Montiel, M.: Testing cognitive theories by creating a pattern-based probabilistic algorithm for melody and rhythm in jazz improvisation. Psychomusicol.: Music Mind Brain **23**, 243–254 (2013)
10. Owens, T.: Charlie Parker: Techniques of Improvisation. University of California, Los Angeles (1974)
11. Pressing, J.: Improvisation: methods and model. In: Sloboda, J.A. (ed.) Generative processes in music (Paperback., pp. 129–178). Oxford University Press, Oxford (1988)

Tuning Systems Nested Within the Arnold Tongues: Musicological and Structural Interpretations

Gabriel Pareyon

Abstract This contribution introduces the concept of musical harmony as a geometric, physical mirror of human biologic proportionality. Although this idea is rather ubiquitous in many aspects and epochs of music theory, mathematical direct modelling is relatively a novelty within the field of dynamical systems. Furthermore, a hypothesis of atomic-molecular harmonicity is provided in order to explain how biologic proportionality is physically biased to perform harmonic patterns eventually codified by culture. This hypothesis is grounded on the topological properties of carbon, and its mapping and embedding within the characteristic geometry of music; from the graphene-Tonnetz analogy, to the map of musical harmony using the Arnold tongues analogy. The topological features of carbon are, then, conceived as crucially influential for the rising of human language and music, and for the development of an associated Euclidean intuition.

1 Theoretical-Philosophical Framework

It is a common place to state that in mathematics a dynamical system is a concept where a function describes the time dependence of a point in a geometrical space. Conceiving such a function uniquely as a set of geometry is a notion gradually less common in musicology, as a borrowing from pure mathematics; this notion provides the theoretical framework for this proposal. Therefore the geometric properties of the studied set will be of key importance, while less attention is paid to the notion of "time dependence" (a concept widely investigated by other authors including [3, 20]).

First of all, it is necessary to note that this proposal does not conceive mathematics merely as a product of biology and culture, but as a precondition for biology and culture modelling self-referential complexity. Thus, at the end, mathematics can evolve through biology and culture.

G. Pareyon (✉)
CENIDIM-INBA, Torre de Investigación, Piso 7, CENART, Av. Río Churubusco 79, 04220 Coyoacán, D.F., Mexico
e-mail: gabrielpareyon@gmail.com

© Springer International Publishing AG 2017
G. Pareyon et al. (eds.), *The Musical-Mathematical Mind*,
Computational Music Science, DOI 10.1007/978-3-319-47337-6_23

Since we assume that the mathematical mind evolves from electrochemical exchanges and electromagnetic fluctuation in neurons and clusters of neurons with potentials triggered following power laws, a hypothesis can be formulated, pointing out a common ground for the description of (a) the participation of a *carbon electrical field* in *Self-referential Abstract Thought* (SAT), (b) the emergence of Euclidean geometric notions from SAT, (c) the empathy of these Euclidean notions in respect to music (equally involved with SAT), and (d) the emergence of "carbonic self-similarity" in music and in language.

1.1 Why Carbon?

Understanding the capital role of carbon participating in the emergence of language, and specifically in the emergence of logic and mathematical language, is a matter of puzzling out appropriate analogies for their corresponding contexts. Particularly because our method of analogy (read *proportionality*) in this case meaningfully connects logic, music and mathematics in their very foundations.

Unlike most of chemical elements tending to perform linear consecutive bonds with other elements, carbon may perform regular and progressively regular-variate compositions with many extraordinary properties, including periodic tiling (i.e. graphene) and variations upon this tiling in two dimensions and emerging complexity in three dimensions (e.g. fullerene manifolds).[1]

A central hypothesis in this proposal is that carbon plays a crucial role organizing nucleobases, so the probabilistic of this organization leads to self-referential, nonlinear carbon circuits (for a first introduction to this concept, see: [18]). The evolution of this organization may also lead to clustering selective circuits with specialized functionality, and ultimately to neural clusters with features of electrochemical coordination and synchronization.[2]

From a physiological holistic perspective, we reject conceiving neurons, axons and the nervous system as from the old-fashion mechanical paradigm. In contrast, we believe there is no need to segregate the nervous system from the whole body, but to conceive the body as a wholeness coordinated through evolution and individuation at the same time. Accordingly, recent biomedical literature [4, 13, 15] allows us to interpret that typical *harmonic*—in geometric sense—patterns of carbon are evidence of this coordination in the form of self-similar biorhythms.

Now, assuming that there is enough empirical and systematic evidence supporting the concept of an organizational role of carbon in neural circuits, and therefore

---

[1] We should remark that, despite its two-dimensional nature, graphene has three phonon modes (LA, TA, in-plane modes with linear dispersion relation; and ZA with quadratic dispersion relation), a fact closely related to the electroacoustic properties of carbon migrating from two to three dimensions.

[2] A clear introductory explanation in [15]: "Coordination of the activity within and between the brain's cellular networks achieved through synchronization has been invoked as a functional feature of normal and abnormal temporal dynamics, the integration and segregation of information, and of the emergence of neural rhythms."

Fig. 1 Schematic representation of a carbonic basic plot. *Straight lines* represent precondition for the Euclidean straight line intuition; vertices represent precondition for the Euclidean point. This scheme represents the basic motif for illustrating the *Hypothesis of Self-Similar Euclidean Axiomatics* (HSA)

in the brain/mind relationship, a more explicit development is required to provide a convincing explanation to connect such an evolution of organic chemistry, with the emergence of mathematical intuition. To this purpose we invoke the *Hypothesis of Self-Similarity in Euclidean Axiomatics* (henceforth HSA), inferred from a self-similarity nesting within biological recurrence in terms of a carbon self-structuring behaviour. According to this hypothesis (already suggested in [14, pp. 133–137, 250–252, 481–484]), a recurrent multi-scalar self-organization of carbon circuits must *repeat* its main physical features, from the atomic to the cellular levels, and then projected to the ecologic-social patterns. Thus, basic intuitions of geometry and spatiality should somehow reflect the *genesis* of this structural principle as their own axiomatic grounds. This is how in Fig. 1, with the scheme of a basic two-dimension carbon structure, straight lines may represent precondition for the Euclidean straight line intuition, and their vertices represent precondition for the Euclidean point. The participation of this sort of *minimal structural bonding*, directly related to the basic forms of organic carbon and to the topology of neural networking, should be meaningful to the emergence of an abstract, *self-reflecting logic* (referred to as SAT, above), in its turn constituting the axiomatic grounds for Euclidean intuition.

　　Whether, according to HSA, Euclidean intuition strongly depends on high-abstract though influenced by carbonic self-similarity, musical high-abstract thought also should reflect its own *carbon footprint*. This is evident as statistical behaviour in generalized, massive samples of music (an idea first suggested by [19]), revealing self-similar patterns of carbon expressed by the so-called *fractional noise* 1/*f*, as documented in [14].[3] There are, as well, structural-geometric evidences for a *musical version* of HSA. The most striking ones are the analogies of the two typical crystal structures of carbon in two dimensions: the graphite simple-hexagonal, and the face-centred diamond-cubic. Both kind of structures are strictly analogous to tonal music self-structuring: the graphite simple-hexagonal in relation to the Tonnetz (the Euler-NeoRiemannian honeycomb lattice that characterizes the tonal functions), and the cubic one firstly described by [8] (later elaborated in [16]). Three dimension analogies

---

[3]See: [14] pp. 242–244, for a general introduction to the concept of *fractional noise* 1/*f* and its interpretation within music theory; and pp. 250–251 for its relationship with carbon.

of carbon also include the harmonic torus [12, 105] (with further development in [1, 21]), by its analogy with fullerene's carbon tubes coupled as torus. This conception may accept forced coupling as physical emulation of harmonic fields, as studied in dynamical systems applied to music [14, 354–367], as well as "special" segmentation (*zigzags*) and self-containment as it occurs in fullerenes [7, 48–50].

Of course, an explanation is required to make intelligible how non-linearity of carbon self-structuring does influence speech and music. This explanation strongly depends of the thermodynamic approach to a specific carbon structure, and something similar would correspond to music and speech analysis. In other words, non-intuitive axiomatics and emergent complexity need particular focus involved with its particular non-linearity. A good example of speech structural complexity at this level is originally proposed as non-linearity of words and phrasing change, by [9, 10]; and its parallel in music leads to the Arnold tongues as explained below, in Sect. 2.

Another necessary explanation is related to the fact that the Tonnetz and its three dimensional projection through the torus of phases, are allegedly "exclusive" of Western European music, i.e. "tonal music" (TM). However we must take into account that TM mapped in the Tonnetz and the torus is just one of the infinite sets of scales able to be mapped in hexagonal lattices (an intuition already visible in [5]).[4] Another quite different story, is that two-dimensional embedding of TM— besides its quality to fit in a maximally efficient hexagonal space—also accomplishes a reasonable set of aesthetic properties, directly related to psychoacoustic human preferences (see [11]). No doubt, this features make TM *special*, particularly in terms of maximal evenness as explained in [6], although *not so special* in terms of roughness and timbre-tone richness,[5] which are aspects of music at least as valuable as *harmony* in its most classical-music sense. Indeed, the simplest *carbonic analogy* representing this dichotomy is pictured in Fig. 2, where a hexagonal lattice-motif represents maximal evenness harmony, and an "amorphous" lattice—altered from the previous—represents seminal self-variation of a carbon structure (here directly related to the evolution of a biorhythmic pattern).

1.2 Cardiorespiratory Performance and Its Inheritance in Music

The HSA does imply verbal language, since speech comes from a common evolution intertwined with music and mathematics. The biorhythmic influence of carbon geometries in music and verbal language, occurs in at least three different processes;

---

[4]Carrillo suggests an infinite harmony embedded within a geometry of series of square roots of 2. During the last twenty years of his life, he attempted to represent this idea by a hexagonal lattice (original manuscripts and blueprints nowadays at the Carrillo Museum, San Luis Potosí, SLP, Mexico).

[5]In this relationship I obviously include the psychoacoustic shades between scales of (micro)tones and timbral roughness, always taking into account instrumental variables in terms of sound color.

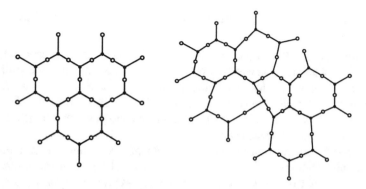

Fig. 2 *Left* Periodic, regular geometry representing a crystal structure of carbon. *Right* Amorphous variation of the latter. *Dots* in both figures represent the ionic composition of the molecular arrangement. Scalar versions of the periodic structure are visible in spatial bio-economies (honeycombs, cacti, photosynthetic cells, etc.). A combination of crystalline and amorphous varieties are common in most of living organisms. In many senses, the characteristic geometries of music in its different parameters are analogous to these *polycrystalline* compositions

even when they differ widely in their surfaces, these geometries produce shared patterns in music and speech: (*a*) cardiorespiratory rhythm (and its analogies), with *self-structuring synchronizing behaviour* (recent literature on the topic includes [4, 13]), featured by oxygenation processes and release of carbon dioxide; (*b*) functional participation of carbon structures in psychoacoustic systems from the middle ear to the most complex brain electrical processing; and (*c*) coordination of these two processes with other cardiovascular constraints that, through bioelectric empathy, tend to phase synchronization (a discovery reported in [17] in the context of *ethnomusicology*, although of deep interest to a universal musicology).

In human cardiorespiratory performance, according to [13] "the heart can act as a pacemaker for respiration". This is a mechanism for *synchronization*, which in physical terms does mean that heart and lungs, and the cardiovascular system tend to adjust pressure and electric potentials within a same harmonic system with constant variation and re-adjustment. Actually [13, p. 5] proposes a diversity of *tunings* (although not exactly using this word) of cardiovascular human synchronization that behaves as a system of harmonic couplings (in its physical sense). Whether brain oxygenation strongly depends on this process of synchronization, [11] provides arguments to hypothesize that Hebbian synaptic plasticity (the adaptation of neurons in the brain during learning and contrasting processes) shares the same kind of proportionality. In few words, music would be an expression of empathy and coordination of a *complex selfness*, a connection and articulation of endorhythms and exorhythms oriented by carbon signals at different levels.

2 Arnold Tongues: Self-similarity in Music and Physiology

As a common approach regarding cardiorespiratory and cardiovascular synchronization [4, 13], and pointing out Hebbian synaptics in the context of music cognition and
perception [11], mathematical modelling through dynamical systems is the foremost
method for the description and explanation of human body harmonicity. Usually this
approach pays special attention to the Arnold tongues in order to portrait human,
complex biorhythms in coordination.

The Arnold tongues models the circle-map of two coupled oscillators, typically one with fixed-periodic rotation, and the second one with incommensurate-
aperiodic rotation. Roughly speaking, Arnold tongues provides a description on how
Pythagorean simple ratios (ratios with small numbers, like $\frac{1}{2}$, $\frac{1}{3}$, $\frac{1}{4}$...) behave as *attrac-
tors* (high numerical hierarchies) among ratios progressively with longer numbers.
A simplified version of the Arnold tongues appears in Fig. 3 (first published in [2]),
obtained from the equation shown below, where θ is to be interpreted as polar angle.
The horizontal low border of the scheme represents the infinitely dense set of $\mathbb{R}$ (thus
a zoom-in at any point of this line would show many other *harmonic* intervals nesting
within bigger intervals); vertical axis represents the coupling strength (K) and the
horizontal one represents the bare winding number (Ω) in the circle map.

$$\theta_{i+1} = \theta_i + \Omega - \frac{K}{2\pi} sin(2\pi\theta_i).$$

Although the Arnold tongues have been suggested in [14, 354–371] as dynamical means for organizing self-similarity scaling in music, this approach to musical
complexity also serves to explain how endorhythms contribute to perform and elab-
orate musical structuring and meaning. An idea prevailing in many musicians, from
pre-Classicism to nowadays theorist and composer Erv Wilson (1928–), is that the
endorhythms of human, animals, plants and other living organisms are somehow
"musical" or "pre-musical" features of an evolutionary society understandible as a
wholeness, although also visible as groups of individuals. The Arnold tongues clearly
grasp this relationship between general and particulars.

The 19th century concept of Farey tree—a simple arrangement of numerical self-
structuring—is closely related to many other self-referential structures studied by
modern mathematics. Introducing a differed approach to successions within these
arrangements may lead to emerging patterns with "new" features and very diverse
behaviour. As a mathematical abstraction, the Arnold tongues fulfil this approach, as
it is analogical to essential self-structuring patterns in living organisms codification
(e.g. basic recursive genetics) and more sophisticated self-structuring grammars in
verbal and non-verbal communication.

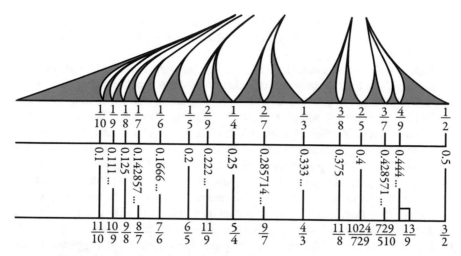

Fig. 3 Arnold tongues ranging from $\mathbb{Q} \frac{0}{0}$ to $\frac{1}{2}$. The *upper row* of numbers corresponds to the absolute values of the tongues in $\mathbb{R}$ (here mapped to the interval [0,0.5]). The *middle row* shows the same quantities as digit representation, and the lower row matches the continuation of the interval [0,1], here mod$\frac{1}{2}$ in order to emphasize the self-similar property of the harmonic intervals nested within the tongues. Notice, among other music intervals, $\frac{3}{2}$, the perfect fifth (the Greek *diapente*), matching at the end of the widest tongue (the white area at the rightmost part of the tongues). Clearly following a harmonic hierarchy, we see $\frac{4}{3}$, the perfect fourth; $\frac{5}{4}$, perfect major third (or fifth harmonic); $\frac{1024}{729}$, Pythagorean diminished fifth; $\frac{6}{5}$, perfect minor third; $\frac{11}{10}$, Ptolemaic second (or neutral second); $\frac{7}{6}$, septimal minor third; $\frac{729}{510}$, Pythagorean tritone; $\frac{8}{7}$, septimal major second; $\frac{13}{9}$, tredecimal diminished fifth; $\frac{9}{8}$, major perfect tone; $\frac{10}{9}$, minor perfect tone or low tone, and $\frac{11}{10}$, neutral second or Ptolemaic second. Notice that a zoom-in between any of these intervals will display subsequent harmonic hierarchies nested among the infinite intervals contained within the tongues lower border

3 Self-contained *Histories of Harmony* and the Ear-Brain-Mind Complexity

Quite obviously, the music intervals listed in Fig. 3 (see caption) do not correspond exclusively to a same cultural tradition of musical tuning or musical scaling, but to many cultural concepts of harmony, nesting within the same self-similar structure. [11] provides interesting evidence to interpret the electric behaviour of the central auditory nervous system, by its analogy with the dynamical structure of the Arnold tongues. This implies that beyond the Neo-Pythagorean numerical interpretation of this model, the Arnold tongues are meaningful to psychoacoustics—an important difference in regard to the linear modelling of music. In this sense, a good question is how the Arnold tongues also do map the self-similar carbonic noise 1/f in terms of potential scaling (i.e. as an Arnold self-similar set mapping a set of carbon power laws). A periodic behaviour of carbon, such as its cyclic stability property of *aromaticity* acting as a quasi-periodic system forcing other geometric or connective arrangements in a same set of atoms, may be part of a hypothesis in this direction. However, up to now little is known about this relationship in the context of pshychoacoustics.

From a distinct viewpoint, if we see the Arnold tongues rather as a set of *musical probabilities*, then we may interpret that the musical parameter measured by the tongues will behave as a *proportionality of aggregates*, a concept already developed by [10] in the context of speech sentence's variation of lengths. This sort of variation does not affect exclusively a general law for musical motif and phrase lengths, but to their related grammars, since, according to analytical musicology, the shape and structure of "little" particles do affect the shape and structure of "big" structures (due to a general property of proportionality in music). In fact, music can be described as a set of collections of quasi-periodic recursive practices, from tuning to motivic recursion, from rhythmic pulsation to metre and phrasing, and so on, through its arrangements of proportionality. This notion also provides a systematic treatment for parametric bifurcations in musical complexity, as suggested by the Arnold tongues inner systems of hierarchical bifurcation; this clearly involves transitions from periodic to quasi-periodic solutions for a wide range of parameters in musical recursion in terms of musical practice.

Whether the Arnold tongues contains the infinite collection of harmonic rational intervals (as well as its correspondence with the irrational ones, since its mapping arises from a periodic-rational rotation coupled with an aperiodic-irrational set of infinite intervals), musical practice operates as a selective process guided by high-rank hierarchies of intervals. This concept connects Neo-Pythagorean music theory and practice with the *axiom of determinacy* (Mycielski-Steinhaus, 1962), where every game of a certain type of musical proportionality is determined by a subset of the Arnold's probabilistic hierarchies. Furthermore, this perspective allows us to interpret the Kripkean worlds—originally *semantic possible worlds*—mentioned in [20], as multiparametric (i.e. not only semantic) possible worlds of self-organized music: a (deterministic) chaotic cascade of multiparametric intervals mappable in Arnold.

Our hypothesis on mathematical intuition (above labeled HSA) seems to be useful to understand how, by pure analogy, Erv Wilson reinvented the Stern–Brocot tree—a developed Farey sequence—which he called the "Scale Tree", adding the convergence of diagonals comprising the Novaro's Triangles in his *natural harmony* lattice.[6] Even the (Peirce)–Schenker–Lerdahl cognitive constraints of musical systems and hierarchical self-structuring, reasonably fit within this tree model, which is not necessarily triadic or *n*-adic, but *n*-layered, *n*-hierarchical (or rhizomatic), dense and self-contained.[7] Accordingly, the embedding of recursive grammars within themselves (e.g. chromaticism within diatonicism, and *Diatonicism*; let this concept be

---

[6]I owe special thanks to Kraig Grady for his personal communication [2015], emphasizing the nesting of the Novaro's triangles within the Wilson reinterpretation of the Stern–Brocot tree.

[7]Here we suggest that the tree diagrams built by Charles S. Peirce, Heinrich Schenker and Fred Lerdahl, although quite distinct by their methods, belong to a same tradition of *analogical hermeneutics*, a concept formalized in recent times by philosopher Mauricio Beuchot (1950–).

useful for an algebraic-geometric definition, and not especially for the naif one), can be associated to a multi-scalar dynamics of music, including musical pragmatics.

Since the identification of initial conditions of a system is usually a basic start point of any dynamical system, the dynamical modelling of music must somehow be of an analogical (i.e. proportional, synecdochic) nature. From the viewpoint of this proposal, the initial conditions of a musical system must be explicitly related to *initial rules* (i.e. probable relationships within a hierarchical universe), such as well-formedness of a pitch scale, or rhythmic proportionality (motivic structure, metre, phraseology). This still valid for any function within a musical grammar regardless its order of complexity, and also is closely related to the emergence of musical Gestalt, in its turn crucial for the meaning of a "grammar in context" leading to music pragmatics.

4 Conclusions

The intuition of nested-complex self-similar harmony seems to be a constant feature of distinct cultures. In this context, the HSA is suitable to provide a philosophical answer to the correspondence between mathematics and physical intuition. Even the most abstract musical phraseology still closely related to the quasi-periodic synchronization inherited from the rhythmics of cardiorespiratory brain oxygenation. Then, it seems to be not a trivial coincidence that the Tonnetz lattice diagram of tonal music resembles the graphene atomic-scale honeycomb lattice: both constitutes a two-dimension maximal economy distribution of carbonic, functional geometry. It is hard to deny that, beyond metaphors, bird vocalization and human speech and songs are *shaped by carbon*. Quite obviously, words and symbols are much more complex that carbon crystals burning in entropy. But we should not ignore that this apparently innocuous behaviour is a *physical guide* to animals and plants that inherit the organizing patterns of carbon, from which the basic structures of language emerge.

The meanings and applications of the HSA may impact music, linguistics, information theory and artificial intelligence, since the implementation of carbonic rhythms in multilevel synchronization processes derive into algorithms that do not simply emulates cognition and expressiveness, but specifically contributes to explain what cognition and expressiveness are. From a philosophical appreciation, this theory, if valid, requires exhaustive research on the nature of carbon, its geometric meanings, its relationship with a general topology and its influence in biology and in human intelligence; if invalid, either a sharper approach is necessary to explain the striking analogies between carbon patterns and biorhythmics so elegantly packed in music.

References

1. Amiot, E.: A survey of applications of the Discrete Fourier Transform in Music Theory, in this volume
2. Bak, P.: Commensurate phases, incommensurate phases and the devil's staircase. Rep. Progr. Phys. **45**, 587–629 (1982)
3. Beran, J.: Statistics in Musicology. Chapman and Hall/CRC, Boca Raton (2004)
4. Bernardi, L., Porta, C., Casucci, G., Balsamo, R., Bernardi, N.F., Fogari, R., Sleight, P.: Dynamic interactions between musical, cardiovascular, and cerebral rhythms in humans. Circulation **119**, 3171–3180 (2009)
5. Carrillo, J.: Leyes de metamorfósis musicales. Talleres Gráficos de la Nación, México, DF (1949)
6. Clough, J., Douthett, J.: Maximally even sets. J. Music Theory **35**, 93–173 (1991)
7. Deza, M.-M., Sikirić, M.D., Shtogrin, M.I.: Geometric Structure of Chemistry-Relevant Graphs: Zigzags and Central Circuits. Springer India, New Delhi (2015)
8. Estrada, J., Gil, J.: Música y teoría de grupos finitos (3 variables booleanas). UNAM, Mexico City (with an introduction in English) (1984)
9. Hřebíček, L.: Fractals in language. J. Quant. Ling. **1**(1), 82–86 (1994)
10. Hřebíček, L.: Persistence and other aspects of sentence-length series. J. Quant. Ling. **4**(1–3), 103–109 (1997)
11. Large, E.W.: A dynamical systems approach to musical tonality. In: Huys, R., Jirsa, V.K. (eds.) Nonlinear Dynamics in Human Behavior, SCI 328, pp. 193–211. Springer, Berlin (2010)
12. Mazzola, G.: Geometrie der Töne. Birkhäuser, Basel (1990)
13. McGuiness, M., Larsen, P.: Arnold tongues in human cardiorespiratory systems. Chaos **14**(1), 1–6 (2004)
14. Pareyon, G.: On Musical Self-Similarity: Intersemiosis as Synecdoche and Analogy, ISS, Acta Semiotica Fennica 39, Helsinki/Imatra (2011)
15. Pérez-Velázquez, J.L., Guevara-Erra, R., Rosenblum, M.: The Epileptic Thalamocortical Network is a Macroscopic Self-Sustained Oscillator: Evidence from Frequency-Locking Experiments in Rat Brains. Sc. Rep. 5, art. 8423 (2015)
16. Tymoczko, D.: The geometry of musical chords. Science **313**(72), 76–79 (2006)
17. Vaughn, K.: Exploring emotion in sub-structural aspects of karelian lament: application of time series analysis to digitized melody. Yearb. Tradit. Music **22**, 106–122 (1990)
18. Voss, R.F.: 1/f noise and fractals in DNA-base sequences. In: Crilly, A.J., Earnshaw, R.A., Jones, H. (eds.) Applications of Fractals and Chaos: The Shape of Things, pp. 7–20. Springer, Berlin (1993)
19. Voss, R.F., Clarke, J.: 1/f noise in music: music from 1/f noise. J. Acous. Soc. Am. **63**(1), 258–263 (1978)
20. Vriezen, S.: Diagrams, Games and Time, in this volume
21. Yust, J.: Schubert's harmonic language and fourier phase space. J. Music Theory **59**(1), 121–181 (2015)

Wooden Idiophones: Classification Through Phase Synchronization Analysis

Gabriel Pareyon and Silvia Pina-Romero

Abstract Idiophone instruments are classified through different methods, ranging from the Hornbostel–Sachs system of musical instrument classification, to time series organized according to features of frequency spectra and time span. We propose an alternative method for analyzing and classifying idiophones according to their timbral complexity, measuring timbral-body continuum and phase-synchronization degree. In order to simplify our exposition, we choose the teponaztli as a model of wooden idiophone, because of its structural unicity and its potential complexity through *extended* musical performance (e.g. through the instrument's individual or group timbral experimentation). We start exposing organological and cultural topics on the teponaztli, and then we discuss its harmonic and musical implications. Finally we explain our experimental development, and discuss the implications for musicological research and eventual new musical output.

1 Introduction

1.1 A Model of Wooden Idiophone Instrument

We study the teponaztli as a undecomposable system and a model of wooden idiophone, since it has no separated sections and it is made of a single piece of carved hard wood (see Fig. 1). A slit drum native to Mexico, the teponaztli typically exhibits two "tongues" or wooden stripes in the form H, struck by two sticks—each for one hand, usually played by a single musician—respectively producing two different pitches.

G. Pareyon (✉)
CENIDIM-INBA, Torre de Investigacion, CENART, Piso 7, Av. Rio Churubusco 79, 04220 Coyoacán, D.F, Mexico
e-mail: gabrielpareyon@gmail.com

S. Pina-Romero
División de Electrónica y Computación, CUCEI – Universidad de Guadalajara, Blvd. Tlaquepaque 1421, 44430 Guadalajara, Jalisco, Mexico
e-mail: slvpina@gmail.com

© Springer International Publishing AG 2017
G. Pareyon et al. (eds.), *The Musical-Mathematical Mind*,
Computational Music Science, DOI 10.1007/978-3-319-47337-6_24

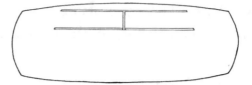

Fig. 1 Line drawing of a contemporary, round teponaztli from central Jalisco (Mexico), carved from a mesquite tree (*Prosopis velutina*) with typical stripes or *tongues* on the *top*, in the form of an H. Other views and details of the same instrument appear in Figs. 2 and 3, below

After [1, 2] it is commonly accepted that the teponaztli is "tuned" to a harmonic interval that for the *Western* (i.e. Western European) *modern* ear, falls nearby a "fourth" or "fifth" with relative roughness,[1] sometimes unfocused because of the variable quality of the instrument's wood, as well as the variability of cuts made when carving the instrument. We believe, however, this "harmonic" description is inadequate because it imposes an exogenous conceptualization over meaningful, original features of the instrument.

1.2   Cultural Implications on the Harmonic Model

Teponaztli's carvers and musicians empirically believe that it is a mistake attempting to fix a Western tuning for this instrument, by the simple argument that such a tuning was locally nonexistent before the European occupation of Mexico (started in 1519). An important conceit in our research, is the claim that such argument is obvious, and there is a stronger, physical—but also cultural—argument to reject any harmonic Western convention adapted to the teponaztli.

First of all, it is necessary to point out that the teponaztli's organological features are directly related to ancient, ritual Mesoamerican practices. These practices were, and still are closely related to regional native languages. Particularly one may trace a complex association of the Nahuatl language's prosody and metrics, with the ritual performance of teponaztli music [3]. Secondly, the use of sacred woods to build this instrument—in its turn a sacred interpreter of Mexican cosmology—is closely related to the conceptualization of a system of four cardinal directions and its center; this 4 + 1 system is represented by five trees (often depicted as a Ceiba tree abstraction) equally distributed with a space in between, as the analogy of *universal space*. A vertical/horizontal representation of this spaciality connects the planes of the Underworld (*Xibalbá* in Mayan, *Mictlan* in Nahuatl), with the terrestrial realm, and the skies. These abstract five trees and their branches and roots are constituted of a double duality (the four extreme trees and their ramifications of double dualities and

---

[1]Indeed, many teponaztlis' tunings may fit within the range of "thirds", "fourths" or "fifths", but in strict sense this terminology does not represent the diversity of timbral-pitch organization in this specific case.

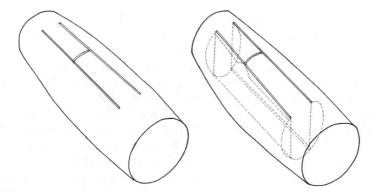

Fig. 2 *Left* upper three-quarters view of the same instrument Fig. 1. *Right* the same drawing including the design of the instrument's acoustic hollow, in *dashed lines*

their symbolic sub-structures), and a sort of *axis mundi* representing centrality and axiology.[2] As a conclusion derived from [4, 99–101] and [5, 220–230], we believe that this spatial complexity represented as a complex tree-branching, has its correspondence in every field of symbolic existence, prominently including the sonority of the teponaztli as a *forest of teponaztlis* (*Teponazcuauhtla* in Nahuatl language); a metaphor that suggest an infinite multiplicity of pitch and timbral features analogous to the *ramifications of double dualities and their symbolic sub-structures*.

The physical calculations and development of a chromatically tuned set of log drums was first put forward by [1, 2], conceiving the traditional teponaztli as a "primitive stage" of a work in progress. This positivistic, evolutionary perspective corresponds to an epoch when—at least in Mexico—the Western chromatic-diatonic system was accepted as the most "developed" system for tuning music (even Julian Carrillo's innovations building new scales and instruments were originally inspired by *well temperament*). In contrast, we believe that the original context and ancient symbolism of the teponaztli may provide us with valuable clues on an unexplored, special system of *harmony*—perhaps better explained as *harmonic-timbrality*—, mirrored in the Mexican tree-branching complexity model.

1.3   Idiophone Timbral Continuum Study and Classification

The Hornbostel–Sachs system classifies the teponaztli within class 111.231, i.e. *directly struck idiophone, individual percussion tube*. However this mechanical-morphological classification has little to say about the timbral variety within a same class. Specialized literature on percussion timbre [6, 7] emphasize a *psychophysical multidimensionality* of timbre, and even [8, 153–155] identifies musical percussive

---

[2]"Tamoanchan is the axis of the cosmos and the set of the cosmic trees", in [4, pp. 19–20, 99–101].

analysis and processing as relevant as its analogies in human speech. Thus, the timbral features of idiophones are at least as important as its organological descriptions; although the latter does not necessarily contribute to a cultural appreciation of musical timbre and to its semantic/syntactic/pragmatic connotations.

Aware that timbral classification may reflect human subjectivity, our approach is indeed rich in terms of comprehensively linking subjectivity of instrumental recognition—e.g. in terms of *color*—with perceptual and cultural contexts.[3] Therefore, and keeping in mind our instrumental and cultural model, the research on idiophone's contextualized production and performance cannot omit the instrument's timbral complexity, timbral continuum, and associated expressiveness.

The teponaztli and other comparable idiophones are actually used and poorly classified as a *two-pitched, non-melodic* percussions. This classification arises from the comparison with "melodic" and "multi-pitched" instruments, paying scarce attention to the subtle timbre-pitch shades of the idiophones. We believe the analogy *microtonality–microtimbrality* would be relatively helpful from a 20th century theoretical approach in order to study these instruments; but this analogy would be too weak for our *continuum timbral* perspective. We rather seek for perceptual intermitency of this analogy after resonance and acoustic synchronization. Within this context we propose a timbral continuum study and classification particularly emphasizing degrees of instrumental synchronization (as explained below in Sect. 2), and ear-instrument synchronization (as further research).

It is worth to mention that rich symbolism may be expressed by the instrument maker, carving reliefs and motifs in the surface of the teponaztli (e.g. see Fig. 4, left). Obviously these carvings may affect and modify the instrument's timbral features; but on the other hand the gross shape of the instrument is irrelevant to the relation pitch/timbre, whilst keeping the same volume and *isospectral manifold* (i.e. the same acoustic sets of eigenvalues) [10]. A comparison between Figs. 3 and 4, respectively with round and square shapes, may be illustrative in this sense.

1.4 Synchronous Motion of a Continuous Oscillatory Medium

Given that the teponaztli is made of a single piece, we can think of it, both, as a continuous oscillatory medium, and as a coupled oscillatory system that, when forced (i.e. hit), may synchronize its phases from the attack to the end of the resonances. Our hypothesis is that the evolution of this synchronization reflects key features of the instrument's timbre and morphology.

It has long been noticed that a wide variety of coupled oscillatory phenomena synchronize, for example the lighting of fireflies, the singing of frogs, and pendulums

---

[3]In [9, 168]: "Timbre spaces reflect human perception and are not necessarily optimal from the viewpoint of partitioning the space into separable classes. However, they can reveal the acoustic properties that enable computing perceptual similarities between instrument classes."

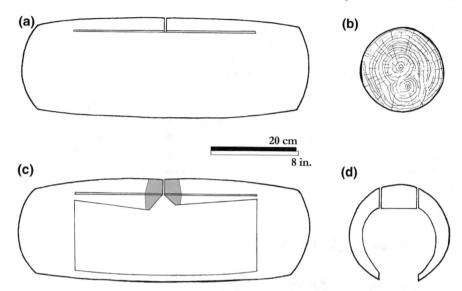

(a)

(b)

20 cm

8 in.

(c)

(d)

Fig. 3 Different views of the same instrument (shown in Figs. 1 and 2): **a** Longitudinal view of the instrument; **b** External side view with the woody plot of the original trunk; **c** Longitudinal cut of the instrument, showing details of the tongues cuts and their *inner shapes*. The *grey middle areas* represent the highest resonance areas in the tongues' tips; **d** Transverse view of the instrument's *middle cut*

Fig. 4 *Square-shaped* teponaztli from Jaral del Progreso, Guanajuato (Mexico), made of white ash (*Fraxinus americana*). *Left* upper view. *Right* bottom view. The proportion of the image in respect to the instrument is preserved from Fig. 3. In this instrument the acoustic tongues exhibit a carved round tip, morphologically suggesting the nodal area for the instrument's better resonance

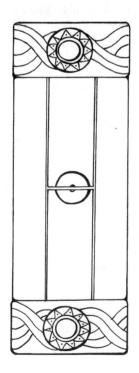

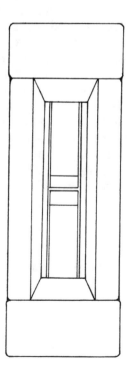

of clocks [11, 12]. Synchronization can occur regarding different features of the oscillatory phenomena, such as phase or frequency. In this work we refer to phase synchronization. The phase of an oscillator at any given time is a quantity that increases 2π in each cycle and corresponds to the fraction of a cycle which has elapsed, relative to an arbitrary point.

A particular kind of phase synchronization, when one of the oscillators leads the oscillation of the overall system is known as *resonance* [11, 13]. Specifically, our work focuses on damped resonance but the framework applies to the more general case of synchronization. Our contribution is the analysis of the timbral features of the teponaztli in the context of its phase synchronization. Eventually, we propose the generalization of this analysis for a variety of idiophones.

2 Experimental Development

2.1 Experimental Set up

Our experimental set up is described in this subsection. For each teponaztli, three sets of three pulses were analyzed, each set of pulses using the exact same set up and protocol except for the drumstick, which changed from *soft*, *medium*, and *hard*. For the first pulse the larger tongue was fitted with a cardioid microphone while the other one was fitted with a transducer microphone; then, the larger tongue was hit once. For the second pulse, the microphones were exchanged and this time the smaller tongue was hit. Finally, for the third pulse, both tongues featured transducers and both were hit once simultaneously. This was repeated for each kind of drumstick.

Each pulse was treated separately, but in all three cases, both, the recording of the sound produced by the vibration of the hit tongue, and the recording of the vibration of the tongue obtained via a transducer, are periodic time series from which it is possible to obtain the respective phases. The recordings were cut starting at the attack and ending when oscillations stopped.

To extract the phase a complex signal is constructed via the Hilbert transform as follows; for each time series $s_j(t)$ with $j = 1, 2$, which correspond to each of the tongues of the teponaztli, we generate a complex signal (1):

$$\zeta_j(t) = s_j(t) + i s_j H(t), \tag{1}$$

where $s_{jH}(t)$ is the Hilbert transform as in (2), that is:

$$H(t) = \pi^{-1} P.V. \int_{-\infty}^{\infty} \frac{s(\tau)}{t - \tau} dt, \tag{2}$$

where $P.V.$ refers to the principal value, and from which Eq. (3) is obtained,

$$\zeta_j(t) = A_j(t) \, e^{i \, \varphi_j(t)}. \tag{3}$$

Equation (3) yields the functions for instantaneous amplitude and phase, $A_j(t)$ and $\varphi_j(t)$, respectively. Once the phases are obtained, the synchronization index is computed. To do so, we use the stroboscopic approach and conditional probability. In this context, the synchronization index is a number between 0 and 1, where 1 is *complete synchronization* and 0 represents *total lack of synchronization*. Each recording is divided in thirty windows of the same length, and a synchronization index is calculated for each of them. Our proposal is to explain how the evolution of the synchronization index characterizes each instrument. In order to calculate each index, both phase functions are mapped around a circle by taking them modulus 2π. A parameter a is selected and the interval $[0, 2\pi]$ is divided in a subintervals, I_i with $i = 1, \ldots a$, of size $\frac{2\pi}{a}$ which cover the circle; a partial synchronization index is obtained for each of these subintervals. More specifically, the synchronization index, λ_i, represents the probability of having the phase of one of the oscillators in a certain subinterval I_i, given that the phase of the other oscillator is at that same interval, this is:

$$\lambda_i = P(\varphi_2(t) \in I_i \, | \varphi_1(t) \in I_i), \tag{4}$$

with t inside the time window in question. Once the phase functions are mapped around a circle, the values of the phase of one of the oscillators (the instrument tongues) are counted and recorded every time the value of the phase of the other oscillator falls inside a given interval I_i. Let M_l be the number of occurrences of the phase of the leading oscillator that fall inside the I_l interval, and let v_l with $l = 1, \ldots, M_l$ be a vector containing the corresponding values of the phase of the other oscillator at those specific times; then the synchronization index for the *l-th* interval is

$$\lambda_l = \left| \frac{1}{M_l} \sum_{j=1}^{M_l} e^{i \, v_i} \right|. \tag{5}$$

2.2 Results

The synchronization index provides useful information for the characterisation of a teponaztli. We can see in Fig. 5 that the evolution of the synchronization index differs from instrument to instrument, but remain similar when trying different drumsticks on a single teponaztli (see Fig. 6). This indicates that is the features of the instrument, rather than the way it is set in motion, what is reflected on this index.

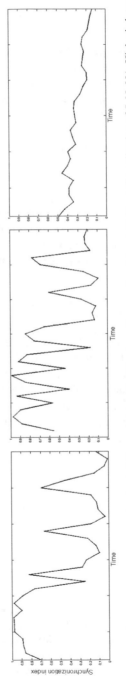

Fig. 5 Synchronization index evolution for three different teponaztlis: Chechen, Guerrero2 and Izcuintli (from the collection of J. N.-N.). High pitch tongue hit with soft drumstick

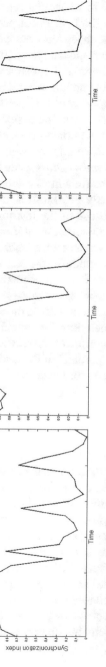

Fig. 6 Synchronization index evolution for a single teponaztli (Chechen) hit a soft (*left*), medium (*center*) and hard (*right*) drumstick

2.3   Discussion

Resonance in idiophones is due to molecular tension-distension struggling forces after an initial perturbation (i.e. the instrument percussion). As a capital phenomenon in physics and in music, resonance has been largely investigated as a main topic of acoustics. In fact, the description and analysis of the teponaztli's resonance is a current procedure assessing the features of the instrument. However, the data extracted from the resonance patterns is not enough to explain the whole teponaztli's acoustic behaviour, insofar as it may imply local (i.e. partial) tension-distension processes leading to synchronicity. As [11] notes, the systematic perturbation of an acoustic device—in this specific case an idiophone—does not necessarily imply synchronicity, but simple resonance (even when this resonance may be somehow "complex").

It is obvious that *simple resonance* occurs when hitting an idiophone. But using a *good quality* teponaztli (i.e. an instrument that does not damp its own resonances by its own material structures), directly streaking one of its tongues may imply indirectly streaking its second tongue, because of the instrument acoustic self-structuring by the same, compact wood piece (the instrument itself).

There is evidence that the resonance of a teponaztli's tongue (low or high) may be partially damped by the resonating body, and even by the acoustic features of the second tongue, "fighting" for its own acoustic partials. Thus, one may speak of synchronization in a specific teponaztli whose struck tongue behaves *adjusting*—increasing or diminishing—its main partials to the synchronic influence of the whole resonating body, meaningfully including the second tongue. In this case *meaningfully* does intend that the harmonic spectrum of the struck tongue may harmonically reflect acoustic properties (resonances, partials) of the second tongue, and even of the whole instrument as a self-coupled oscillatory system.

3   Conclusions

The synchronization analysis of the teponaztli is clearly useful for classifying timbral nuances of the idiophone timbral continuum. Typically rich-timbre teponaztlis—as well as other idiophones non-industrially produced, and therefore non-homogeneous pitch/timbre idiophones—may produce a timbral continuum difficult to be described by musicians. Synchronization analysis provide clues for understanding how an idiophone like this produces a timbral continuum as a shade of phase self-synchronizing, ranking from null-synchronization to full-synchronization, being the former in direct relation to sound opaqueness and harmonic roughness; and the latter in direct relation to sound sharpness, rich in harmonics and therefore in musical color brilliance. Thus, the *elusiveness* of the teponaztli's timbral continuum classification would no longer be a musicological problem, but its very quiddity in its own philosophical and cultural context.

In an epistemological context, and considering that the teponaztli was used for centuries as a didactic artifact for explaining geometry and arithmetic (a means neglected and even prosecuted during the European colonization of Mexico), one may open the following question: Is the scrutiny on the nature of the *fuzzy epistemics* of the teponaztli an opportunity for discussing on the epistemics of modern (*open*, fuzzy logic of) mathematics? The same question in other words: May our mathematical thought be subject of hidden colonialist strategies and therefore mathematics are somehow affected by its socio-historical context? In any case this contribution wants to provide matter to this discussion, with the specific case of complexity that can be found in many idiophone instruments whose harmonic patterns could be ignored or cut by cultural biases.

Acknowledgements We are grateful to José Navarro-Noriega, for his generous support providing us with statistical information extracted from his personal, unique collection of teponaztlis, located in Aculco, Mexico City.

References

1. Castañeda, D., Mendoza, V.T.: Instrumental precortesiano, Instrumentos de percusión, vol. 1. Museo Nacional de Arqueología, Historia y Etnografía, Mexico City (1933)
2. Castañeda, D., Mendoza, V.T.: Los teponaztlis en las civilizaciones precortesanas. In: Anales del Museo Nacional de Arqueología, Historia y Etnografía, 4th epoch, vol. 8, no. 2, Apr.–Jun., pp. 5–80, Mexico City (1933)
3. Leon-Portilla, M. (ed.): Cantares mexicanos, 3 vols., IIH – IIFL – UNAM, Mexico City (2011)
4. López-Austin, A.: Tamoanchan y Tlalocan. FCE, Mexico City (1994)
5. Dehouve, D.: El imaginario de los números entre los antiguos mexicanos. Publicaciones de la Casa Chata, CIESAS, Mexico City (2014)
6. Bell, R.: PITCH: The percussion instruments timbral classification hierarchy. In: ICMC Proceedings, vol. 1994, pp. 481–482 (1994)
7. Brent, W.: Physical and perceptual aspects of percussive timbre, University of California, San Diego, 2010 (PhD diss.)
8. FitzGerald, D., Paulus, J.: Unpitched percussion transcription. In: Klapuri, A., Davy, M. (eds.) Signal Processing Methods for Music Transcription, pp. 131–162. Springer, Berlin (2006)
9. Herrera-Boyer, P., Klapuri, A., Davy, M.: Automatic classification of pitched musical instrument sounds. In: Klapuri, A., Davy, M. (eds.) Signal Processing Methods for Music Transcription, pp. 131–162. Springer, Berlin (2006)
10. Kac, M.: Can one hear the shape of a drum? Am. Math. Month **73**(4), 1–23 (1966)
11. Pikovsky, A., Kurths, J.: Synchronization: A Universal Concept in Nonlinear Sciences. Cambridge University Press, Cambridge (2003)
12. Rosenblum, M.G., Pikovsky, A.S., Kurths, J.: Synchronization approach to analysis of biological systems. Fluctuat. Noise Lett. **4**(1), L53–L62 (2004)
13. Rosenblum, M.G., Pikovsky, A.S., Schäfer, C., Tass, P., Kurths, J.: Phase synchronization: from theory to data analysis. In: Moss, F., Gielen, S. (eds.) Handbook of Biological Physics, pp. 279–321. Elsevier Science (2001). Neuro-informatics, Chap. 9

A Fuzzy Rule Model for High Level Musical Features on Automated Composition Systems

Iván Paz, Àngela Nebot, Francisco Mugica
and Enrique Romero

Abstract Algorithmic composition systems are now well-understood. However, when they are used for specific tasks like creating material for a part of a piece, it is common to prefer, from all of its possible outputs, those exhibiting specific properties. Even though the number of valid outputs is huge, many times the selection is performed manually, either using expertise in the algorithmic model, by means of sampling techniques, or some times even by chance. Automations of this process have been done traditionally by using machine learning techniques. However, whether or not these techniques are really capable of capturing the human rationality, through which the selection is done, to a great degree remains as an open question. The present work discusses a possible approach, that combines expert's opinion and a fuzzy methodology for rule extraction, to model high level features. An early implementation able to explore the universe of outputs of a particular algorithm by means of the extracted rules is discussed. The rules search for objects similar to those having a desired and pre-identified feature. In this sense, the model can be seen as a finder of objects with specific properties.

1 Exposition

Algorithmic composition systems (ACs) that create music by means of formalizable methods, are now well-understood and documented [6, 9]. Examples have followed the trend of the formalization of thought and technology development, starting around 1000 AC with the Guido D'Arezzo method for the generation of melodies from text material, based simply on the mapping of vowels over pitches. Later in the "musical dice game" of the 18th century, the outputs were generated based on a combinatorial structure. Players rolled a dice to choose, for every temporal unit, a bar from a table until completing a short piece. Although playing was trivial, the system required

I. Paz (✉) · À. Nebot · F. Mugica · E. Romero
Departament de Llenguatges i Sistemes Informátics, Soft Computing Research
Group, Technical University of Catalonia, Barcelona, Spain
e-mail: ivnpaz@gmail.com
URL: http://www.semimuticas.org

© Springer International Publishing AG 2017
G. Pareyon et al. (eds.), *The Musical-Mathematical Mind*,
Computational Music Science, DOI 10.1007/978-3-319-47337-6_25

expertise in music composition to consider the harmonic and voice-leading aspects of the combinations. In 1956 the Illiac suite was the first completely computer-generated composition. The piece was composed by Lejaren Hiller and Leonard Isaacson using a Monte Carlo method which outputs were musical notes. Over the 20th century, compositional methods, like twelve tone technique (that assured all 12 notes of the chromatic scale sounded as often as one another, and so, preventing the emphasis of any one note), and serialism (that uses a series of values to manipulate musical elements) used structural parameters and logical conditions. Xenakis' theory of sieves, is an application of logical combinations for the generation of musical structure. With the development of techniques of machine learning composition systems are becoming more and more complex. For example, "Autocousmatic" [4] is an algorithmic system for electroacoustic music composition that incorporates machine-listening processes within the design cycle. In this way, the system is able to assess the "worthiness" of intermediate files from which the final output will be created. The formal structure is drawn from models based on the analysis of an exemplar corpus of pieces. A similarity-measure is used to decide between candidates for the final mix.

Despite the extensive work developed on ACs, its capacity for accurately produce high level musical features (like coherence, emotion or personality), is still object of discussion. Until this moment, the representations obtained for high level musical features were side effects of the machine learning research. This has motivated the emergence of new paths for explicitly work on this subject. For example, "affective algorithmic composition", which seeks to develop systems for the selective adjustment of emotional responses using parameterizations of musical features [5]. In his work "Musical form and algorithmic composition" [3] argue that "algorithmic music often seems stuck in a static moment form, able to abruptly jump between composed sections but unable to demonstrate much real dramatic direction." Models for musical form that takes into account the psychological perspective of the listener to a great degree have not been well explored. Then, whilst it is possible to generate sections of a composition by means of the different ACs, to achieve greater expressiveness, it is still necessary to implement strategies to capture the relations between sections, hierarchical layers (possibly by relating low and high level features) and evolution of the tension.

On the search for designing systems capable to consistently produced outputs perceived by the listener with specific characteristics, two approaches have been addressed. On one hand, some systems have followed the explicit rule paradigm, using conditions (many times expressed as rules) given by experts. On the other hand, other system have used machine learning tools to learn patterns and produce compositional models with the learnt parameters. However, the obtained representations do not takes into account musical considerations and then, are more suitable for tasks like pattern recognition. In other words, the fact that machine learning processes have effectively captured high-level features to a great degree is still object of discussion. Moreover, designed systems have not extensively incorporated perception and semantic of the form of generated music, including the psychological sensation of the listener. Attempts to do this often use modules designed for the evaluation and adjustment of the outputs based on pre-established symbolic domain [4], like the fitness functions of genetic algorithms to modify the outputs until the desired ones are obtained. As a consequence of this lack, in practical implementations, the composers have to proceed by analyzing the outputs one by one, in order to select the best outputs among all the possible ones. This happens either in automated or assisted composition, in statistical models, and even in algorithms for stochastic synthesis (see for example, [8]). This means that the composer has to create all the possible outputs and then, listen as much as possible in order to select those with the desired expressiveness, style, etc. This process is performed by using expertise on the model or by sampling techniques. However, the universe of all possible valid outputs of such kind systems normally range over thousand or even million, and so, it is impossible for the composer to explore it manually. In the present work we introduce a methodological approach for modeling high level musical features in the outputs produced by ACs. The model takes advantage of fuzzy quantifiers for modeling high level features. The objective is to climb in the semantic level by relating high level features with its representation at beat or raw level data. A remarkable point is that the design of the system requires the whole example to be listened before the evaluation, and so different levels of perception in the time and form domain are taking into account. The rest of the work is structured as follows. Section 2 present and discussed the model. Section 3 presents a general discussion of the results and elucidate possible further work.

2 Development

Given the subjective nature of the high-level features, the use of fuzzy systems to work with them sounds feasible. Fuzzy systems are based of fuzzy logic theory, they allow us to work with objects that are approximate or subjective rather than exact. In the context of the modeling of musical features, fuzzy systems can help to work without been restricted to pre-established structures for the evaluation modules of the outputs. Then, human expertise and its associated psychological perspective can be included within the design (or the evaluation) cycle. This allows the system to extract the musical representation of the expert experience as it is perceived and then, to translate it in terms of combinations of low level variables. We have to point out that the perception of the features constitute a subjective evaluation of the listened outputs, and so it is subjected to context and cultural experiences. Fur this reason it could hardly vary form one listener to other. However, it is possible, for example, to consider different opinions that may be pondered according with the degree of experience. A general structure of the system design is showed in Fig. 1.

The system is composed by several modules, each of which can be independently treated. The first module is the algorithmic composer or generative system. It, either belongs to one of the stablished paradigms or a hybrid one.[1] From the system, a sam-

---

[1]Different algorithm classes enable specific approaches to musical structure generation. The most popular are Statistical models including Markov models, generative grammars, transition networks,

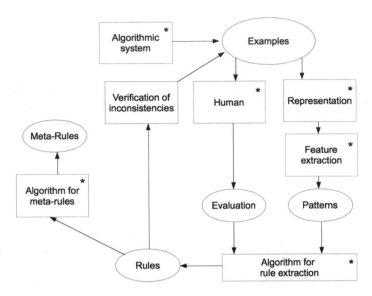

Fig. 1 The system architecture and process flow of the proposed approach. *Squares* denoted the different modules of the system and *ovals* denote data

ple of pieces (patterns) are extracted. Then, the sample is evaluated by human experts by using linguistic variables in order to categorize if they have or not the desired feature. This process can be performed for short pieces, patterns, phrases or motifs, to explore the system's capacities over different scales of length and hierarchical levels. It has to be noticed that, as the whole output is listened before the evaluation, it takes into account the different levels of organization. For example, if dealing with rhythmic patterns it will consider the different arranges of contrast and repetition as well as its locations present in the whole output. Or, if dealing with melody sequences the evaluation will consider the transitions of adjacent notes, but also relations among notes at different points, as well as its evolution over time. In other words, this kind of evaluation allow us to include the different semantical levels and the subjective human perception of the outputs. The representation module structure the data in the most efficient manner for analysis and feature selection. The feature selection module performs the feature extraction/dimensionality reduction and prepare the data for the algorithm of analysis. The extraction algorithm takes the linguistic evaluation and the set of patterns as input set. The extraction algorithm is the module that analyzes the episodical behavior of the system. i.e. the occurred patterns and its association evaluation, and it is able to abstract a model for its description. Algorithms been currently investigated are of the type of logical rules extraction [10]. The extracted rules are used to explore and classify the universe of legal outputs of the original

---

(Footnote 1 continued)
chaos and self-similarity, genetic algorithms, cellular automata, neural networks, as well as other artificial intelligence methods.

algorithmic system looking for inconsistencies until a terminal criteria for the quality of the obtained rules is meet. In order to perform a higher level analysis we include an algorithm for the construction of meta-rules, that take the extracted rules as input and perform the extractions of new rules of higher level. It should be noticed that this module does not starts until the iterations of the previous steps finish. Also, the extracted rules used to explore and classify the universe of legal outputs can also be used for the implementation of generative frameworks.

2.1 Implementation

For the first implementation of the methodology we used a generative model, based on probability templates, for beat patterns generation in the context of UK garage proposed by [2]. UK Garage is characterized by a high tempo, around 140 bpm, with a triplet swing groove, and the '2 step' feel [2]. In the usual 4/4 metric at the second and fourth beats we found a snare, and we can found at the first and third beats a kick. A simplified model of the style can be codify in a regular grid of 16 points per bar (swing variations can be incorporated during performance). This points correspond with each of the sixteenth notes. The beat patterns are modeled in templates as being played of three independent sixteen bits streams, for instruments kick, snare and hihat. These streams codify the information presence/absence for the instruments. For simplify the data representation in the analysis the original algorithm was restricted to considered only cases when one instrument strikes at the same time. Although this can appear as a great simplification, produced patterns are still valid cases of the original, i.e. it is a subset of the space of states. The set of probabilities (independent for each strike and taken from 0.0 to 1.0) are:
    //kick
[0.7, 0.0, 0.0, 0.1, 0.0, 0.0, 0.0, 0.2, 0.0, 0.0, 0.0, 0.0, 0.0, 0.0, 0.0, 0.3]
    //snare
[0.0, 0.0, 0.0, 0.0, 1.0, 0.0, 0.0, 0.5, 0.0, 0.2, 0.0, 0.0, 1.0, 0.0, 0.0, 0.3]
    //hihat
[0.0, 0.0, 1.0, 0.0, 0.0, 0.0, 1.0, 0.7, 0.0, 0.0, 1.0, 0.0, 0.0, 0.0, 1.0, 0.7]

This template can produce 128 valid outputs. From these, we produced short examples composed by 8 bars length. The first 4 bars correspond with a particular output of the system and the following 4 bars with another. We represent each output in an array containing 32 sixteenth notes, 16 from the first pattern and 16 from the second.
[1,2,3,4,5,6,7,8,9,10,11,12,13,14,15,16,17 18 19 20 21 22 23 24 25 26 27 28 29 30 31 32]

At each entrance for representing the different values that the variables can take we adopt the following convention: kick = 1, snare = 2, hi-hat = 3, and silences = 4. The cardinality of all possible outputs taken in this way is 16384 (128 times 128). It should be notice that all this cases including the extreme cases are valid outputs. From this set we extracted a sample to be evaluated by an expert. However, as different outputs have different probabilities, first all probabilities were calculated, and associated with each one of the 128 possible cases. Then, we create a sample to explore the space. We were interested in being sure that even the rare patterns, those with lower probability (probabilities range between 0.000101 and 0.070560), represent interesting outputs. After the exploration, we decided that all the space was interesting enough. By taking this into account, patterns were selected in a manner that every possible variation (class) in the training data were equally represented. From this set we create all possible combinations (400) among patterns. Then, combinations of each pattern with itself were removed. This was due that, for this experiment, we were interested in patterns constituted by different parts. So, the resulting (380) patterns will constitute our universe. Based on this set we trained and tested our model. Although the size of this set could appear small in comparison with the complete test set (of 16384 patterns), the amount of time needed for listen and evaluate the pieces makes the use of bigger sets impractical. However, the algorithm's restrictions, seeking to emulate a particular musical style, suggest that it is possible to obtain good results with few but well labeled data (as will be shown below).

The training set was evaluated by experts focussing in the perception of the "transition quality" between parts, considered as a high-level feature. We define the transition quality as "how good patterns match together" when they are perceived by a particular listener, in terms of the quality of the individual patterns and by considering the contrasts and repetition sensation between the parts. More precisely, this notion corresponds with a "motif coherence" (or coherence between rhythmic cells) of musical patterns. The evaluation of the transition will depend on the contrast and repetition points between parts A and B and on the moments on which those contrast and repetitions points are situated in the structure. For example, suppose that A and B are rhythmical motifs composed by 4 bars each one, as in our case. In the context of occidental music (at which the UK garage belongs), it could be considered that if the 4th bar of A coincides with the 4th bar of B, it will produce a greater sensation of coherence than if the 2nd bar of A coincides with the 4th bar of B. The evaluation was made by using linguistic variables (fuzzy quantifiers [7]) defined as low, medium or high. The evaluated training set was processed with the fuzzy inductive reasoning methodology and the algorithm for logic rules extraction (FIR + LR_FIR). A detailed explanation of both algorithms can be found in [1, 10]. Each time, the rules were used for the classification of the training set, looking for inconsistencies, and then, for the classification of the universe created with the sample patterns. After the definitive training set were selected, several tests were performed using other training sets. Some of them were chosen randomly, and others by sampling the possible patterns considering its probabilities (e.g. those with higher probability). It is remarkable that in the performed tests, the classification of the training set throws no inconsistencies, and that the proportion of classified patterns remains. In the same sense, after processing the rest of the universe (test data), the proportion of patterns that can be classified by rules is also similar among tests. And the order in the number of inconsistencies (in the test set 380 patterns) remains around ten percent. The rules are expressed in the IF THEN form. The numbers of the variables, for example V24 or V32, correspond to its location at the 32 sixteenth notes. The values in antecedents are 1 = kick, 2 = snare, 3 = hihat, 4 = silence. In consequences (system's output V33) are 1 = low transition quality, 2 = medium transition, and 3 = high transition quality. The expressions "Spec" and "Sens" stand for the usual specificity and sensitivity measures for the rules. The set of extracted rules are show below. They describe (or encode) the combination of variables that, according with the model, will produce a transition perceived as "low, medium and high".

LOW rules

IF V24 - 4 AND V32 - 1 THEN V33 - 1 Spec = 1 Sens = 0.33
IF V16 - 1 AND V24 - 4 THEN V33 - 1 Spec = 1 Sens = 0.18
IF V16 - 2 AND V17 - 4 AND V24 - 2 THEN V33 - 1 Spec = 1 Sens = 0.078
IF V16 - 1 AND V26 - 2 AND V32 - 2 THEN V33 - 1 Spec = 1 Sens = 0.078
IF V8 - 3 AND V24 - 3 THEN V33 - 1 Spec = 1 Sens = 0.078
IF V10 - 2 AND V24 - 2 AND V26 - 2 AND V32 - 4 THEN V33 - 1 Spec = 1 Sens = 0.078
IF V10 - 2 AND V16 - 2 AND V32 - 4 THEN V33 - 1 Spec = 1 Sens = 0.078
IF V8 - 4 AND V26 - 2 AND V32 - 4 THEN V33 - 1 Spec = 1 Sens = 0.078
IF V16 - 2 AND V24 - 2 AND V26 - 2 THEN V33 - 1 Spec = 1 Sens = 0.059
IF V16 - 3 AND V24 - 3 AND V32 - 1 THEN V33 - 1 Spec = 1 Sens = 0.059
IF V8 - 4 AND V20 - 4 AND V32 - 3 THEN V33 - 1 Spec = 1 Sens = 0.059

MEDIUM rules

IF V4 - 4 AND V24 - 2 AND V32 - 2- 3 THEN V33 - 2 Spec = 1 Sens = 0.2
IF V10 - 4 AND V24 - 2 AND V32 - 2 THEN V33 - 2 Spec = 1 Sens = 0.14
IF V10 - 4 AND V17 - 1 AND V32 - 4 THEN V33 - 2 Spec = 1 Sens = 0.11
IF V8 - 4 AND V20 - 1 AND V32 - 3 THEN V33 - 2 Spec = 1 Sens = 0.11
IF V16 - 1 AND V24 - 2 AND V32 - 3 THEN V33 - 2 Spec = 1 Sens = 0.086
IF V16 - 2 AND V24 - 2 AND V32 - 2 THEN V33 - 2 Spec = 1 Sens = 0.086
IF V8 - 1 AND V24 - 2 AND V32 - 3 THEN V33 - 2 Spec = 1 Sens = 0.086
IF V8 - 1 AND V16 - 1 AND V32 - 4 THEN V33 - 2 Spec = 1 Sens = 0.057
IF V16 - 3 AND V24 - 4 AND V32 - 4 THEN V33 - 2 Spec = 1 Sens = 0.057
IF V8 - 2 AND V24 - 3 AND V32 - 4 THEN V33 - 2 Spec = 1 Sens = 0.057
IF V10 - 4 AND V24 - 4 AND V32 - 4 THEN V33 - 2 Spec = 1 Sens = 0.057
IF V16 - 1 AND V26 - 4 AND V32 - 4 THEN V33 - 2 Spec = 1 Sens = 0.057
IF V8 - 4 AND V24 - 1 AND V26 - 2 AND V32 - 3 THEN V33 - 2 Spec = 1 Sens = 0.057

HIGH rules

IF V16 - 2- 3 AND V24 - 1 AND V32 - 1 THEN V33 - 3 Spec = 1 Sens=0.38
IF V24 - 1 AND V26 - 2 AND V32 - 1 THEN V33 - 3 Spec = 1 Sens=0.34
IF V10 - 4 AND V24 - 1 AND V32 - 1 THEN V33 - 3 Spec = 1 Sens=0.28
IF V16 - 3 AND V24 - 1 AND V26 - 2 THEN V33 - 3 Spec = 1 Sens=0.24
IF V8 - 2 AND V24 - 1 AND V26 - 2 THEN V33 - 3 Spec = 1 Sens=0.17 1
IF V8 - 1 AND V16 - 1 AND V24 - 3 AND V32 - 3 THEN V33 - 3 Spec = 1 Sens = 0.069
IF V24 - 1 AND V32 - 2 THEN V33 - 3 Spec = 1 Sens = 0.069

With these rules we explored the universe of all possible outputs. The results after the classification were: the total amount of patterns that satisfied at least one rule were 10656. From this, 5512 were low, 4080 one medium, and 2432 one high. Crossed patterns were: one rule low and one high = 192. One rule low and one medium = 1112. One medium and one high = 88. Finally patterns that satisfy rules low, medium and high = 24. In order to validate the model, we implement a "blind review" process of the classified outputs. Around 60% of the outputs were classified and 14% perceived with the desired property. However, when reviewing the misclassified outputs it was clear that the rules were expressing a partial property of the pattern and that more information is needed. This could be improved by using another level of analysis that can be performed by inspection or by using an algorithm for extraction of meta rules (rules of rules). However, it is important to take into account that as the rules are more specific they tend to describe particular patterns, and so, even though the outputs are the desired ones, the search could be limited.

3 Recapitulation

Although the system is in its primary implementation, the approach sounds promising. The obtained representation classified around 60% of all the possible patterns (with A + B structure) produced by the original system. This is a good result if we consider that the system is labeling instances unlabeled before. Furthermore, the analysis of all possible patterns considered with the same probability, allowed us to find patterns classified as high, that otherwise would have been difficult to obtain with the original model (the smallest associated probability is 0.0001). Remember that the decision of considering all the patterns with the same probability was taken when exploring the less probable patterns with the original probability distribution and patterns considered as "highly" interesting were found. Early programmed versions for pattern generation have been tested in the context of live performances with good results. This implementations were designed by filtering out the low and medium outputs. Other versions oriented to "desk composition" can include other options for selecting the patterns according to rules, for example by choosing those rules supported by great amount of instances for the main parts, and less supported rules for short passages.

Acknowledgements We kindly thank the support of the Faculty of Sciences of the National Autonomous University of Mexico (UNAM).

References

1. Castro, F., Nebot, A., Mugica, F.: On the extraction of decision support rules from fuzzy predictive models. Appl. Soft Comput. **11**(4), 3463–3475 (2011)
2. Collins, N.: Algorithmic Composition Methods for Breakbeat Science ARiADA No. 3, May 2003

3. Collins, N.: Musical form and algorithmic composition. Contemp. Music Rev. **28**(1), 103–114 (2009)
4. Collins, N.: Automatic composition of electroacoustic art music utilizing machine listening. Comput. Music J. **36**(3), 8–23 (2012). Fall 2012 Massachusetts Institute of Technology
5. Duncan, W., Kirke, A., Miranda, E., Roesch, E.B., Slawomir, J., Nasuto, S.: Towards affective algorithmic composition. In: Luck, G., Brabant, O. (eds.) Proceedings of the 3rd International Conference on Music & Emotion (ICME3), Jyväskylä, Finland, 11th - 15th June 2013 (2013)
6. Fernández, J.: AI methods in algorithmic composition: A comprehensive survey. J. Artif. Intell. Res. **48**, 513–582 (2013)
7. Kosko, B.: Fuzzy cognitive maps. Man-Mach. Stud. **24**, 65–75 (1986)
8. Luke, S.: The Stochastic Synthesis of Iannis Xenakis. Leonardo Music Journal, vol. 19, p. 7784. Edited by Leonardo/The International Society for the Arts, Sciences and Technology, and published by The MIT Press (2009)
9. Nierhaus, G.: Algorithmic Composition. Paradigms of Automated Music Generation. Springer, Wien (2009)
10. Paz-Ortiz, I., Nebot, À., Mugica, F., Romero, E.: Using the Fuzzy Inductive Reasoning methodology to improve coherence in algorithmically produced musical beat patterns. In: Museros, L. (ed.) Artificial Intelligence Research and Development, vol. 289. IOS Press (2014)

The Musical Experience Between Measurement and Computation: From Symbolic Description to Morphodynamical Unfolding

Mark Reybrouck

Abstract Music and mathematics have a lot of common grounds. They both involve processes of thought, but where mathematics is concerned basically with symbols without any physical connection to the world, music has sound as its major category. Music, in this view, is characterized by a sonorous articulation over time, which can be described in physical terms. Yet, it is possible to conceive of music also at a virtual level of imagery and to carry out symbolic computations on mental replicas of the sounds. The major aim of this contribution, therefore, is to explore some basic insights from algebra, geometry and topology, which might be helpful for an operational description of the sounding music. Starting from a conception of music as a formal system, it argues for a broadening and redefinition of the concept of computation, in order to go beyond a mere syntactic conception of musical sense-making and a mere symbol-processing point of view.

1 Introduction

Music is sonic matter shaped in time. As such it calls forth two levels of description: the physical description of the sonorous unfolding and the level of sense-making by the listener. Many attempts have been made to bring together these two levels of explanation—ranging from philosophical studies of time to psychological contributions that are grounded in empirical research—but several dividing lines still hamper the development of a coherent approach. They can be summarized as the objective/subjective and discrete/continuous dichotomy. Both dichotomies, however, are not opposed to each other but are related to some extent in the sense that the flow of sensory impressions, which is continuous, is interrupted by the perceiver in order to make sense [16]. This phenomenological claim is quite important: it illustrates the transition from continuity to discretization and makes possible the allocation of discontinuous points in time—conceived as temporal windows—, which can be focal or extended in time, as elaborated in Husserl's philosophical discussion of the inner

M. Reybrouck (✉)
University of Leuven, Blijde Inkomststraat 21, PO Box 3313, 3000 Leuven, Belgium
e-mail: Mark.Reybrouck@kuleuven.be

© Springer International Publishing AG 2017
G. Pareyon et al. (eds.), *The Musical-Mathematical Mind*,
Computational Music Science, DOI 10.1007/978-3-319-47337-6_26

253

consciousness of time [10]. It provides an operational description of the sounding music in terms of now moments, extended now moments and relational networks that apply to actual now moments and their relationships. As such, there is a third dichotomy, to be taken also into account, namely the in time/outside of time distinction, which is closely related to the perceptual/computational dichotomy. The actual now moment, in fact, is time-bound and is characterized by perceptual immediacy with actual sounding stimuli being presented to the senses. The relational network, on the contrary, goes beyond perceptual immediacy by anticipating future elements and by recollecting previous ones in memory. As such, it is possible to transcend the inexorable character of time and to carry out symbolic computations on mental replicas of the sounds, which can proceed outside of time in a kind of virtual simultaneity, somewhat related to Saussure's distinction between syntagmatic continuity and associative relationships in language: the former is "in praesentia", relating two or more elements that are equally present in an effective series; the latter unites elements "in absentia" in a virtual mnemonic series [7, p.171].

2 Experience and Computation: Internal and External Semantics

Music, in its broadest definition, is a collection of vibrational events, which have the potential of being structured, either in an analog-continuous or discrete-symbolic way, somewhat analogous to the distinction between the bottom-up and the top-down approach to auditory processing. The former is continuous and proceeds in real time; the latter proceeds at a level of mental representation by applying discrete symbolic labels to the sounds, allowing a description either in experiential or computational terms depending on whether the processing is time-consuming (continuous-analog) or proceeding in a much more economic way by reducing temporal unfoldings to single representations with an all-or-none character which lean themselves to symbolic computations (discrete-symbolic).

Computations, further, take as a starting point a set of elements to operate upon with as basic idea formal symbol manipulation by axiomatic rules with a complete conceptual separation between the symbols and their physical embodiment. They are by definition implementation-independent, as exemplified most typically in computer programs, which handle discrete symbols and discrete steps in rewriting them to and from memory to sequences of rules in an axiomatic-deductive way. "Steps", according to Pattee, can be defined by a measurement process and "symbols" as records of a measurement. A programmable computation, then, can be described in physical terms as a formal dynamical system that is internally constrained to regularly perform a sequence of simple measurements that are recorded in memory. The time of measurement, moreover, has no coherence with the time of the dynamics, which means that the sequence of computational steps is rate-independent. Formal systems, in other words, must be free of all influence other than their internal syntax.

To have meaning, however, they must be informally interpreted, measured, grounded or selected from the outside, which involves a transition from rate-independent programmed computation to a rate-dependent dynamic analog with measurements proceeding in real time [13].

Dealing with music, accordingly, can be understood in terms of symbolic modeling and computation somewhat in line with the syntactization of semantics, which began in the 1930s with the "logical semantics" of Carnap [4] and the "model-theoretic semantics" of Tarski [19], and which is accomplished by completely encoding the world, so that symbols are seen in relation to a completely logical-symbolic structure, postulating merely sets of possible worlds and world-states without having to specify any sets of observables or having to verify any truth values with respect to the external world. If the symbols are without relations to the external world, they can be conceived in terms of internal semantics; if they establish a relation to the outer world, they should be explained in terms of external or real semantics [1].

Music, as a sounding art, cannot be described exhaustively in terms of internal semantics. Both measurement and computation are related to the musical experience but they differ with regard to the level of abstraction and distance vis-à-vis the sounding stimuli. The distinction is somewhat related to Langacker's division between "peripherally connected" and "autonomous cognitive events"—with the former referring to sensations which are directly induced by stimulation of the senses and the latter to corresponding images that are evoked in the absence of such stimulation [11, p.12]—and Jackendoff's distinction between "lower" or more "peripheral levels" and "higher" or more "central levels" of structure [28, p. XX]. The lower levels interface most directly with the physical world and highlight the interactional and experiential approach to musical sense-making; higher levels represent a greater degree of abstraction, integration and generalization with respect to sensory input and are dealing with the sounds at a symbolic level of functioning.

3 Measurement and Symbolic Play

The notion of measuring device was introduced by Hertz [9] who pointed out the possibility of linking particular symbol states to particular external states of affair. A measurement, in his view, is produced by measuring devices that interact with their environment and provide a pointer-reading or pointer-sign of an observable that functions as the initial condition of a formal model for predicting the value of a second one. It reflects the particular interactions of the measuring apparatus with the external world, thus providing the initial conditions to carry out predictive arithmetic and/or logical calculations on the pointer-signs, which are completely rule-governed and syntactic in character [2, 3]. As such, the role of symbolic play must be considered here, as formal computation is carried out on the symbolic counterparts of the observables, and not on the observables themselves. It is a major component of symbolic functioning, which has a theoretical elaboration in the concept of "internal model" of the outer world and which reminds us of the older concept of epistemic rule system with its epistemic generalizations of *homo sapiens*, *homo faber* and *homo ludens*, each of which can be considered in terms of automata: homo sapiens as a "perception machine" (selection and classification), homo faber as an "effector machine" and homo ludens as a "playing automaton" [29]. The latter, in particular, is of paramount importance for symbolic functioning at large. It calls forth the introduction of intermediate variables between perception and action [31, 34] and raises the functioning of the rule system to a level that transcends the reactivity of causal stimulus-reaction chains. Symbolic play, for short, makes it fitted for goal-directed behavior that involves deliberate planning and mental simulation at the level of imagery, which is typically the hallmark of the homo ludens as a playing automaton. It stresses the possibility to perform internal dialogues and to carry out symbolic computations on mental replicas of observables. In order to do so, however, the "player" must have at his/her disposal a symbolic repertoire for doing the mental arithmetic.

Computations, thus, are considered mainly from a symbol-processing point of view. There is, however, a broader conception of computation, which can be handled in terms of modeling or predictive computations [20]. Computation, in this view, embraces the whole range of mental operations that can be performed on symbolic representations of the sounds. They are exemplified in elementary mathematical activities, which can be reduced to the logico-mathematical operations of classifying, seriating, putting in correspondence and combining, which were defined already by Piaget [33] as abstractions of concrete operations such as collecting, ordering and putting things together, all of which can be subsumed under the concept of symbolic play.

4 Music as an Algebraic Structure: The Concept of Musical Space

The computational approach to music offers the possibility to carry out symbolic computations on mental replicas of the sounds. It takes as a starting point a set of elements upon which to operate, and which can be labeled symbolically as discrete things with unit character. These elements can be of any length, ranging from discrete focal points to larger temporal events, which can have a continuous representation as well. It is useful, therefore, to conceive of them not only as discrete symbolic units but as functions of time, combining the quantal aspect of discrete labeling with the continuous aspect of temporal extension [35]. Their delimitation can be described in algebraic terms, conceiving of them as variables, taking as a starting point a "domain" that represents the sonic universe, which can be defined as a virtual infinity of possible combinations of individual vibrational events [22, 36]. These sounding elements, further, can be described in terms of mathematical sets as collections of elements, without any constraints as to their character and number and without any order or relation being defined on them. As such, it is possible to transcend the rather limited scope of existing musical systems, with a reduction to the delimitation of elements in terms of structural variables (sets of discrete pitches in the frequency domain, discrete durations in the temporal domain, a limited set of intensity levels and timbres) in favor of a multidimensional approach that allows a quantitative description of sound characteristics (amplitude envelope, amplitude and phase spectrum, spectral envelope, harmonicity, formants, noise level, etc., see [39]) that raises them to the level of time-varying acoustic events.

As such, it is possible to provide an operational description of sound as exemplified in the spectrographic description of sound as a function of time [6, 22, 23]. The pioneering work of Schaeffer [28] on musical objects is still important here. By defining 33 criteria of sonorous characterization—with 19 of them referring to the dynamic aspect, 9 referring to timbre and only 5 referring to pitch—, he argued that the articulation of sound cannot be reduced to spectral and durational aspects, but that it has to be supplied with a dynamic description as well. This can be represented graphically as a sonogram or spectrogram, i.e. a system of coordinates that combines the spectral, dynamic and melodic features of the sound.

The broader concept of sonic universe is very fruitful here. As a generic category, it encompasses all kinds of subuniverses, such as music in general or its basic building blocks, allowing to conceive of musical elements and configurations as subsets of a more encompassing sonic universe, which can be described in algebraic and geometric terms. It is possible, in fact, to reduce the sonic universe to its arithmetical substrate, and to conceive of it as a musical space, consisting of a set of points, each of them corresponding to a number. Spaces, indeed, are networks within which points can be fixed by giving them some numbers, called coordinates, allowing a matching of a geometrical space and a corresponding number space. The starting point for an objective description, therefore, is the sounding music and its numerical encoding. Musical figures can be delimited in this space and may be considered as configurations of points, which can move from one configuration to another. The geometric space that figures as a framework for these transformations has to take account of this and must be chosen according to some criteria (e.g. every possible point must have an allocation in the space, and every transition from one configuration to another must be possible). This calls forth a dynamic conception of geometric space, as exemplified in Leibniz' conception of space as a method of knowledge [21, p. 270]. In this conception, space is not enclosed in itself, but is considered a relative concept with three major moments, namely multiplicity, continuity and coexistence. This dynamic definition of space has been very influential for the conception of geometrical space as a collection of points. Especially the concept of continuous space can be formulated in an elegant way by introducing sets of points.

Musical space, accordingly, can be conceived as a collection of elements to be described formally as an algebraic structure, i.e. a non-empty structured set together with a collection of (at least one) operations and relations on this set. As such, it calls forth both set theoretical and algebraic methodology. The central problem, however, is the definition of the elements, since musical space and time have to be integrated in the definition, together with set theoretical, geometrical and algebraic points of view.

5   Musical Space as Topological Space

Musical space can be defined as a collection of points that constitute the domain (the arguments) upon which predication processes can be applied. The results of these processes are propositions that assign some general term to individuals. Predication, however, does not apply to points without extension, but to units that are recognizable as such. At a formal level these units are systems of isolated points in one or more dimensions, somewhat comparable to the point-events of physics which are determined by three space coordinates and one time coordinate [15, p.183]. Each point of space has a "world-line" that corresponds to the flow of time at a particular moment or during an interval of time. As such, it must be possible to construct a mathematical model for the description of the physical domain (the sonorous universe) from which the units may be recruited, and to give a numerical description of them. This is possible, by conceiving of the space in which collections of points can be delimited as a metrical space (S, d), namely as a set S on which a metric or distance function (d) is defined with the real number d (a, b) being called the distance from a to b [12]. It allows an expansion from a metrical to a topological space, conceiving of geometrical figures as collections of points that can be subject to transformations. Set-point topology, in fact, is concerned with notations of continuity and relative position, regarding geometrical figures as collections of points with the entire collection often considered a space.

Applied to music, this means that every musical structure can be defined by selecting sets of points. Musical configurations, in this view, are point-sets that can be transformed into other configurations, and this in a gradual or rather abrupt way. In applying such transformations to sets of points, further, the configurations mostly are left invariant with respect to at least some properties, which are called topological invariants. The sets, however, must be structured, allowing the mapping of each element of set A onto set B, with elements of A being the domain and elements of B being the co-domain. Most interesting, further, are operation preserving mappings, that preserve the structure of the algebraic system, as is the case with mappings or homomorphisms that generate a transformed image of the original structure (the domain) in the image set (the range) and provide a numerical basis for identification and transformation algorithms [18]. A special case of images, furthermore, are those images that are generated within the same topological space and which can be conceived as functions (f: A→ B or f(x) = y). There is, however, one problem, which is related to the special position of music as a temporal art: the arguments of the function are themselves functions of time, and transformations of musical figures are to be conceived as transformations of functions. Musical space, therefore, is essentially a virtual space, becoming a function space when sonorous articulation is going on. The unfolding of music, then, can be described at two levels: the sonorous articulation as a dependent variable (function) of the time as the independent variable, with the points of the time continuum being the arguments and the points of the spectral configuration of the sound being the images.

6   From Static Description to Morphodynamical Unfolding

Music can be considered as a collection of sound/time phenomena, with the structure being defined at several temporal scales, ranging from the processing of focal now moments of a few milliseconds, over temporal units or events in the range of 2–3 s to large-scale temporal spans [14, 42, 43]. The pioneering work of Schaeffer [17] is still important here. Elaborating on the concept of sonorous object, which he defined as an intentional unit, constituted in our consciousness by our own mental activity and being interpreted as sound-in-itself ([37, p. 263], see also [26]), he drew a distinction between three levels of description: the large-scale context in which the sonorous object may be included, the level of the sonorous object and the internal substance of the sonorous object, zooming in and out in order to focus on the continuous acoustic substrate, which is divisible down to the size of a single point in time as well as on the overall level of continuity [37, p. 503]. It was one of his central claims that all sounding events can be defined and classified as sonorous objects, which led him to the elaboration of a morphology and typology of sonorous objects on the basis of the criteria of "sustaining" and "articulation" of the energy.

As such, we can conceive of music in "morphological" terms, which genuinely combines a discrete with an analogous description of the sound. The sonogram has proved to be useful for meeting some of these requirements [27]. The spectrographic description of the sound as a function of time, however, is likely to be more interesting [6] as exemplified in Cogan's conception of spectral morphology [22, 23]—where sonic morphologies may resemble one another, may be transformations of one another and may oppose one another—and in related elaborations of spectromorphology [40] and acousmatic morphology [24]. Such a morphological way of thinking is challenging. It provides a description of typical patterns of temporal unfolding as well as a description of their sounding articulation, and has received impetus from other areas of research, to mention only the morphological and morphodynamical procedures for delimiting morphological lexicons as proposed by Petitot [32] and Thom [41]. Such lexicons consists basically of elements which are defined as being dependent upon an interplay between stability and instability in order to deliver the fundamental perceptual effects of invariance and discretization (for musical applications, see [25, 30]). Once the elements are delimited, it is then possible to carry out syntactic operations on them in an attempt of knitting them together in a relational network and to consider the transition from discrete particulars to an organized piece. The idea has been advocated already by Schoenberg [38] who defined musical form as being constituted by elements that operate like a living organism. The essential and necessary requirements to the creation of an understandable form are "logic" and "coherence" and these are based on "internal connectedness".

The organization of the "macrostructure", however, proceeds mostly out-of-time. It entails computational work, which is basically syntactic in nature, and wich reduces the sounding music to its formal-symbolic counterparts. There is, however, another approach, which is more fruitful, especially with regard to the morphodynamical unfolding of the sonorous articulation through time, and wich is related to the concept of continuous knowledge representation. To quote Desain and Honing [8]:

"In general it appears that representations of a continuous nature can improve the flexibility of representational systems considerably. They sometimes yield a level of performance that is not obtained by their discrete counterparts. Continuity has been underrated for too long now, both from a technical viewpoint —in many cases considering a discrete representation a harmless simplification—, and from musicological and psychological perspectives which, more or less, overstressed the importance of discrete categories."(pp. 15–16).

References

1. Cariani, P.: Some epistemological implications of devices which construct their own sensors and effectors. In: Varela, F., Bourgine, P. (eds.) Towards a Practice of Autonomous Systems, Proceedings of the First European Workshop on Artificial Life, pp. 484–493. MIT Press, Cambridge, MA (1991)
2. Cariani, P.: Symbols and dynamics in the brain. Biosystems **60**(1–3), 59–83 (2001)
3. Cariani, P.: Cybernetic systems and the semiotics of translation. In: Petrilli, S. (ed.) Translation, pp. 349–367. Rodopi, Amsterdam (2003)
4. Carnap, R.: Logische Syntax der Sprache. Springer, Wien (1934)
5. Carnap, R.: On the character of philosophic problems. Philos. Sci. **1**(1), 5–19 (1934)
6. Dannenberg, R.B., Desain, P., Honing, H.: Programming language design for music. In: De Poli, G., Picialli, A., Pope, S., Roads, C. (eds.) *Musical Signal Processing*, pp. 271–315. Swets & Zeitlinger, Lisse (1997)
7. de Saussure, F. (1969[1916]). Cours de linguistique générale. Paris: Payot
8. Desain, R., Honing, H.: Music, Mind and Machine. Studies in Computer Music, Music Cognition and Artificial Intelligence. Amsterdam, Thesis Publishers (1992)
9. Hertz, H.: Principles of Mechanics. Dover, New York (1956 [1894])
10. Husserl, E.: The phenomenology of internal time consciousness. In: Heidegger, M., trans. Churchill, J.S. (eds.). Indiana University Press, Bloomington, IN (1964)
11. Langacker, R.: Foundations of Cognitive Grammar, vol. 1. Stanford University Press, Stanford (1987)
12. Lipschutz, S.: General Topology. McGraw-Hill, New York (1965)
13. Pattee, H.: Artificial life needs a real epistemology. In: Moran, F., Moreno, A., Merelo, J., Chacon, P. (eds.) Advances in Artificial Life, pp. 23–38. Berlin, Springer (1995)
14. Pöppel, E.: A hierarchical model of time perception. Trends Cogn. Sci. **1**(2), 56–61 (1997)
15. Reichenbach, H.: The Philosophy of Space and Time, pp. 144–145. Dover Publications, New York (1958)
16. Ricoeur, P.: Hermeneutics and the Human Sciences. Cambridge University Press/Editions de la Maison des Sciences de l'Homme, Cambridge (Paris) (1981)
17. Schaeffer, P.: A la recherche d'une musique concrète. Seuil, Paris (1952)
18. Smith, D., Eggen, M., & St. Andre, R.: A Transition to Advanced Mathematics. Belmont, Wadsworth (1986)
19. Tarski, A. (1933). The Concept of Truth in Formalized Languages. In Tarski, A. (1983). Logic, Semantics, Metamathematics. Papers from 1923 to 1938. (Trad. J. H. Woodger. Ed. J. Corcoran) (pp. 152–278). Indianapolis: Hackett Publishing Company
20. Bel, B., Vecchione, B.: Computational musicology. Comput. Humanit. **27**, 1 (1993)
21. Cassirer, E.: Leibniz' System in seinen wissenschaftlichen Grundlagen. Georgs Olms Verlachbuchhand, Hildesheim (1962)
22. Cogan, R.: New Images os Musical Sound. Harvard University Press, Cambridge (1984)
23. Cogan, R., Escott, P.: Sonic Design: The Nature of Sound and Music. Prentice-Hall, Englewood Cliffs (1976)

24. Desantos, S.: Acousmatic morphology: an interview with François Bayle. Comput. Music J. **21**(3), 11–19 (1997)
25. Dufourt, H.: Musique et psychologie cognitive: les éléments porteurs de forme. In: McAdams, S., Deliège, I. (eds.) La musique et les sciences cognitives, pp. 327–334. Pierre Mardaga, Liège - Bruxelles (1989)
26. Godøy, R.I.: Gestural-sonorous objects: embodied extensions of Schaeffer's conceptual apparatus. Organ. Sound **11**, 149–157 (2006)
27. Helmuth, M.: Multidimensional representation of electroacoustic music. J. New Music Res. **25**, 77–103 (1996)
28. Jackendoff, R.: Consciousness and the Computational Mind. MIT Press, Cambridge (1987)
29. Klaus, G.: Kybernetik und Erkenntnistheorie. Deutscher Verlag der Wissenschaften, Berlin (1972)
30. McAdams, S.: Contraintes psychologiques sur les dimensions porteuses de la forme. In: McAdams, S., Deliège, I. (eds.) La musique et les sciences cognitives, pp. 257–280. Pierre Mardaga, Liège - Bruxelles (1989)
31. Paillard, J.: L'intégration sensori-motrice et idéo-motrice. In: Richelle, M., Requin, J., Robert, M. (eds.), Traité de psychologie expérimentale. vol. 1, pp. 925–961. Presses Universitaires de France, Paris (1994)
32. Petitot, J.: Perception, cognition et objectivité morphologique. In: McAdams, S., La Deliège, I. (eds.) Musique et les sciences cognitives, pp. 242–256. Pierre Mardaga, Liège - Bruxelles (1989)
33. Piaget, J.: Biologie et connaissance. Essai sur les relations entre les régulations organiques et les processus cognitifs. Gallimard, Paris (1967)
34. Reybrouck, M.: Musical imagery between sensory processing and ideomotor simulation. In: Godøy, R.I., Jørgensen, H. (eds.) Music. Imag., pp. 117–136. Swets & Zeitlinger, Lisse (2001)
35. Reybrouck, M.: Music cognition, semiotics and the experience of time. ontosemantical and epistemological claims. J. New Music Res. **33**(4), 411–428 (2004)
36. Reybrouck, M.: Music as environment: an ecological and biosemiotic approach. Behav. Sci. **5**, 1–26 (2015)
37. Schaeffer, P.: Traité des objets musicaux. Editions du Seuil, Paris (1966)
38. Schoenberg, A.: In: Strang, G. (ed.) Fundamentals of Musical Composition. Faber and Faber, London (1970)
39. Serra, M.-H.: Introducing the phase vocoder. In: Roads, C., Pope, S., Piccialli, A., de Poli, G. (eds.) Musical Signal Processing, pp. 31–90. Swets & Zeitlinger, Lisse (1997)
40. Smalley, D.: Spectromorphology: explaining sound-shapes. Organ. Sound **2**(2), 107–126 (1997)
41. Thom, R.: Modèles mathématiques de la Morphogenèse. Bourgois, Paris (1980)
42. Wittmann, M.: Time perception and temporal processing levels of the brain. Chronobiol. Int. **16**, 17–32 (1999)
43. Wittmann, M., Pöppel, E.: Temporal mechanisms of the brain as fundamentals of communication – with special reference to music perception and performance. Musicae Scientiae, Special Issue, pp. 13–28 (1999–2000)

Generic Additive Synthesis. Hints from the Early Foundational Crisis in Mathematics for Experiments in Sound Ontology

Julian Rohrhuber and Juan Sebastián Lach Lau

Abstract Motivated by an investigation of the historical roots of set theory in analysis, this paper proposes a generalisation of existing spectral synthesis methods, complemented by the idea of an experimental algorithmic composition. The background is the following argument: already since 19th century sound research, the idea of a frequency spectrum has been constitutive for the ontology of sound. Despite many alternatives, the cosine function thus still serves as a preferred basis of analysis and synthesis. This possibility has shaped what is taken as the most immediate and self-evident attributes of sound, be it in the form of sense-data and their temporal synthesis or the aesthetic compositional possibilities of algorithmic sound synthesis. Against this background, our article considers the early phase of the foundational crisis in mathematics (*Krise der Anschauung*), where the concept of continuity began to lose its self-evidence. This permits us to reread the historical link between the Fourier decomposition of an arbitrary function and Cantor's early work on set theory as a possibility to open up the limiting dichotomy between time and frequency attributes. With reference to Alain Badiou's ontological understanding of the praxis of axiomatics and genericity, we propose to take the search for a specific sonic situation as an experimental search for its conditions or inner logic, here in the form of a decompositional basis function without postulated properties. In particular, this search cannot be reduced to the task of finding the right parameters of a given formal frame. Instead, the formalisation process itself becomes a necessary part of its dialectics that unfolds at the interstices between conceptual and perceptual, synthetic and analytic moments, a praxis that we call *musique axiomatique*. Generalising the simple schema of additive synthesis, we contribute an algorithmic method for experimentally opening up the question of what an attribute of sound might be, in a way that hopefully is inspiring to mathematicians, composers, and philosophers alike.

J. Rohrhuber (✉)
Institute for Music and Media, Robert Schumann Hochschule, Georg-Glock-Str. 15, 40474 Duesseldorf, Germany
e-mail: julian.rohrhuber@musikundmedien.net

J.S. Lach Lau
Conservatorio de Las Rosas, Morelia, Mexico
e-mail: lachjs@gmail.com

© Springer International Publishing AG 2017
G. Pareyon et al. (eds.), *The Musical-Mathematical Mind*,
Computational Music Science, DOI 10.1007/978-3-319-47337-6_27

1 Spectres of Accumulation

Adding up two numbers, adding up many numbers – this appears to be a most unquestionable and intuitive activity. Cutting a number in two, grouping its parts, is no less self-evident. The concept of natural number itself suggests a definite idea of accumulation, and thereby serves as a blueprint for other domains. Be that as it may – anyone who has ever worked with sound knows that understanding what happens in adding up and mixing, separating and analysing, is really far from trivial. The addition of one element may cancel out another, one part may interfere with, or may recontextualise others, become indistinguishable or irrecognisable – or may, for no apparent reason, suddenly turn out as entirely separable, untouched from the whole it coexists with. This is finally the reason why harmonic and rhythmic relationships have never ceased to provide an interesting and endless topic for investigation. It could also be part of the reason why it is so difficult to specify sound.

The simplest correlate in the realm of elementary arithmetic (we could call it Pythagorean) is the fact that addition entails multiplication: in general it is an undecidable question, for example, whether an unknown even number will result in a prime when adding one to it. In other words, the properties of a sum are non-trivial, and they are so already in the truly elementary case. For the inventory of mathematical entities that has ever grown and shifted over its history, such as infinitesimals, functions, sets, groups, categories, it doesn't become much easier and, even more, the very notion of addition becomes a matter that needs, dependent on the subject matter, a separate justification.

Adding up, taken in full generality, does indeed entail both mathematical and philosophical challenges. In particular, in absence of an immediately given continuous grounding, the consequences of "making the next step" may be unforeseeable in the most general sense. Alain Badiou notes:

> To understand and endure the test of the additional step, such is the true necessity of time. [...] There is nothing more to think in the limit than in that which precedes it. But in the successor there is a crossing. The audacity of thought is not to repeat 'to the limit' that which is already entirely retained within the situation which the limit limits; the audacity of thought consists in crossing a space where nothing is given. We must learn once more how to succeed. [1, 81f.]

2 The "Birth Place of Set Theory" and Its Potential Relevance to the Ontology of Sound

Spectrum and *multiplicity* are historically related concepts. The 19th and early 20th century attempts to gain a better understanding of the concept of function is one of the most telling in this respect. Let's briefly recapitulate[1]:

---

[1] In the historical description, we largely follow [7], as well as Cantor's collected papers [4].

The idea of 'being the function of something', of a linear continuum in particular, was at the source of the concept of function, which entailed ideas of dependent change (derivatives) and cumulative volume (integrals) that made it possible to ask questions about the specific properties and laws of functions. Thereby, the infinite series, the possibility to understand a function as a sum of other functions, became one of the most indispensable as well as problematic devices in the then emerging branch of mathematics, real analysis. But this idea of a 'spectrum' of a function also led to the radical rethinking of its domain, the continuum.

From a very general point of view, one can say that the idea of prismatic composition/decomposition exposes the possibility of looking at one and the same thing from different perspectives. Its effectiveness lies in the fact that some perspectives reveal properties that could never have been understood from any other. This is also what explains the ontological gravity of the spectrum: if its partials are mere *devices* to approach the whole of an intuitively continuous shape, what does it mean if, for some points, their sum does not converge to a single number? Or if, for some spectra, a rearrangement of their terms leads to a different result? Has one chosen the wrong 'alphabet' to form the 'words' of a given relation?

The decomposition of a function into trigonometric functions had its beginnings in the problem of understanding the movement of a plucked string, and because of the potential of the Fourier Series for calculating the 'image' of any function whatsoever, over the 19th century, harmonic analysis became a paradigmatic medium for the understanding of functions. Of course not only of mathematical functions in general, but also of sound. Even if the qualities of sound may escape immediate understanding, once the partial is assumed to be intuitive and self-evident, should it not be possible to finally access the totality of all possible sounds in one spectral world image? Should not the knowledge of the principal dimensions of sound allow access to every one of its instances?

Even though additive synthesis and harmonic analysis sufficiently approach completeness in many cases and can thus be helpful indeed, the harmonic spectrum is by far not as productive as one may think in solving and of finding and understanding unknown sounds. As it turns out, the difficulty remains in the interrelation between the coefficients and of finding the law that describes them best. The case of transients (or discontinuities) illustrates that the sum of partials does not converge well to some wave forms, and that the rules according to which it does are not helpful for understanding and, by implication, for finding interesting variations.

Over the course of the 19th century, establishing alternative and operatively adequate perspectives on the properties of functions, the non-trivial domain of partials and the limits of their series stabilised a process that slowly eroded the intuitive geometric image of a function. In the face of so called "monstrous" functions (today they are less dramatically called "pathological"), many obvious concepts had to be reviewed, an important one of them being the hitherto rather unsuspicious identity between the continuous and the differentiable. Essentially, the early "crisis of intuition"[2] was an ontological one: should those monsters be admitted as properly

---

[2]The "crisis" of intuition was called the "Krise der Anschauung" in the German discourse [17].

existing, even though they contradicted the most basic spatial intuitions and could not be clearly visualised? Inspired by his senior colleague in Halle, Eduard Heine, from 1869 onward, Georg Cantor endeavoured to extend the possibility of representing (and thus making sense of) arbitrary functions in terms of infinite sums of trigonometric functions. He succeeded in showing that the series is unique (and its coefficients thus irreplaceable by another set of parameters) even for functions for which infinitely many points fail to converge to a single number. Sums of harmonic oscillations can indeed represent extremely discontinuous functions.

In subsequent years, the mathematical devices that Cantor developed in the course of these proofs were to become the impulse for his development of transfinite numbers, and were to motivate his conception of actual infinite and transfinite sets: accepting infinite series of rational numbers as properly existing entities (rather than mere approximations), allowed him to convey access to the extremely rich, but also disputed, structure of the continuum. In such a way, what is now called the Fourier Series is the entry point into modern set theory. Ernst Zermelo, who in 1932 edited the collection of Cantor's papers, writes:

> In the concept of "higher order derivations" of a point set, we thus should behold the proper nucleus, and in the theory of trigonometric series the birth place, of the Cantorian "set theory" (p. 102).

Considering the significance that set theory and harmonic analysis has for each of these two fields respectively, making sense of this transitory moment should be of interest for those who work at the intersection between mathematics and sound. So, how do we understand this fact from the perspective of sound? As Alain Badiou has emphasised, Cantor's affirmation of the transfinite is an essential step in the history of ontology, because it departs from the idea of the "unity of being as such"—the continuum, rather than being a lawless or tensionless matter that serves as a medium of inscription for the arbitrary cuts enacted by thought, turns out to be a cloven, abstracted and non-unifiable landscape of structures. The idea is not, however, a total rule of orderless noise over each local part that renders it unintelligible. Monsters, even though counter intuitive, always constitute some new laws.

An aspect, or property, that cannot be described with the given means, a subset that is therefore indiscernible from its background with the means given in this horizon, has been called "generic". According to the reconceptualisation of Paul Cohen's notion by Badiou, the generic set is

> neither a known or recognized multiple, nor an ineffable singularity, but that which detains in its multiple-being all the common traits of the collective in question: in this sense, it is the truth of the collective's being. [2, p.17]

It is in this sense that mathematical monsters are *generic*: they have no proper place in the given order, so, if one chooses to accept their existence nevertheless, they make it necessary to find a new analytical apparatus instead of relying on the *generality* of the existing one. Such a process cannot proceed from a full understanding, a transparent intuition of a space for a free unfolding of self-evident laws. Finding an appropriate description, conversely, requires an incomplete process of experimentation and conjecture, which in the following we shall call *partial understanding*.

We are now in the position to ask: how can we find new laws of sound and how can we enter a process of partially understanding their consequences? As one possible step in this direction, we propose a generalised, or better *generic*, form of additive synthesis that is inspired by the so far discussed "birth place" of set theory.

3 The Epistemic Value of Base Functions

In general, Fourier's most celebrated contribution is widely applicable because it provided a method (the Fourier Transform) to calculate coefficients for each partial that works in many cases. It also serves as an intuitive model of breaking a complex signal into more accessible parts. The cosine function (or its equivalents), parameterised in phase and amplitude, effectively is a *coordinate system* that gives access to every point in the space of possible (and thus arbitrary) functions.

Since its discovery, many other functions that serve as 'equally general' basis functions for linear combinations have been found, e.g. the *Chebychev polynomials* (1854) and *Spectral modeling* [16]. Perhaps most influential today is the application of the uncertainty principle from quantum physics to sound by Dennis Gabor, with its information theoretical approach, that explicates possible trade offs between frequency and time representation [6, 13, 14] in the form of acoustical quanta. Among others, Gabor's ideas inspired *wavelet analysis* [10], which uses distributions of suitably windowed and translated partials in order to render the decomposition more adequate to certain sound qualities. Even in the ideal lossless case, however, each method still may convey or obscure given properties. As the authors of yet another decomposition, namely *Chirplet Transform* (introduced for radar image processing) argue, that

> [e]ach of the chirplets essentially models the underlying physics of motion of a floating object. Because it so closely captures the essence of the physical phenomena, the transform is near optimal for the problem of detecting floating objects.[3]

A decompositional basis is an observational *paradigm*: the choice of a coordinate system determines how an object can be understood, and the very coordinates themselves constitute which properties, or aspects, become apparent and what kind of transformations are thinkable. By consequence, despite the universality of the Fourier series, its partials may be more or less well suited to construct or understand a given waveform, its decomposition being more or less able to convey its hidden inner logic. Hence, a more general perspective on the idea of a 'spectrum' may be practically helpful, and ontologically necessary.

---

[3]They continue with acoustic examples: "Besides applying it to our radar image processing interests, we also found the transform provided a very good analysis of actual sampled sounds, such as bird chirps and police sirens, which have a chirplike nonstationarity, as well as Doppler sounds from people entering a room, and from swimmers amid sea clutter" [11].

4   Generic Additive Synthesis

Most transforms mentioned so far have inspired specific methods in sound synthesis. Chebychev polynomials, for example, are typically used in waveshaping [9], wavelet-like sound functions in granular synthesis and microsound [5, 13]. The difference between analysis and synthesis is *gradual*: just as each method has a distinct sound character, it equivalently has its own domain of sonic investigation.

The general method underlying all of the above is the *additive in a broader sense*: fixing a number of simple functions (pseudo-partials) that can be transformed in some systematic way, and then combining them together (by pseudo-addition). So one starts with a list of functions that obey a common law, then combines them in some systematic way, usually regulating by coefficients 'how much' a certain partial contributes to the whole.

Expressed as a function of time, such generic additive synthesis can be written in terms of partials g and combinator G:

$$f(t) = \overset{n}{\underset{i=1}{G}} g_i(t, c_i) \tag{1}$$

where g_i represents a *partial* (each different, depending on i). Every partial is a function of time t, and takes a coefficient c_i, conveniently in a way that only if $c_i \neq 0$, the partial contributes.[4] Finally, G is the *combinator*, a generalised map[5] that joins n partials into $\mathbb{R}^m$, in a way that entirely depends on the method chosen.

The basic schema at work here is an *interweaving of two perspectives*: the partial function describes the 'horizontal' dependence on time t (e.g. the shape of a harmonic oscillation), as well as the 'vertical' dependence on the partial number i (e.g. the frequency). By consequence, only a minor shift is needed for both c_i and g_i to be undestood as a function of i ('vertical order') and t ('horizontal order'). Thus it is sometimes adequate to treat the generic spectrum as factored into a new basis function $g_i^\times(t)$:

$$f(t) = \overset{n}{\underset{i=1}{G}} g_i^\times(t) \tag{2}$$

Apart from its temporal evolution, each of the n partials is determined by its place i in the spectrum, and $g^\times$ thereby is the name of the crossing point between the

---

[4]In the general case, these partials need not be linearly independent, and the coefficient need not be unique for a given resulting function. It is convenient, however, if we know a coefficient that cancels the contribution of the respective partial (typically zero). This means that depending on the combinator G, we need different scaling functions for each partial. With an explicit generalised scaling function, and a neutral element e with regard to G (usually, the neutral element, i.e. zero for addition and one for multiplication), we can write: $f(t) = \overset{n}{\underset{i=1}{G}} c_i g_i(t, 1) + (1 - c_i)e$

[5]In all 'conventional' series, the combinator is just the iterated addition. $G = g_1(t, c_1) + g_2(t, c_2) + \ldots g_n(t, c_n)$, or conveniently $\sum_{i=1}^n g_i(t, c_n)$, where usually $g_i(t, c_i) = c_i g_i(t)$. In the general form, however, a combinator is thought of as any interpretation of '+', thus any form of 'one more'.

specifications of partial and spectrum. Rather than a fixed space and a variable set of coordinates, both are here on the same level, and may equally be subject to variation.[6]

Factoring the other way round, the combinator and the partials can be seen as a single function that takes a sequence of coefficients, a *generic spectrum*:

$$f(t) = G^\times(t, s) \tag{3}$$

As it is the case with conventional additive synthesis, each instance of a generic spectrum s is an ordered tuple of coefficients $\langle c_1, c_2 \ldots c_n \rangle$. We shall come back to this formulation later.

Before we discuss some consequences, a note on terminology. In related methods like additive synthesis, the basis function is assumed to be known—it is the 'type' of the dimensions of the space, and, for a given function, it is really the coefficients that are unknown. Movement is understood as a movement through a fixed space. In this narrow sense, a distribution can be taken as *general* in so far as it completely and uniquely represents any arbitrary function given in another well-defined domain— the main task is to find the right coefficients. In the broader sense of decomposition proposed here, however, there is no given basis function with reliable properties, and thus no 'type' given in advance.[7] Instead of being general, it is *generic*.[8]

The task is now to show how the two schemata of generic additivity become productive under the specific conditions of algorithmic sound synthesis.

5   Musique Axiomatique

It is well known that the immediacy of the visualisation of a wave form or a spectrum is misleading: sound can be very difficult to specify. That is, the relation between some formal or causal description of a sound and its aesthetic or even physical consequences is non-trivial.

In such cases, the classical method is to make a clear divide between what is given in advance (e.g. the instrument or synthesis method) and what is subject to variation (e.g. the score or parameters). The instrumentation is then, first of all, the choice

---

[6] Operations on the spectrum will in this case be operations on the mapping $i \to c_i$. Because both coefficient and partial are dependent on the same i, the two terms (1) and (2) can be used exchangingly.

[7] This general schema does not lead to any method to calculate the coefficients for a given case and neither does it guarantee that it is orthogonal, unique, and linearly independent. But as we shall see more clearly in the next section, these properties need not be secured in advance where no type can be given anyhow.

[8] We are aware that the term *generic* may lead to misunderstandings, in particular due to the existing terminology in topology. We use the term to mark a distance from the idea of 'generalisation', following Alain Badiou's and Paul Cohen's concept of a *generic set*, as briefly explained in the last part of section 2. We have to leave open to what degree the precise ramifications of this concept remain adequate to its origin.

of a suitable relation between those two parts, the given and the unknown. Or, in the context of our present discussion, one can say that it is the search for a relation between a given basis function and an unknown set of coefficients.

The above mentioned foundational discourses in 19th and early 20th century mathematics not only brought about the discovery of new subject matters, but also affected the relation between known and unknown: while in the classical understanding, the *axiom* was to be understood as that which is self-evident and indubitably true, it increasingly became that of a posit, a starting point, even a counterintuitive precondition necessary for a certain fabric of investigation. Questioning the self-evidence of the continuum was one of them. Since then, axiomatic thought has become a back and forth movement between conditions and consequences rather than simply a construction from first premises.

In such a movement, formal languages have attained the role of a medium, pretty much like that of measurement instruments in a laboratory. And while today algorithmic proof systems slowly enter mathematical reasoning, high level programming languages are already a well established medium for sound synthesis and algorithmic composition. Having a common language for instrument and score has decisively blurred their distinction. Being able to modify code at runtime (*interactive programming* or *live coding*) further allows us to reconsider the temporal distinction between precondition and consequences. Therefore we are well equipped to embark on an experimental praxis of modern axiomatics that neither denies the sensual and situational qualities of sound nor the possibility of its mathematical and algorithmic formalisation—a praxis which we like to call *musique axiomatique*. Here, there is no need to keep the order between first devising a fixed synthesis method and then looking for the appropriate parameters. Rather, it becomes the very principle for interweaving algorithms that unfold in time and algorithms that specify their mutual relations, so that the path to finding a new sound moves back and forth between the rewriting of the one or the other.

6   Experiments in Partial Understanding

The existing and widespread decompositions—what Mazzola has called "omnibus-decompositions" [12, 899]—obey constraints that are necessary to address specific domains. These domains are inhabited by certain properties, in particular the complementary pair of frequency spectrum and points in time. This is why the laws of such decompositions can be seen as epistemological consequences of the ontological structure of the sound that they investigate. If we want to investigate other domains, by experimenting and reasoning, we may find other decompositions which are adequate to them, in particular implying properties that do not have to be 'located' in time and frequency as with the others. Axiomatics in the modern sense is, as we have seen, not the positing of self-evident properties; here, it means the *search* for a basis function and its logic of combinations.

So how to start such an investigation, how to set up a generic additive synthesis experiment? How to 'proceed'? Here we can only mention a few elements that serve as one of many possible starting points, keeping in mind that the aim is to develop a partial understanding—in the double sense of the word—of the procedures involved. In the experiments so far, we have worked with the SuperCollider programming language, which—given the necessity of dealing with multidimensional signals and arbitrary functions—is most suitable for the task at hand.[9]

6.1 A Comparison of Two Examples

Here are two very simple examples. The first is a sum of harmonically related cosines (multiples of $110\,\text{Hz}$) whose coefficients are a composite modulo function:

$$c_i = 1/((i \mod 7) + (i \mod 8) + (i \mod 11) + 1)).$$

```
// generic additive synthesis with a sine basis function and addition
(
Ndef(\g, {
    var combinator = { |a, b| a + b }; // just binary sum here
    var c = { |i| 1 / ((i % 7) + (i % 8) + (i % 11) + 1) };
    var g = { |i|
        SinOsc.ar(110 * i) * c.(i); // basis function ("dimension")
    };
    var z = (1..30); // number of operands
    var set = z.collect { |i| g.(i) }; // sequence
    // combine and scale output:
    set.reduce(combinator) ! 2 * (1 / z.size)
}).play;
)
```

In the second example we instead have a *product* of pulsed frequency modulated cosines, where $c_i = 1/i$.

```
// generic additive synthesis with a simple spectrum
// but a more complicated basis function and a product combinator
(
Ndef(\g, {
    var combinator = { |a, b| a * b };
    var c = { |i| 8 / i };
    var g = { |i|
        var cn = c.(i);
        var y1 = SinOsc.ar(120 * i, SinOsc.ar(cn * 10 * i) * (1/i));
        var y2 = LFPulse.kr(cn, 0, SinOsc.ar(cn * i, i, 0.2, 0.3));
        y1 * y2 * cn + 1
    };
    var n = (1..12); // number of operands
    var set = n.collect { |i| g.(i) }; // sequence
    LeakDC.ar(set.reduce(combinator) * (0.01 / n.size)).tanh ! 2
}).play;
)
```

From a conventional point of view, these two examples combine very different synthesis methods. The main difference lies in the function of each partial and the method of combining them. They implement the same structure in so far as both define the three components—partial, spectral coefficient, and combinator—separately, and then combine them according to the schema of generic additive synthesis.

---

[9]Note that in the SuperCollider signal semantics, the time parameter t is usually factored out: UGens are essentially arrows, similar to the description given by Hughes [8].

6.2 Comments

A few observations and remarks from the experiments so far:

1. From conventional additive synthesis we expect that a large number of partials is necessary. This is often not so with a different basis function. In such cases, we can say that the series converges almost 'too quickly'. Looking closer, the situation is this: the inherently polyphonic character of generic additive synthesis becomes interesting because of the *interference* between the partials: adding two waveforms may well result in cancellation or other unexpected but characteristic effects. For example, in the low frequency range and with sparse functions, the resulting sounds resemble percussion ensembles. Thus, it is sometimes useful to start with the minimal case in which only two partials are combined. This minimal constellation can then be extended by finding new laws for both the coefficients and the basis functions (i.e. the intersection between horizontal and vertical features). Here, partial understanding implies a search for spectral basis functions *in conjunction with* its parametrisation law.
2. The resulting function need not be used directly as an audio output signal. It may well be sonified by different means, e.g. by modulating a parameter of a carrier wave.
3. Allowing a certain distance from the predominant idea of the 'preset', axiomatic composition does not need to always externalise the parameters. This is the justification for the unusual inclusion of the coefficients into the partial in Eq. (2). At any stage the spectrum can again be factored out again (1), moving to and fro between the first two equations.
4. In many cases, the norm of what it means to have found a solution cannot be given in general (this is somewhat unsurprising as it applies to music in particular). One basic method of algorithmic composition responds to this challenge by superposing the *algorithmic description* as much as possible with its *temporal unfolding*, and thus with its perceptual and aesthetic qualities. As a program is by definition a future process, this superposition is necessarily incomplete. By consequence, rewriting code at runtime makes it necessary to delimit the relation of changes in the description to changes in the process. Proxies are an approach to solve this problem [3, 15]. Partial understanding means here to understand the relations between a *partially* changed description and its corresponding *partial* change in sound.

7 One More Step: Two Meanings of 'Concatenating Combinators'

A generic combinator G can consist of any ordered sequence of operations. Having defined the operation of 'addition' in the most generic sense—namely of a binary operation of a 'next step'—suggests cases where the operands are composed, rather than accumulated. In other words, the result of g_i, g_j, and g_h is not any more e.g. $g_i + g_j + g_h$ or $g_i g_j g_h$, but instead $g_i(g_j(g_h))$. In the simplest case, this can be written as:

$$f(t) = g_1(t, c_1) \circ g_2(t, c_2) \cdots \circ g_n(t, c_n) \tag{4}$$

Here, the sum operator becomes the function composition operator,[10] and the coefficients c_i of the spectrum determine the contribution of each partial in a series of nested function applications.

Thereby, e.g. a kind of spectral modulation, 'concatenative phase modulation', can be formulated. In the example that follows, each partial takes the previous one as phase input, and each partial's carrier frequency depends on its index i in the series. The spectrum is slowly modulated by linear triangular oscillators.

```
// simple case of concatenative phase modulation
(
Ndef(\g, {
    var combinator = { |a, b| a <> b };
    var c = { |i| LFTri.ar(1 / i, 1 / i).range(-1, 1).max(0) };
    var g = { |x, i| SinOsc.ar(i * 40, x * 2) };
    var n = (1..22); // number of operands
    n.inject(0, { |x, i| // inject is also known as left fold. Base case is 0 here
        g.(x, i) * c.(i) + (x * (1 - c.(i))) }) * 0.1
}).play;
)
```

But note that function composition is indeed only the first of two possible interpretations of a concatenating combinator. The second interpretation one might call spectrum composition. It changes from an internal to an external perspective of concatenation: instead of combining a sequence of elementary partials, it concatenates a sequence of spectra.

For this, we re-expose the *generic spectrum* $s = \langle c_1, c_2 \ldots c_n \rangle$ (see Eq. 3), consisting of the coefficients of each partial, in the form of $g_i(t, c_i)$. Treating the coefficients as m extra parameters of G, the spectrum can itself become a *time varying argument* of a function $G^\times(t, s)$. Because the original combinator G can in principle map any number of partials into any number of 'channels' in $\mathbb{R}^m$), we can interpret the output (codomain) of one as a spectrum (domain) of the other.

This requires to consider the combined signal as a set of functions. A sequence of G is then 'horizontally' combined by concatenation:

$$f(t) = G_1^\times \circ G_2^\times \circ G_3^\times \cdots \circ G_n^\times \tag{5}$$

Such a string of concatenated generic terms $G_n^\times$ essentially represents an ordered set of mappings between generic spectra. In terms of sound synthesis, we simply have an m-channel signal chain, where each node maps one spectrum to the next. The mappings can be conveniently arranged so that they form a monoid: they can be combined arbitrarily, because each output can serve as input for any other. The

---

[10] 'One more step' here simply means 'one more f unction applied'. Note that this is a case where the order in which the partials are combined influences the outcome (the operation of function composition is in general noncommutative). Furthermore, the coefficient scaling function is a little more complicated: a coefficient of zero must result in the identity function $f(x) = x$, when applied to a partial g.

composition operation could, in turn, also be expressed by a second order combinator, and a corresponding second order spectrum that encodes the contribution of each operand. Instead, we have devised a domain specific language[11] that is useful for experimenting with heterogenous mappings that do not follow from a single definition by variation. In favour of a final resume, we leave this topic to future discussion.

8   A Final Note on the Ontology of Sound

Formally, our proposal is indeed very minimal, little more than a spectral skeleton. We hope, however, that the historical and conceptual analysis has oriented it in such a way that it inspires new ideas at the intersection between mathematics, philosophy, and music.

Generic additive synthesis results in sounds that are on the verge between singularity and plurality. It starts from the multiple without presupposing unity, arising from a common law without presupposing that the result will cover a given domain completely. Being much less specific than other forms of 'additive' synthesis, it comes with no guarantees of completeness, and, paradoxically perhaps, enforces a much more specific treatment. Intertwining an observational paradigm (consisting of a decompositional basis function and the combinator map) and the law that parametrises a singular sound object, this synthesis method makes a good example, but only one example, of *musique axiomatique*.

In many contemporary treatments of Fourier analysis, a strong opposition is made between frequency and time perspective, where the frequency and phase spectrum are shown to be insufficient with regards to capturing the discontinuous structure of the time evolution. The spectrum is an illegitimate 'eternalist' rationalisation of the anomalies of noise. It is interesting, however, that historically, the harmonic decomposition had precisely the opposite role, namely to provide a way to find and convey ever larger sets of discontinuous points in the seemingly smooth continuum. The experience of an insufficiency of the Fourier series may thus be merely the result of the projections of an infinite series to a finite one, and from the difficulty of actually finding the laws that allow us to understand the spectrum of a given function. In this sense, the experimentation with alternative basis functions assumes the role of opening up new methods for conveying a mix of the continuous and the discontinuous, and escaping the false choice between immediacy and eternity. More than that, perhaps, it permits a focus on the particularly difficult problem of choosing the right partial: as we have seen in the experiments so far, generic additive synthesis is not so much a question of convergence at a high number of partials anymore—it is less a matter of the *limit*, as it is a matter of finding the adequate *successor*.

---

[11] The concatenative language *Steno* is embedded in SuperCollider. See https://github.com/telephon/ Steno. For examples of generic additive synthesis, see: https://github.com/musikinformatik/ Generic-Additive-Synthesis

In truth, the ordinal limit does not contain anything more than that which precedes it, and whose union it operates. It is thus determined by the inferior quantities. The successor, on the other hand, is in a position of genuine excess, since it must locally surpass what precedes it. As such – and this is a teaching of great political value, or aesthetic value – it is not the global gathering together 'at the limit' which is innovative and complex, it is rather the realization, on the basis of a point at which one finds oneself, of the one-more of a step. Intervention is an instance of the point, not of the place. [2, Appendix 3, p. 451]

Sound is a domain that matches this description surprisingly well.

Acknowledgements In the process of experimenting with generic additive synthesis in a multi-channel laboratory environment, the inspiring contributions by Hans W. Koch and Florian Zeeh were essential. We would also like to thank Guerino Mazzola for his ideas on frequency modulation in the present context. This paper would have lacked much of what we like about it without the continuing exchanges with Gabriel Catren, Maarten Bullynck, Renate Wieser, Tzuchien Tho and Alberto de Campo. The clarity of James McCartney's programming language design choices made it easy to develop these ideas. Last but not least, Frank Pasemann and Till Bovermann have given extremely valuable comments on the terminology and formalisation used – it goes without saying that we take full responsibility for remaining errors.

References

1. Badiou, A.: Number and Numbers (Le Nombre et les nombres). Des Travaux/Seuil (1990). Translation into English 2005 by Robin Mackay
2. Badiou, A.: Being and Event. Continuum International Publishing Group, London (2007)
3. Bovermann, T., Rohrhuber, J., de Campo, A.: Laboratory methods for experimental sonification. The Sonification Handbook. Logos Publishing House, Berlin (2011)
4. Cantor, G.: Gesammelte Abhandlungen Mathematischen und Philosophischen Inhalts. von Julius Springer, Berlin (1932)
5. de Campo, A.: Microsound. In: Wilson, S., Cottle, D., Collins, N. (eds.) SuperCollider Book, pp. 463–504. MIT Press, Cambridge (2008)
6. Gabor, D.: Acoustical quanta and the theory of hearing. Nature **4044**, 591–594 (1947)
7. Grattan-Guinness, I. (ed.): From the Calculus to Set Theory, 1630–1910. An Introductory History. Princeton University Press, Princeton (1980)
8. Hughes, J.: Generalising monads to arrows. Sci. Comput. Program. **37**, 67–111 (2000)
9. Le Brun, M.: Digital waveshaping synthesis. J. Audio Eng. Soc. **4**(27), 250 (1979)
10. Hemandez, E., Weiss, G.: A first course on wavelets. In: Studies in advanced mathematics. CRC Press LLC, Boca Raton, London, New York, Washington, D.C. (1996)
11. Mann, S., Haykin, S.: The chirplet transform: a generalization of Gabor's logon transform. In: Vision Interface '91. Communications Research Laboratory, McMaster University, Hamilton Ontario (1991)
12. Mazzola, G.: The Topos of Music. Geometric Logic of Concepts, Theory, and Performance. Birkhäuser Basel, Zürich (2002)
13. Roads, C.: Microsound. The MIT Press, Cambridge (2004)
14. Rohrhuber, J., de Campo, A.: Waiting and uncertainty in computer music networks. In: Proceedings of ICMC 2004: the 30th Annual International Computer Music Conference (2004)
15. Rohrhuber, J., de Campo, A., Wieser, R.: Algorithms today - notes on Language design for just in time programming. In: Proceedings of International Computer Music Conference, pp. 455–458. ICMC, Barcelona (2005)
16. Serra, X.: A System for Sound Analysis / Transformation / Synthesis based on a Deterministic plus Stochastic Decomposition. Ph.D. thesis, Stanford University, Stanford, California (1989)

17. Volkert, K.T.: Die Krise der Anschauung. Studien zur Wissenschafts-, vol. 3. Sozial - und Bildungsgeschichte der Mathematik. Vandenhoeck & Ruprecht, Göttingen (1986)

Dynamical Virtual Sounding Networks

Edmar Soria, Roberto Cabezas and Roberto Morales-Manzanares

Abstract This work will present a method for algorithmic music composition based on concepts from graph theory and non deterministic finite state automaton. The core formulation lies on the construction of a basic mathematical formal set structure over music rhythmic elements with two arithmetic operations: sum and multiplication. This structure allows to generate a whole compositional structure where mathematical functions can be directly related, or interpreted as musical rhythmic generators. We present then a brief scheme of a proposed algorithmic music composition system which we call Automaplex and its implementation in programming language Supercollider.

1 Introduction

Graph theory and non-deterministic finite state automata are both well known tools within the algorithmic composition and electroacoustic music. They offer a powerful and broad frame to develop and create artistic scoped ideas with a relatively simple implementation possibility. In this work we first establish formal theoretical foundations of the way we propose to mathematically manage traditional rhythmic elements from a set theory perspective at a very primary level; the one that is involved about significance. We will define sets for rhythm elements and for their further vectorial representation, as well as addition and multiplication operations over these sets. We will also define the concepts of *associated rhythm function*, *grouping operators* and *algebraic rhythm structures* which will be the underlying basis of the further practical

E. Soria (✉) · R. Cabezas
Music Technology Graduate Program, UNAM, Calle Xicotencatl 126,
04100 Coyoacán, Del Carmen, D.F., Mexico
e-mail: kenshi.shinobi@gmail.com; esoria.sonicart@gmail.com

R. Cabezas
e-mail: bbeto@gmail.com

R. Morales-Manzanares
Facultad de Música, DAAD, Universidad de Guanajuato, 36000 Guanajuato, Mexico
e-mail: robeto.morales@gmail.com

© Springer International Publishing AG 2017
G. Pareyon et al. (eds.), *The Musical-Mathematical Mind*,
Computational Music Science, DOI 10.1007/978-3-319-47337-6_28

implementation. The concept of *automaplex* will be defined as an algorithmic system that hybrids foundations of graph theory and finite state automata to interpret, manage and organize raw data from any data series expressed as ordered pairs. Finally, a couple of practical examples are presented using this concept *automaplex*. The first one outputs rhythm structures and the second one works as sound processing system constructed by modules.

2 Basic Definitions

In order to fulfill the computational requirements of this model and its further practical applications, some basic formal definitions need to be done. This also shapes the model not only as a practical one-case application, but as a more general theoretical framework. Let $\mathcal{A}$ be the set of all the known individual music rhythmic figures which will be called the *rhythm source set*. Let $\alpha(a) : \mathcal{A} \to \mathcal{V} \subset \mathbf{R}$ be the *value function*, the one which relates each rhythmic element to its corresponding time value in abstract musical terms, according to the condition shown in Fig. 1.

As the reader can note in the Fig. 1, we have defined the *quarter note* as a reference element within this set and so each one of the other elements will have values according to it. We call $\mathcal{V}$ the *rhythm value space*.

Rhythmic forms can be seen as grouping structures of individual notes. The basic elements that creates the minimum rhythmic structure is the *slur*. When we join two individual rhythmic elements we form this basic structure which is strongly related to traditional music notation but this idea can be extended for algorithmic composition purposes. In this sense, the *primary rhythm set* could be seen as the alphabet source.[1] If $\mathcal{A}$ is indeed this alphabet source, consequently, a second set $\mathcal{B}$ could be formed as the gathering of all the possible combinations of elements $a \in \mathcal{A}$, which can be

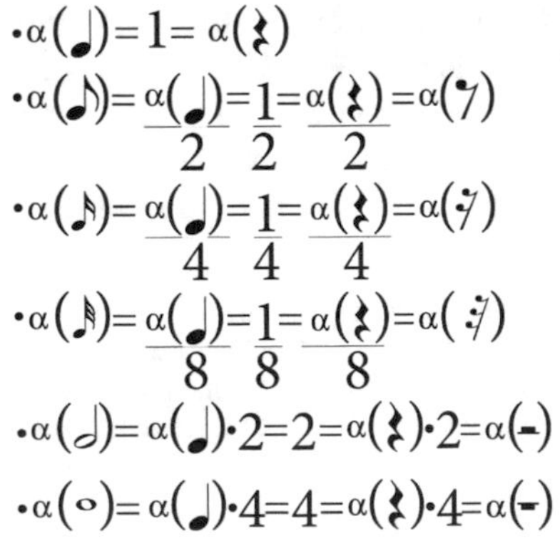

Fig. 1 Rhythm Value Function

$$\cdot\alpha\left(\textstyle\rule{0pt}{0pt}\right)=1=\alpha\left(\rule{0pt}{0pt}\right)$$

$$\cdot\alpha\left(\rule{0pt}{0pt}\right)=\alpha\left(\rule{0pt}{0pt}\right)=\frac{1}{2}=\alpha\left(\rule{0pt}{0pt}\right)=\alpha\left(\rule{0pt}{0pt}\right)$$

$$\cdot\alpha\left(\rule{0pt}{0pt}\right)=\alpha\left(\rule{0pt}{0pt}\right)=\frac{1}{4}=\alpha\left(\rule{0pt}{0pt}\right)=\alpha\left(\rule{0pt}{0pt}\right)$$

$$\cdot\alpha\left(\rule{0pt}{0pt}\right)=\alpha\left(\rule{0pt}{0pt}\right)=\frac{1}{8}=\alpha\left(\rule{0pt}{0pt}\right)=\alpha\left(\rule{0pt}{0pt}\right)$$

$$\cdot\alpha\left(\rule{0pt}{0pt}\right)=\alpha\left(\rule{0pt}{0pt}\right)\cdot2=2=\alpha\left(\rule{0pt}{0pt}\right)\cdot2=\alpha\left(\rule{0pt}{0pt}\right)$$

$$\cdot\alpha\left(\circ\right)=\alpha\left(\rule{0pt}{0pt}\right)\cdot4=4=\alpha\left(\rule{0pt}{0pt}\right)\cdot4=\alpha\left(\rule{0pt}{0pt}\right)$$

---

[1]From an automata theory perspective, this means that the set of all the available symbols within the system can be combined among them to create chains or words.

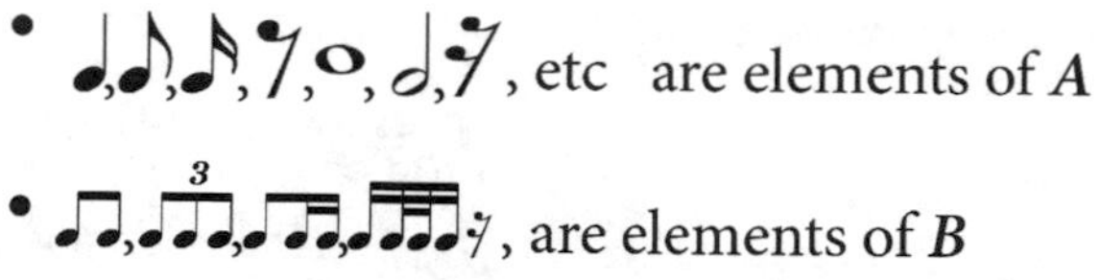

Fig. 2 Rhythm Value Function

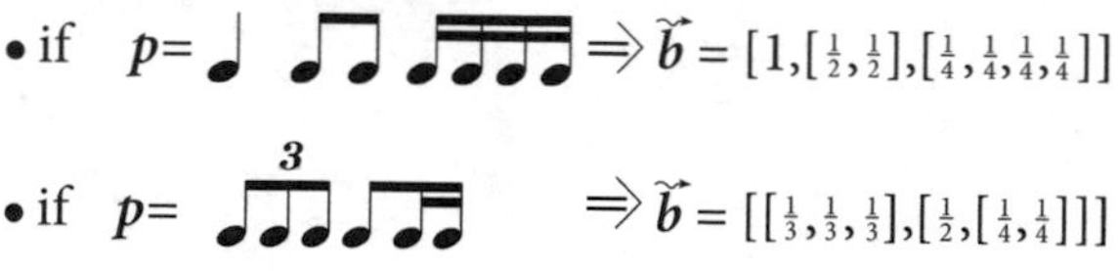

$\bullet$ ♩,♪,♫,𝄾,o,♩̇,𝄾̇ , etc   are elements of A

$\bullet$ ♫,♬,♬,♬𝄾, are elements of B

Fig. 3 Representation of rhythmic elements

$\bullet$ if $p=$ ♩ ♫ ♬ $\Rightarrow \vec{b} = [1,[\frac{1}{2},\frac{1}{2}],[\frac{1}{4},\frac{1}{4},\frac{1}{4},\frac{1}{4}]]$

$\bullet$ if $p=$ ♬ ♫ $\Rightarrow \vec{b} = [[\frac{1}{3},\frac{1}{3},\frac{1}{3}],[\frac{1}{2},[\frac{1}{4},\frac{1}{4}]]]$

interpreted as the language of any automata; we call B: the *generated rhythm set*. It can be easily seen that $A \subset B$ (Fig. 2).

As we have stated before the main aim of this work is to establish formal foundations for computational practical applications; and so, array or *vectorial* representation for elements of B are more than useful. This representation can be stated with this set:

$B^{\leadsto} = \{b^{\leadsto} : b^{\leadsto} = [z_1, \ldots, z_m], z_i = \alpha(b), i = \overline{1, n}, b \in B\}$

Figure 3 shows a few examples of this array representation for elements of B.

We now define two basic operations over sets A and B. Let $+ : A \times A \rightarrow \mathbf{R}$ and $* : A \times A \rightarrow \mathbf{R}$ be the *sum operation* and *multiplication operation* respectively; both of them compute the time value of two individual rhythmic elements according to the usual context (Fig. 4). The Fig. 5 shows a few examples of an application of these operations. Note that their function within the musical context is to provide a measure of duration of any possible rhythmic structure.

With these sets and these operations defined, we can go one step beyond and apply these ideas to general mathematical functions. Let $f(x)$ be a one variable function,

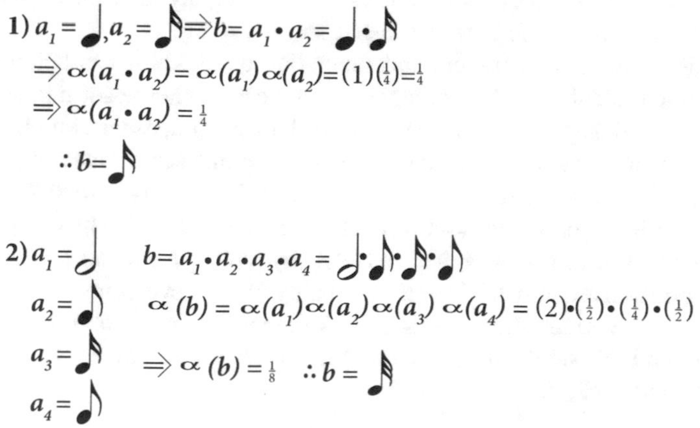

1) $a_1 =$ ♩ $, a_2 =$ ♪ $\Rightarrow b = a_1 \cdot a_2 =$ ♩ $\cdot$ ♪
$\Rightarrow \alpha(a_1 \cdot a_2) = \alpha(a_1)\alpha(a_2) = (1)(\frac{1}{4}) = \frac{1}{4}$
$\Rightarrow \alpha(a_1 \cdot a_2) = \frac{1}{4}$
$\therefore b =$ ♪

2) $a_1 =$ ♩ $b = a_1 \cdot a_2 \cdot a_3 \cdot a_4 =$ ♩̇ $\cdot$ ♪̇ $\cdot$ ♪̇ $\cdot$ ♪
$a_2 =$ ♪ $\alpha(b) = \alpha(a_1)\alpha(a_2)\alpha(a_3)\alpha(a_4) = (2) \cdot (\frac{1}{2}) \cdot (\frac{1}{4}) \cdot (\frac{1}{2})$
$a_3 =$ ♪ $\Rightarrow \alpha(b) = \frac{1}{8}$ $\therefore b =$ ♪
$a_4 =$ ♪

Fig. 4 Multiplication operation

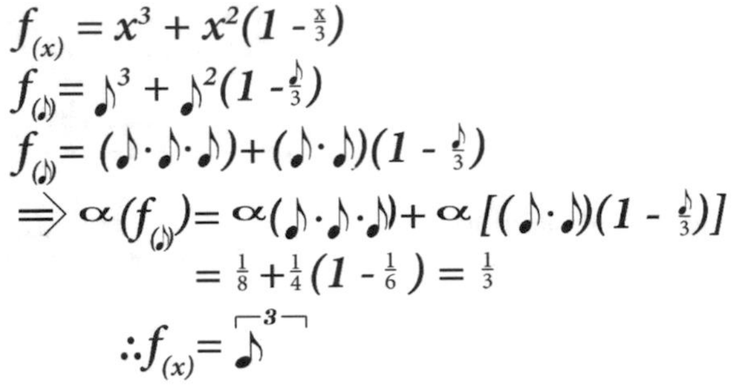

Fig. 5 Associated rhythm functions

$$f_{(x)} = x^3 + x^2\left(1 - \tfrac{x}{3}\right)$$

$$f_{(\flat)} = \flat^3 + \flat^2\left(1 - \tfrac{\flat}{3}\right)$$

$$f_{(\flat)} = (\flat \cdot \flat \cdot \flat) + (\flat \cdot \flat)\left(1 - \tfrac{\flat}{3}\right)$$

$$\Rightarrow \alpha(f_{(\flat)}) = \alpha(\flat \cdot \flat \cdot \flat) + \alpha\left[(\flat \cdot \flat)\left(1 - \tfrac{\flat}{3}\right)\right]$$

$$= \tfrac{1}{8} + \tfrac{1}{4}\left(1 - \tfrac{1}{6}\right) = \tfrac{1}{3}$$

$$\therefore f_{(x)} = \overset{\ulcorner 3 \urcorner}{\flat}$$

we can define the *associated rhythm function* $\mathcal{F}(x)r : \mathcal{B} \rightarrow \mathbf{R}$. For example, let $f(x) = x^3 + x2(1 - \tfrac{x}{3})$ be a third degree polynomial function, then, the *associated rhythm function* would look like that shown in Fig. 6.

As the reader can note, what is really happening in the last process is what actually happens with any function; an input is transformed into an output by an specific process. In this case, the *associated rhythm function*, converts any element from both, the *rhythm* or *the complex set* and transforms it into another rhythmic element which is defined ultimately by the duration. All this allows us not only to establish a formal foundation in our model but to create an intrinsic aesthetic meta-language, which derives in the convergence of music and mathematics at the very root level; the realm of the sign and its significance.

We can now develop the concept of a *Tie-slur operator* which will allows us to define and generate rhythmic structures according to a predefined grouping:

$$\phi : \mathcal{B}^k \times \mathbf{N} \rightarrow \mathcal{B}^{\leadsto}$$

such that $\phi(x, n) = [x_1, x_2, \ldots, x_k] \urcorner (n)$, for $x = (x_1, x_2, \ldots, x_n) \in \mathcal{B}^k$ and $\Sigma_{i=1}^n \alpha(x_i) = n$. Note that this operator allows to create rhythmic structures as complex as desired within a predefined numeric slur grouping.

There is a nice Supercollider implementation of RTMs list notation by Mike Laurson which calculates subgrouping of rhythmic structures according to an initial numeric slur. Although this is a very useful implementation, as the complexity of the structure rises up -specially in nesting cases- the complexity of the notation grows proportional and so its usage from the client side. With this alternative notation it is possible to incorporate that useful notation and also to avoid the complexity we have talked about. In this way, any musical rhythmic phrase can be described in vectorial terms. In order to continue the formality of this development we will define a *phrase* as any rhythmic structure such that is composed for any number of elements of $\mathcal{B}^{\leadsto}$. We call this set the *phrase set* and it will be denoted by $\mathfrak{P}$. This both ideas are represented in Fig. 6.

Fig. 6 Vectorial
representation of rhythmic
elements

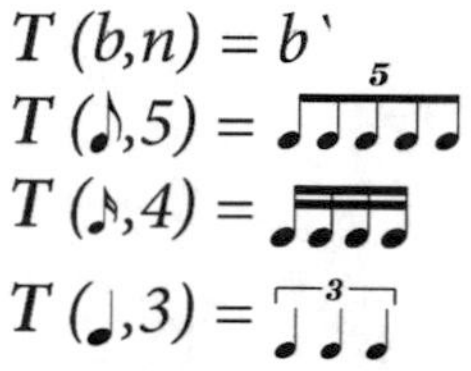

$\bullet\, b =$ 𝅘𝅥𝅮𝅘𝅥𝅮𝅘𝅥 $\Rightarrow b = [\frac{1}{2},\frac{1}{1}]$

$\bullet\, b =$ $\overline{}^{\,3}\overline{}$ 𝅘𝅥𝅮 𝄾 𝅘𝅥𝅮 $\Rightarrow b = [\frac{1}{3},\frac{1}{3}^{\cdot},\frac{1}{3}]$

$\bullet\, b =$ 𝄾 𝅘𝅥 𝅗𝅥 𝄾 𝅘𝅥𝅮 $\Rightarrow b = [1,1,2,\frac{1}{2}^{\cdot},\frac{1}{2}]$

Fig. 7 Tie grouping operator

$$T\,(b,n) = b\,^{\backprime}$$
$$T\,(𝅘𝅥𝅮,5) = \overset{5}{\text{𝅘𝅥𝅮𝅘𝅥𝅮𝅘𝅥𝅮𝅘𝅥𝅮𝅘𝅥𝅮}}$$
$$T\,(𝅘𝅥𝅮,4) = \text{𝅘𝅥𝅮𝅘𝅥𝅮𝅘𝅥𝅮𝅘𝅥𝅮}$$
$$T\,(𝅘𝅥,3) = \overset{3}{\text{𝅘𝅥𝅘𝅥𝅘𝅥}}$$

We now define the *Tie-grouping operator* $\mathcal{G} : \mathcal{A} \times \mathbf{N} \to \mathcal{B}$ which is a very useful computational representation of the traditional grouping of rhythmic notes. As it can be seen, the domain takes a single representative element from $\mathcal{A}$ and a natural number. In this way, the *Tie-grouping operator* creates traditional rhythmic groupings from an arithmetic perspective and it can be seen as a particular case of the *Tie-slur operator*. The Fig. 7 shows some examples about this concept.

2.1 Algebraic Rhythmic Structures

We talked previously about how we can make the language of math and rhythmic notation to converge by usual math functions and we computed the time value for a couple of examples for that. Now we can go one step beyond and define *math-rhythm functions* to create complex algorithm structures. The basic idea here is to interpret math functions at their primal level and to apply that to the realm of music rhythm, so we can be able to define in the same way *math-rhythm functions* $f : \mathcal{B} \to \mathcal{B}$.

The reader should remember that we defined at the beginning of this paper the quarter note as the unity or reference point element. With this concept in mind, it can be seen from the figure that the way the *math-rhythm functions* work is based precisely in the rhythm value of each element. For example, the first case can be interpreted as this: *take one rhythm element, and input it to the function, then it will output a phrase containing that element and two elements with half the rhythm value of the original.* In this way, we can define any function we want and we will get algorithmic rhythmic structures that will preserve the shape of the original function in a nested way (Fig. 8).

Fig. 8 Math rhythm
functions

3   Automataplex

As a particular case implementation we developed an algorithmic system in programming language Supercollider for data mapping onto sonic parametric processing
based on foundations of *non*-deterministic finite automata and graph theory.

There is a technique in image processing called morphological thinning which is
used to remove selected foreground pixels from binary images. This is particularly
useful for skeletonization and for tidying up the output of edge detectors by reducing
all lines to single pixel thickness. For this paper an image of a plant was analyzed by
this process using Mathematica Software to generate an adjacency matrix as defined
within the graph theory.

There is an issue with this matrix as it is output from Mathematica, and it is
that sometimes the size may be as large, as 1000×1000 with 50% of actual useful
information. This of course generates difficulties of data managment and processing
for aims focused on musical or artistic applications. In order to solve this problem
the system proposed to collect the data in a CSV file filtering all the inactive nodes
and rewriting the file as ordered pair list. Since there is still a large amount of raw
data in this second list, the option proposed to make it useful for compositional and
artistic aims was to develop an algorithm that was able to automatically generate
-given some initial conditions- a set of what can be called *mini complex networks*
with two basic principles of hybridization with finite state automata:

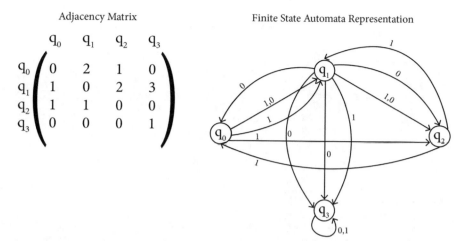

Fig. 9 Incidence Matrix to Automata conversion

1. The class divides the original list according to a **maximum number of states** stated by the user such that the *minicomplex networks* can be generated.
2. Given the maximum number of states, the user enters the *range length* which define the size of the individual incidence matrices corresponding to each of the *directed subgraphs*.

As a result the class outputs a set of *directed subgraphs* where each of the nodes its uniquely related to each of the sates of a finite automata. The Fig. 9 will illustrate this whole idea. Figures 10 and 11 show the code written in Supercollider for this whole process.

As it can be seen each state of the potential finite state automata is connected in the very same way as the nodes of the subgraph are connected among them by the links. Of course, the configuration for each individual finite state automata is given by the previously generated adjacency submatrices.

In the last stage, a set of subgraphs were generated and each subgraph was related to a potencial finite state automata with a one to one relation between nodes and states. Until this step the configuration is constructed but is completely static, so now the inner process needs to be stated. This next algorithm interprets the previous set -or any adjacency matrix- into a non-deterministic finite sate automata or NDFA. The user first defines its alphabet in the very convenient way representing each letter of the alphabet normally with a number, the elements of the alphabet are selected. Here are some examples:

1. (0.5) Will denote an alphabet of six elements: a,b,c,d,e,f.
2. (8, 12) Will denote an alphabet of four elements: i,j,k,l.
3. [0, 4, 9, 10, 23] Will denote an alphabet of five elements: a,e,j,k,z.

```
| PairToMatrix {
    var <>source_matrix, <>states, <>matrixForm, <>final_network, <>converted_list, <>index_matrix,
<>count_matrix,<>range_length, <>processList, <>final_network;

    *new {
        arg source, //Pair list source to be converted //source list needs to be of the type
[[a_1,b_1],..[a_n,b_n]],...[]]
        states,
        range_length;  //Number of columns of the final matrix...this is equivalent to the number of
states of the former
                        //automataplex

        ^super.new.init(source, states, range_length);

    }

    init {
        arg src, sts, range;
        var dummy1, preList, components;

        source_matrix = src;
        states = sts;
        range_length = range;

        preList = source_matrix % states; //divide the list for fitting the number of states.
        processList = List[];

        components = (source_matrix.size/range_length).floor;
        components.postln;

        (components.asInteger).do{arg i;
         dummy1 = preList[(range_length*i)..(range_length*(i+1)-1)];
         processList.add(dummy1);
            };
    }

    execute {
        final_network = List[];
        processList.size.do{ arg i;
        final_network.add(this.convert(i).asArray);
        };
    ^final_network;
        }
```

Fig. 10 PairToMatrix class

Once the user has specified the input alphabet, she -he- has then two options:

1. Let the algorithm to automatically generate the NDFA. In this way, the algorithm
   generates the transition function with weighted probabilities and so the user can
   ask for the final word related to a defined state series or the inverse case; the user
   inputs a word and the algorithm outputs a state series possibilities.
2. Manually define the set of final states and the transition function.

So, summarizing the process we have, in the first stage the user specifies the incidence matrix to process; then, the class automatically generates the non-deterministic
finite automaton according to the defined alphabet and the reinterpretation of the
nodes connected to the matrix as active links finite state automaton. The following
stage is to obtain the chain elements of the alphabet and the trajectory by specify-
ing an initial state and a finite number of transitions. Since the automaton is non-
deterministic, for a given set of initial conditions different chains could be obtained
since the class automatically assigns the transition functions in each state. Thereby,
you obtain the list that represents the final trajectory given the initial state conditions,
which is defined by three elements: the first state, the second state, and the transition

```
Automataplex {
     var <>states_list, <>alphabet, <>states, <>range, <>source, <>path_string,<>connection_source;

     *new{
           //creates an hybrid algorithmic non deterministic finite state automata->complex network
           //user define an alphabet or can use  default one.
           arg alphabet, //user specifies the number of the alphabet items or if wished the block of
letters
           //number of states which define the automataplex
           connection_source; //list


           ^super.new.init(alphabet, connection_source);

     }

     init {
           arg al, connection;
           var alpha;

     connection_source = connection;
     states = connection_source[0].size;

     states_list = List.fill(states,{arg i; (\q_++i).asSymbol}); //fill a list with states labels from q0
to qn

     alpha = List.with(\a,\b,\c,\d,\e,\f,\g,\h,\i,\j,\k,\l,\m,\n,\o,\p,\q,\r,\s,\t,\u,\v,\w,\x,\y,\z);

     alphabet = List[];

     al.do{arg i;                    //create alphabet for automataplex
     alphabet.add(alpha[i]);
          };
     alphabet.postln;
     this.matrix;
     }
```

Fig. 11 Automaplex class

function between them. And so the final list of states and the entire string is obtained. We call this hybrid proposal *Automaplex*.

3.1   *Mapping Data to Sound Realm*

With the whole system previously defined we can map information from a source as abstract as a plant photo to actual sonic parameters or music structures through an hybrid foundation of graph theory and finite state automata. There are of course endless possibilities for this mapping so we will present a couple of examples; one for sound processing and the other for algorithmic rhythmic structure generation.

Lets take one of the subgraphs as an example. As we have seen before we propose to treat all of them as finite state automata where each of the nodes acts as a unique state. Let $\mathfrak{A}t$ be an automaplex with 5 states named $q_0, \ldots, q_1$ where each state is a math rhythm function as shown in the figure:

As it can be seen for each word in the automata, an algebraic rhythm structure is generated and it can grow really complex but always keeping a whole grouping

```
matrix {
range = List[];

connection_source.do{ arg item, id;
item.do{arg in,id_2;
    while({in>0},{
                //[id,id_2,alphabet.choose].postln;
                range.add([id,id_2,alphabet.choose]);
                in=in-1;
            });
}
;

range.size.do{arg j;
2.do{arg i;
    range[j][i] = states_list[range[j][i]];
};
    };

    }

val {
arg pos,state,letter;
var pass_list;
pass_list = List[];

range.do{arg item;
if(item[pos] == state && item[2] == letter,
    {item.postln;
    pass_list.add(item);
        },{});
    };
        ^pass_list;
    }

//given an initial state, looks for all possible states to transtiton through the aplhabet
look {
arg state;
var path2 = List[];

alphabet.do{ arg item;
    path2.add(this.val(0,state,item).asArray);
};
^path2;
    }
```

Fig. 12 Automaplex class

coherence assisted by the predefined math functions for every state (Fig. 12). Since every state can be algorithmically programmed for virtually any math function there are no computational boundaries for defining automaplexs of any number of states and with any *math rhythm functions* (Fig. 13).

The second example proposed is for sound processing. Now each one of the states will work as dedicated processing module and with the same mechanism, it can be possible to algorithmically process an input sound signal. Lets take as an example the same automaplex of the Fig. 14. Suppose now that instead of *math-rhythm functions*, each node is a particular sound process:

Fig. 13 Automaplex class

```
path {

    arg iteration, initial_state;
    var path_f, x,y,w;

    path_f = List[];
    path_string = List[];

    path_f.add(initial_state);

     iteration.do{ arg i;
        x = path_f[i];
        w = this.look(x);
        x = w.flat;
        x = x.reshape(w.size,3);
        y = x.choose;
        if(y.notNil,{
        path_f.add(y[1]);
        path_string.add(y[2]);
            },{});
    };
    ^path_f;
     }

string {
        ^path_string;
        }

s_range {
        ^range;
    }

}
```

1. $q_0 = delay$ with random time parameter in between 50 and 500 milliseconds.
2. $q_1 = comb\ filter$ with random time parameter in between 150 and 800 milliseconds.
3. $q_2 = resonant\ filter$ with random frequency range.

With this configuration any incoming signal source can be processed according to different words input to the automaplex. This system offers the same wide range possibilities and since each module has random values, each output will have consistent sonic differences.

4 Conclusions

Algorithm hybridization is a useful alternative to create more complex and dynamical mapping structures within the algorithmic composition and sound processing scopes. Interpreting the concepts of such theoretical tools at their very basic level allows us to establish convergence between apparently non related frameworks. The system we proposed is a particular example of this hybridization idea and the whole process

```
//automataplex help

//read an CSV File and store it in a List. Each item is a pair of numbers that indicates the nodes that are
connected
(
x = CSVFileReader.readInterpret("/Users/Draco/Downloads/gafo.csv");

x = x.asList;

~activos = List.new;
for(0,x.size-1,{|i|
    for(0,x.size-1,{|j|
        if(x[i][j]==1,{e=[i,j];~activos.add(e)});
    });
});
~activos;
)

(
//PairToMatrix interprets any exhaustive List of pairs of numbers into a complex network set of subgrids each
one representing an incidence matrix of a given number of states and range length;
// create a new object
a = PairToMatrix.new(~activos, //List source
    8, //number of states
    18); //range length

b = a.execute; //get the final general List
b.size; //number of sub networks that compose the general hipernetwork
c = Matrix.with(b[1]);  //get incidence matrix for the first automataplex or for any of other item
)

//Create an individual automataplex. Args are alphabet and source array. number of states automatically matches
with the source array. General Alphabet is standard english alphabet, automata alphabet can be defined in any
desired way by an id array, identifing letters with their position. for example [0,3,5,23] will generate
[a,d,f,z]
//as automata`s alphabet, (0..2) will generate [a,b,c], (5..9) will generate [f,g,h,i] as automata`s alphabet

~a1 = Automataplex.new((0..2),b[1]) //define automata alphabet); //define list source (incidence matrix)

~path1 = ~a1.path(5,\q_1); //get an stochastic path with initial state q_10 and 10 iterations, if an state has
more than one link out, each one of those will have the same probability to be choosen

~string1 = ~a1.string; //this array gives out the string that needs to be addressed for reaching that particular
path.

~path1_1 = ~a1.path(19,\q_3); //get a different path from different initial conditions
~string1_1 = ~path1_1.string;
```

Fig. 14 Automaplex class

reflects a complexity upgrade in the mapping from theoretical source to sonic realm. The math foundations are necessary from a formal and organizational perspective and for develop of future works that are scoped in the same direction (Fig. 15).

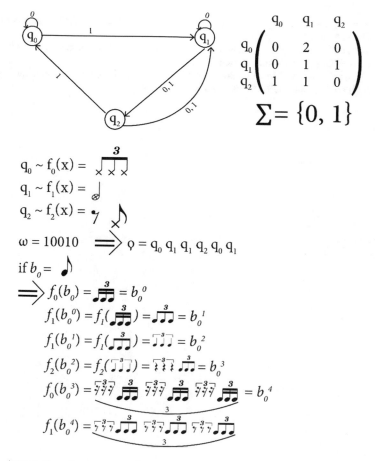

Fig. 15 Automaplex structure example

References

1. Barrat, I., Barthemy, M., Vespignani, A.: Dynamical Processes on Complex Networks. Cambridge University Press, Cambridge (2008)
2. Kaur, K., Sharma, M.: A method for binary image thinning using gradient and watershed algorithm. Int. J. Adv. Res. Comput. Sci. Softw. Eng. **3**(1), 1–4 (2013)
3. Laurson, M., Kuuskankare, M.: From RTM-notation to ENP-score-notation. In: Proceedings of Journées d'Informatique Musicale (2003)
4. Miranda, E.R.: Composing Music with Computers. Focal Press - Routledge, New York (2002)
5. Oliwa, T., Wagner, M.: Composing Music with Neural Networks and Probabilistic Finite-State Machines. Lecture Notes in Computer Science, vol. 4974, pp. 503–508. Springer, Berlin (2008)
6. Szeto, W.M., Wong, M.H.: A Graph-Theoretical Approach for Pattern Matching in Post-Tonal Music Analysis. Department of Computer Science and Engineering, The Chinese University of Hong Kong Shatin, N.T., Hong Kong

7. Van Steen, M.: Graph Theory and Complex Networks: An Introduction. On Demand Publishing, LLC-Create Space, Seattle (2010)
8. Vincent, L.: Morphological algorithms. In: Dougherty, E. (ed.) Mathematical Morphology in Image Processing. Marcel-Dekker, New York (1992)
9. Wagner, F., Schmuki, R.: Modeling Software with Finite State Machines: A Practical Approach (Auerbach Publications). CRC Press, Boca Raton (2006)

Melodic Pattern Segmentation of Polyphonic Music as a Set Partitioning Problem

Tsubasa Tanaka and Koichi Fujii

Abstract In polyphonic music, melodic patterns (motifs) are frequently imitated or repeated, and transformed versions of motifs such as inversion, retrograde, augmentations, diminutions often appear. Assuming that economical efficiency of reusing motifs is a fundamental principle of polyphonic music, we propose a new method of analyzing a polyphonic piece that economically divides it into a small number of types of motif. To realize this, we take an integer programming-based approach and formalize this problem as a set partitioning problem, a well-known optimization problem. This analysis is helpful for understanding the roles of motifs and the global structure of a polyphonic piece.

1 Motif Division

In polyphonic music like fugue-style pieces or J.S. Bach's *Inventions and Sinfonias*, melodic patterns (motifs) are frequently imitated or repeated. Although some motifs are easy to find, others are not. This is because they often appear implicitly and/or appear in the transformed versions such as inversion, retrograde, augmentations, diminutions. Therefore, motif analysis is useful to understand how polyphonic music is composed.

Simply speaking, we can consider the motifs that appear in a musical piece to be economical if the number of types of motif is small, the numbers of repetitions are large, and the lengths of the motifs are long. Assuming that this economical efficiency of motifs is a fundamental principle of polyphonic music, we propose a new method of analyzing a polyphonic piece that efficiently divides it into a small number of types of motif. Using this division, the whole piece is reconstructed with a small number of types of motif like the puzzle game *Tetris* [1] (In tetris, certain

T. Tanaka (✉)
IRCAM, Paris, France
e-mail: tsubasa.tanaka@ircam.fr

K. Fujii
NTT DATA Mathematical Systems Inc., Tokyo, Japan
e-mail: fujii@msi.co.jp

© Springer International Publishing AG 2017
G. Pareyon et al. (eds.), *The Musical-Mathematical Mind*,
Computational Music Science, DOI 10.1007/978-3-319-47337-6_29

291

domains are divided with only seven types of piece). We call such a segmentation a *motif division*.

If a motif division is accomplished, it provide us a simple and higher-level representation whose atom is a motif, not a note, and it will be helpful to clarify the structures of polyphonic music. The representation may provide knowledge about how frequent and where each motif is used, the relationships between motifs such as causality and co-occurrence, which transformations are used, how the musical form is constructed by motifs, and how the long-term musical expectations are formed. This analysis may be useful for applications such as systems of music analysis, performance, and composition.

Studies about finding boundaries of melodic phrases are often based on human cognition. For example, [2] is based on grouping principles of gestalt psychology, and [3] is based on a short-term memory model. While these studies deal with relatively short range of perception and require small amounts of computational time, we focus on global configuration of motifs on the level of compositional planning. This requires us to solve an optimization problem that is hard to solve. To deal with this difficulty, we take an *integer programming*-based approach [4] and show that this problem can be formalize as a *set partitioning problem* [5]. This problem can be solved by integer programming solvers that use efficient algorithms such as the branch and bound method.

2 Transformation Group and Equivalence Classes of Motif

In this section, we introduce *equivalence classes* of motif derived from a group of motif transformations as the criterion of identicalness of motifs. These equivalence classes are used to formulate the motif division in Sect. 3.

Firstly, a motif is defined as an ordered correction of notes $[N_1, N_2, \ldots, N_k]$ ($k > 0$), where N_i is the information for the ith note, comprising the combination of the pitch p_i, start position s_i, and end position e_i ($N_i = (p_i, s_i, e_i)$, $s_i < e_i \leq s_{i+1}$). Next, let $\mathcal{M}$ be the set of every possible motif, and let T_p, S_t, R, I, A_r be one-to-one mappings (transformations) from $\mathcal{M}$ to $\mathcal{M}$, where T_p is the transposition by pitch interval p, S_t is the shift by time interval t ($p, t \in \mathbb{R}$), R is the retrograde, I is the inversion, and A_r ($r > 0$) is the r-fold argumentation (diminution, in the case of $0 < r < 1$). These transformations generate a transformation group $\mathcal{T}$ whose operation is the composition of two transformations and whose identity element is the transformation that does noting. Each transformation in $\mathcal{T}$ is a *strict imitation* that preserves the internal structures of the motifs.

Here, a binary relation between a motif m ($\in \mathcal{M}$) and $\tau(m)$ ($\tau \in \mathcal{T}$) can be defined. Due to the group structure of $\mathcal{T}$, this relation is an equivalence relation (i.e., it satisfies

Fig. 1 Possible motifs of the
first voice of J.S. Bach's
Invention No. 1 In the case
where the maximum number
of notes in a motif is 4)

reflexivity, symmetry and transitivity [6]). Then, it derives equivalence classes in $\mathcal{M}$. Because the motifs that belong to a same equivalence class share the same internal structure, they can be regarded as identical (or the same type).[1]

3 Formulation as a Set Partitioning Problem

A set partitioning problem, which is well known in the context of operations research, is an optimization problem defined as follows. Let N be a set that consists of n elements $\{N_1, N_2, \ldots, N_n\}$, and let M be a family of sets $\{M_1, M_2, \ldots, M_m\}$, where each M_j is a subset of N. If $\bigcup_{j \in X} M_j = N$ is satisfied, X, a subset of indexes of M, is called a cover, and the cover X is called a partition if $M_{j_1} \cap M_{j_2} = \varnothing$ is satisfied for different $j_1, j_2 \in X$. If a constant c_j called a cost is defined for each M_j, the problem of finding a partition X that minimizes the sum of the costs $\sum_{j \in X} c_j$ is called a set partitioning problem.

3.1 Condition of Motif Division

If N_i corresponds to each note of a musical piece to be analyzed and M_j corresponds to a motif, the problem of finding the most economically efficient motif division can be interpreted as a set partitioning problem. The index i starts from the first note of a voice to the last note of the voice and from the first voice to the last voice. $M_j (1 \le j \le m)$ corresponds to $[N_1]$, $[N_1, N_2]$, $[N_1, N_2, N_3], \ldots, [N_2], [N_2, N_3]$, $\ldots$ in this order. The number of notes in a motif is less than a certain limit number (Fig. 1).

This information can be represented by the following matrix A:

---

[1]Although the criterion for identical motifs defined here only deals with strict imitations, we can define the criterion in different ways to allow more flexible imitations, such as by (1) defining an equivalence relation from the equality of a shape type [7–10] and (2) defining a similarity measure and performing a clustering of motifs using methods such as k-medoid method [11] (the resulting clusters derive an equivalence relation). In any case, making equivalence classes from a certain equivalence relation is a versatile way to define the identicalness of the motifs.

$$A = \begin{bmatrix} 1\ 1\ 1\ 1\ 0\ 0\ 0\ 0\ 0\ 0\ 0\ 0\ \cdots \\ 0\ 1\ 1\ 1\ 1\ 1\ 1\ 1\ 0\ 0\ 0\ 0\ \cdots \\ 0\ 0\ 1\ 1\ 0\ 1\ 1\ 1\ 1\ 1\ 1\ 0\ \cdots \\ 0\ 0\ 0\ 1\ 0\ 0\ 1\ 1\ 0\ 1\ 1\ 1\ 1\ \cdots \\ 0\ 0\ 0\ 0\ 0\ 0\ 0\ 1\ 0\ 0\ 1\ 1\ 0\ \cdots \\ 0\ 0\ 0\ 0\ 0\ 0\ 0\ 0\ 0\ 0\ 0\ 1\ 0\ \cdots \\ \vdots\ \vdots\ \vdots\ \vdots\ \vdots\ \vdots\ \vdots\ \vdots\ \vdots\ \vdots\ \vdots\ \vdots\ \vdots\ \ddots \end{bmatrix} \tag{1}$$

where each row corresponds to each note N_i and each column corresponds to which notes are covered by each motif M_j. This matrix is the case where the maximum number of notes in a motif is 4.

Representing the element of A as a_{ij}, the condition that the whole piece is exactly divided by a set of selected motifs can be described by the following constraints, which mean that each note N_i is covered by one of M_j once and only once:

$$\forall i \in \{1, 2, \ldots, n\}, \quad \sum_{j=1}^{m} a_{ij} x_j = 1, \tag{2}$$

where x_j is a 0-1 variable that represents whether or not M_j is used in the motif division. These conditions are equivalent to the condition of partitioning.

3.2 Objective Function

The purpose of motif division is to find the most efficient solution from the many solutions that satisfy the condition of partitioning. Then, we must define efficiency of motif division. We can consider that the average length (the number of notes) of motifs used in the motif division is one of the simplest barometers that represent the efficiency of motif division. Also, the number of motifs and that of the types of motif used in motif division will be efficient if they are small.

In fact, the average length of motifs is inversely proportional to the number of motifs. Therefore, if the number of types of motif (denoted by P) is fixed, the number of motifs will be what we should minimize.

The number of motifs can be simply represented by $\sum_{j=1}^{m} x_j$. This is the cost function $\sum_{j=1}^{m} c_j x_j$ whose c_j is 1 for each j. We adopt this cost function. However, in the next subsection, we introduce additional variables and constraints to fix the number P.

3.3   Controlling the Number of Equivalence Classes

Let C be the set of equivalence classes of motif, which is derived from M, which is the set of all possible motif classes that can be found in a piece (only the motif classes whose number of notes is less than a certain number is included in M). This means that M is derived from $\mathcal{M}$ by a restriction.

Let y_k be a 0-1 variable that represents whether or not one of the members of C_k appears in X (the set of selected motifs), where each element of C is denoted as $C_k (1 \leq k \leq l)$. This means that statement "$y_k = 1 \Leftrightarrow \sum_{j \in C_k} x_j > 0$" must be satisfied. This statement can be represented by the following constraints that use $\sum_{j \in C_k} x_j$, the number of selected motifs that belong to C_k:

$$\forall k \in \{1, 2, \ldots, l\}, \ y_k \leq \sum_{j \in C_k} x_j \leq Qy_k, \tag{3}$$

where Q is a constant that is sufficiently large.

Then, the statement that the number of equivalence classes is P can be represented by the following constraint:

$$\sum_{k=1}^{l} y_k = P. \tag{4}$$

If P is small to a certain degree, the motif division will tend to be simple. However, if P is too small, covering whole piece with few motif classes will be difficult and one note motif will be used too many times. This will lead to a loss of the efficiency of motif division.

Therefore, we should find good balance between the smallness of the objective function and the smallness of P. Because knowing which number is adequate for P in advance is difficult, we will solve the optimization problems for respective P in a certain range. Then, we will determine an adequate number for P, observing the solutions for respective P.

4   Result

We analyzed J.S. Bach's *Invention No. 1* by solving the optimization problem described in the previous section. The maximum length of motif was set as 7. An IP solver *Numerical Optimizer 16.1.0.* and a branch and bound method was used for searching the solution. From the observation of solutions for various values for P, P was set as 13. It took less than one minute to obtain a solution for $P = 13$.

Figure 2 shows the result of motif division. The slurs represent the motifs and the one-note motifs don't have a slur. Figure 3 shows the representatives of 13 motif classes that are used in the motif division.

Fig. 2 The representatives of motif classes that appear in the motif division of J.S. Bach's *Invention No. 1* in the case that $P = 13$. Some flats are replaced by sharps for the purpose of programming

This result tells us many things. For example, 4th, 10th, and 11th motif classes in Fig. 3 are slightly different but can be regarded as the same motif, which corresponds to the subject of this piece. Searching for the domains where the subject doesn't

Fig. 3 The motif classes that appear in the motif division of J.S. Bach's *Invention No. 1* (the number of motif classes was set at 13)

appear, we find that there are three domains whose durations are one and half bars (These are indicated by the big rectangles). The ends of these domains coincide with the places where the cadences exist. Therefore, we could detect three sections of this piece properly.

The last motif class in Fig. 3 is a leap of octave. This motif class appears in all of the cadence domains and is related to the ends of sections. It also co-occurs with 2nd motif class, which is a two-note motif, in the cadence domains. The 12th motif class is a very characteristic one that includes a doted note and a large leap. This motif class only appears before the cadence domains (the two motifs surrounded by the rounded rectangle). We can consider that this remarkable motif class plays an important role that tells listeners the end of the exposition of subject and the beginning of the cadence domain.

The 9th zigzag motif class and the motif classes that are one-way slow movements shown by the arrows in Fig. 2 only appear as the ascending form in the first "2" sections. In contrast, these motif classes appear only as the descending form in the final section. We interpret this contrast means that the ascending form creates a sense of continuation of the piece and the descending form creates a sense of conclusion. Thus, long-term musical expectations seems to be formed by the selections of transformation.

In such ways, motif division is useful to make us understand the roles of motifs and how global musical structures are formed.

5 Conclusion

In this paper, we formulated the problem of motif division, which decomposes polyphonic music into a small number of motif classes, as a set partitioning problem, and we obtained the solution using an IP solver. It was shown that the motif division provides useful information to understand the roles of motifs and how global musical structures are constructed from the motifs.

Future tasks include construction of a program that automatically analyzes global structures utilizing the obtained motifs and automatic composition of new pieces that use the same motifs as the original piece using the result of the analysis program. To create a criterion for determining adequate value of P automatically is also a remaining problem.

Acknowledgements This work was supported by JSPS Postdoctoral Fellowships for Research Abroad.

References

1. http://tetris.com
2. Cambouropoulos, E.: The local boundary detection model (LBDM) and its application in the study of expressive timing. In: Proceedings of ICMC, pp. 290–293 (2001)
3. Ferrand, M., Nelson, P., Wiggins, G.: Memory and melodic density: a model for melody segmentation. In: Proceedings if the XIV Colloquium on Musical Informatics, pp. 95–98 (2003)
4. Nemhauser, G.L., Wolsey, L.A.: Integer and Combinatorial Optimization. Wiley, New York (1988)
5. Balas, E., Padberg, M.W.: Set partitioning: a survey. SIAM Rev. **18**(4), 710–760 (1976)
6. Armstrong, M.A.: Groups and Symmetry. Springer, Berlin (2012)
7. Buteau, C., Mazzola, G.: From contour similarity to motivic topologies. Musicae Scientiae Fall 2000 **4**(2), 125–149 (2000)
8. Mazzola, G., et al.: The Topos of Music: Geometric Logic of Concepts, Theory, and Performance. Birkhäuser, Basel (2002)
9. Buteau, C.: Topological motive spaces, and mappings of scores motivic evolution trees. In: Fripertinger, H., Reich, L. (eds.) Grazer Mathematische Berichte, Proceedings of the Colloqium on Mathematical Music Theory, pp. 27–54 (2005)
10. Buteau, C.: Melodic clustering within topological spaces of schumann's *Träumerei*, Proceedings of ICMC, pp. 104–110 (2006)
11. Bishop, C.: Pattern Recognition and Machine Learning, pp. 423–430. Springer, New York (2006)

Diagrams, Games and Time (Towards the Analysis of Open Form Scores)

Samuel Vriezen

Abstract How to analyze open form scores? Generally, analysis will not be able to proceed by describing the architecture of a sequence of events, which renders most traditional analytical tools of music theory (structural voice leading, harmonic progression, thematic development, etc.) inoperative. But even scores that do not prescribe anything about event order contain an idea about time, and the treatment, or architecture, of time that is implicit in them is part of their musical subject matter. The temporal architecture in such compositions will be a non-linear field, a network of possible performance developments, and this structure, which we will refer to as a "time field", can be studied for its formal properties. What scores in open form express, then, is the character of a time field.

1 On the Concept of "Action Grammar"

A musical piece's take on time does not only express itself as an overarching network. It also expresses itself at every moment of indeterminacy during a performance. The space of musical actions that could occur at any moment, to be referred to as the "action grammar", and the way it is (locally) structured, determine the character of the musical subjectivity involved in actually performing a piece, that is, how it will feel to play the piece and the performer's role in it. In open form scores, then, there is a relationship between the performers' subjectivity and the structure of time itself. The shape of time is virtually present at every moment in performance, contributing to the musical expression as roads not taken hovering in the background to the events that do actually occur. Actions and time co-determine one another.[1] What analysis of open form scores requires, then, are tools that can address this interrelationship. Developing such tools requires letting go of some conventional assumptions about

---

[1] This perspective has been developed by the author in [8]. The terms "time field" and "action grammar" were first used there.

S. Vriezen (✉)
Independent Researcher & Composer, Amsterdam, Netherlands
e-mail: sqv@xs4all.nl

© Springer International Publishing AG 2017
G. Pareyon et al. (eds.), *The Musical-Mathematical Mind*,
Computational Music Science, DOI 10.1007/978-3-319-47337-6_30

299

music. Firstly, music does not coincide with the sounds heard: what constitutes a musical performance also includes the processes, choices and ideas that inform playing, giving shape to the audible. Secondly, as a consequence, we will not assume there to exist a single unified way of parametrizing musical structure. No single format for describing musical events—not even one that could account for the movements of every air molecule in a concert hall during a certain time span—will be assumed adequate to represent music in general. Thirdly, the essential characteristic of time itself will not be held to be its linearity. Instead, linearity of time will be interpreted as purely emergent—it is the way rich potential structures can collapse into single paths.[2] The enactment of such a collapse is precisely the act of performing; it is making time real.

It follows that all proposed structures for describing time fields must have a provisional character. Here, time fields will be assumed to consist of *states* or *situations* of the piece, together with laws governing how states can transition to other states. Our prototype here will be simple, finite, directed graphs. Since we assume no general musical structure, that which the states of a time field refer to will not generally be fixed either, and compositions may in fact imply multiple, interrelated time fields. Each such time field indicates an aspect of the piece in question. Each delineation of situations is one map of what the relevant parameters of the music can be. Any concept of music in general must include all conceivable time fields, which will not be a unified, consistent ("small") collection. These "states" of the piece can refer to the sonic qualities of moments. Examples include key areas and harmonic functions of tonal music, with the transitions between those determining tonal grammar. They also include anything generated in terms of the "combinatorial constraints" and "local morphological constraints" discussed by Michael Winter [9], with the former corresponding to the states and the latter to transitions. And they include the graph-based harmonic progressions of Tom Johnson, such as his compositions based on block design theory, including his sequence of *Networks* for piano.[3] However, the states may also refer to a developing *history* of a performance. Consider, as an example, *ba da duos* by Antoine Beuger. In this work, two performers alternate playing notes; but there is considerable freedom in the durations, and the notes could overlap, indeed almost be played simultaneously, but also be separated by pauses, so that performances that are entirely "staccato" would be possible as well as performances featuring a two-part entirely legato counterpoint, or any mixture of these two extremes. If these issues of phrasing are seen as determining, then the state space would not consist of qualities of a particular moment, but of the entire unfolding phrase. A highly simplified time field could have (from the second note played onwards) three states for the quality of the entire phrase: a "staccato" state of all the notes being separate, a "legato" state of all the notes overlapping, and a "mixed" state with pauses between some notes and overlapping of others. This simplified scheme has only two possible transitions, from both the first two states to the mixed one. Of

---

[2]We will, however, make a case for *sequentiality* and irreversibility as fundamental characteristics of time.

[3]Similar visual representations are investigated in [2, 4].

course, a full determination of the relevant state space for this piece should include much more subtle qualities and gradations.

How are these time fields to be related to the notion of *action grammars*? The transitions between states can be understood to be musical actions. This intuition is supported by mathematical terminology, as it is said that a group or monoid *acts* on a set, and such actions can be pictured in the form of a Cayley graph. Indeed, there exists a tradition of relating group structure to musical form, such as in Xenakis' description of his cello solo composition *Nomos alpha* [11] where the mapping of parametric sets to modes of playing shifts section by section according to a group action. Thus, monoid actions can be linked to time: according to a basic result in topos theory, the category of actions of a monoid M is a topos, of which the logic is classical iff M is a group. That is, if there is an element in M that has no inverse, the topos of its actions will have an intuitionistic logic, implying the existence of Kripkean models for it, and hence temporal progressions.[4] Going back to Xenakis' use of groups, it is striking that in his book *Formalized Music* he introduces his own group structures, as well as the *metabolae* that organize transitions between patterns, under the banner of "music outside time". The present focus will be on works structured on the basis of individual actions, which are not necessarily defined in monoid terms. Instead, we will propose a system for notating actions entirely from a local, bottom-up perspective. This system is inspired by Christian Wolff's composition *For One, Two or Three People*, and the notations will be named "Wolff diagrams" (Fig. 1).

Wolff notation, as it occurs in *For One, Two or Three People*, is a method for notating actions, specifically in their relationship to other actions or sounds that occur, as performed by the other players (or, in the case of solo performance, in the environment). The basic notation specifies very little about the actions themselves, though additional notational elements can of course always be added as needed, to indicate parameters such as pitch, loudness, timbre, and so forth. The basic interest here however is purely in temporal coordination between sounds. Take the example of this figure, from the score of *For One, Two or Three People*:

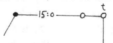

This is to be interpreted as follows: immediately following the end of another sound (in another part), play a short sound, followed by fifteen seconds of nothing (silence), followed by any sound, directly followed by any sound that is however to be precisely coordinated with something happening in another part. The 't' here indicates "a sound made by tapping or touching or tracing or the like". The full score consists of ten pages, with little diagrams like these spread all over the page, to be interpreted by the performer(s) in free order. Much of the fun of the piece derives from the unexpected ways these actions end up lining up and interacting, producing a very specific game of waiting, coordination, reaction, cueing among the players, a highly particular rhythm of interactions.

---

[4]This observation, which I owe to Fernando Zalamea, suggests that, though we have rejected linearity, irreversibility may be essential for time.

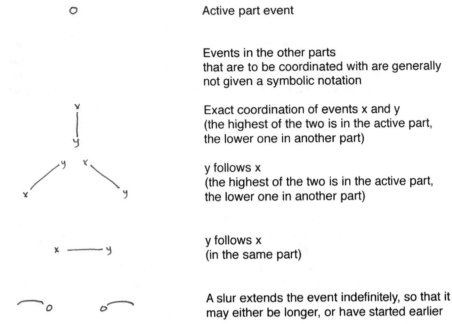

Fig. 1 Wolf Notation, basic symbols

Basic cases that this notation makes possible are given here as a Wolff alphabet (see Fig. 2). These give the notations for indicating that a certain action should coincide with another action in another part, with beginning, middles, or ends coinciding—or with the entire event.

Wolff's style of notation primarily notates *conditions* for a sound, determining under what circumstances a sound can start, what should happen during the sound, and when it may end. These conditions are not unlike the "states" or "situations" of the time fields as described above. An action notated in this way takes as its input the situation of the music, and by adding itself to it, transfers the music into a different situation.[5] Interestingly, these situations are themselves the result of actions, so that Wolff notation suggests a musical form where actions act on complexes of actions. The challenge to performers inherent in the notation is to make an action fit into the developing network, the performance's accumulating history, not unlike the way, in Tetris, a player should fit the falling blocks into the accumulating block

---

[5] Actions are thus very much like moves in a game, suggesting that compositions using Wolff diagrams, or equivalent instructions, can be studied in game theory terms. What is specifically interesting about the games that are made possible by Wolff's notation is that the moves all happen in real-time, potentially simultaneously, requiring a continuous game theory approach. Additionally, certain moves require coordinations that can only be reached at by cueing. That is, the game involves live (real-time) negotiation between the performers about the order of actions and the structure of time and coordination itself. It is in this sense that one can say, as we have done, that performance is "making time real"—doing so in real time.

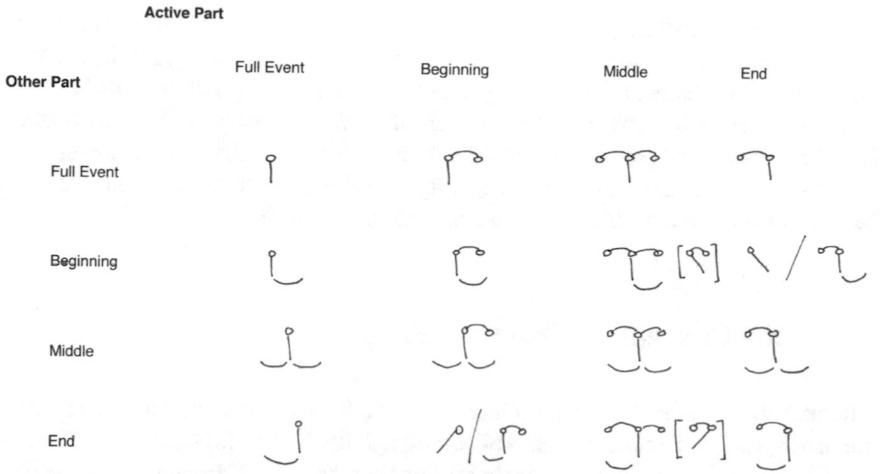

Fig. 2 A Wolff alphabet: notating basic action-coordination figures

structure.[6] These Wolff notations are diagrams to be inserted into an *emergent graph* of the performance, a graph describing the horizontal/vertical network of the events unfolding in time.

We describe Wolff diagrams as things that are to be embedded in a graph that will function as a code for the performance. It will be interesting to compare this set-up with the use of diagrams in gesture theory, as outlined by Guerino Mazzola and Moreno Andreatta in their 2007 paper, "Diagrams, Gestures and Formulae in Music" (see [5]). Vastly simplified, in that paper, diagrams refer to directed graphs, which act as representations of gestures coordinating multiple movements, and which can be embedded in topological spaces describing musical, or more generally physical, phenomena.

Certainly, the "emergent graphs" that we are talking about here could similarly be understood to be something close to a "gesture" coding an entire performance. But compared to the diagrams in Mazzola and Andreatta, the Wolff diagrams that we are describing here are generally *incomplete* graphs, with vertices that remain unconnected, as they indicate coordination with events not given within the diagram itself. One could say that the gestural approach, for which the authors present as paradigmatic the movement of a hand (including the arm and the fingers), stresses an expressive dimension: an external realization of the motion potential of a single body (an e-motion). The paradigm for Wolff is rather the social functioning of a musician, with the expressive dimension being supplemented with a dimension of interaction—and therefore openness and incomplete control. What is so intriguing about Wolff's notation is the way the actions tacitly presuppose one another's structures in order to function at all; which gives a piece like *For One, Two or Three People* its characteristic uncertain rhythm.

---

[6]On this very topic, see Tanaka and Fujii, in this volume.

However, a Wolff diagram-based action grammar and time field graphs can be related in other ways, too. A time field for the piece could be a graph of a tree of possible developments of its playing, much like Kripkean possible worlds. Every action will make possible a certain branch of performances and close off others.[7] Related to this is the notion of states describing a quality of the piece's "history", as described above (one can imagine a quality to correspond to a statement that "will be true" of the actual performance in a particular "world").

2 On the Concept of "Situation Field"

Alternatively, actions can also correlate to moving from some situation of the piece to another; associated graphs are maps of *situation fields*. These different time fields that can be conceivably associated with the composition will have structural relationships amongst themselves. For instance, if a full tree of possible worlds can be described, each of its nodes should correspond to an initial segment of an emergent graph, and the tree itself can be mapped surjectively onto the situation fields. An analysis of the form of the piece should account for the relationship between these different graph types. That is, the temporal architecture of an open form musical piece is a composite of graphs, and they could be many, as there may be multiple ways of measuring what the situations of a piece are, or even of what constitutes an "event"—and therefore, an emergent graph. Indeed, by the assumed non-existence of a general representation of music, the singular reality of the specific performance will not be available to us as a structure that answers to a generalised format; all we have are these composites of time fields.

As a very simple example of this relationship between an action grammar and a multiplicity of time fields, we will look at the ensemble piece *Ensemble* (2008) by the author. The score of this piece is in the form of a prose score, which implicitly determines a rigorous action grammar and a number of relevant time fields.

The idea behind this piece was to write a work in open form, in which every performer would at every moment of the piece be able to exert a maximal influence on its development. In order to complete a performance of *Ensemble*, cooperation is required among all the members of the group, but any member can always choose to "sabotage" the next step and postpone completion. Thus, the piece is structured like a game, and the score gives its rules:

Ensemble

> At least four instruments. No more instruments than would permit every performer to hear every other performer clearly.

---

[7]This concept is related to Đurđevich's *quantum circles theory*, in this volume. See also the concept of *musical recursion* in [6, pp. 94–97, 173–175], for a more in detail explanation; and [6, pp. 207–220] for the concept of "tree of possible developments" in music.

Three movements. In each movement a group of 4 pitches: I – C D F G; II – C# D# E F#; III – D E♭ E F, positioned within the span of one fifth. (The piece may be freely transposed to other pitches and to every octave, as long as every pitch can be played by every instrument).

Each movement starts and ends in silence. At any moment between those two silences 1, 2 or 3 of the pitches are heard. The second silence may only start when each of the 4 pitches has been heard by itself at least once, and when each instrument has played each pitch at least once (in mvt. I), twice (in mvt. II) or three times (in mvt. III).

Play single tones (no figures). Play such that everyone can be heard. Maximum dynamics is *mf* in mvt. I, *p* in mvt. II, *pp* in mvt. III.

These instructions imply that the ensemble, as a whole, has to navigate a field of harmonic possibilities, such that every one of the four pitches in a movement will be heard unison at least once. This means transitions have to be forged from unison situation to unison situation. Situations with at least two notes sounding will function as bridges. However, situations with three notes sounding are admissible as well, which create ambiguity of the direction of harmonic development. These are the uncertainties that the ensemble has to negotiate. Additionally, the harmony needs to be sustained, since there has to be at least one pitch at all times during a movement.

The rules imply the actions available to performers in the piece. Essentially: if one or two notes are sounding, the available actions are to play any of the four notes, and stop at any moment, provided another sound is going on. If three notes are sounding, only one of those three can be played. Furthermore, beginning and ending actions can be specified: following a general silence, any note can be played; and if the conditions for ending the movement have been met, it is possible to have a note followed by general silence. These actions can be notated in terms of Wolff diagrams, using an adapted version of the Wolff notation, in which we do specify the pitch content of events, and add a symbol "ø" to indicate a general silence. As in the Wolff diagrams, the active part is notated on the upper staff, and the environmental conditions that affect the active part to be coordinated with it, on the lower staff. This notation is equivalent to the harmonic conditions outlined in the prose score, but now given from the immanent perspective of individual performers, explicitly stating the space of their available actions (Fig. 3).

These same rules can also be made explicit in terms of time fields, as graphs. To give an example of the multiplicity of the fields involved, let us assume that *Ensemble* will be performed by a quintet. Then all the admissible situations involve any subset (except the empty subset) of the five playing, and any subset of the four notes, except the empty and the full subsets, sounding. There are, then, at least two time fields that are implied by the rules. They are both given here in graph form, with a connection indicating a minimal change: a part or note being added or dropping out. In this case, all these transitions are reversible, so the graphs need not be directed.

The time field for the harmonies is an incomplete *tesseract* (every one of the four pitches can be present or absent, giving a fourth-dimensional structure), using a total of fourteen nodes. Likewise, the time field for the sub-ensemble formations is an incomplete *pentaract*, of thirty-one nodes. The two fields can in fact be seen as

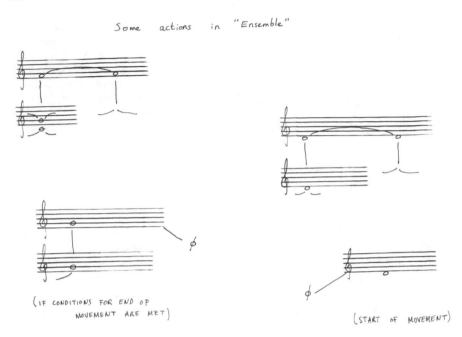

Fig. 3 *Ensemble* (2008) basic actions

contracted versions of a much larger graph of situation transitions, where the nodes include information about both harmony and active sub-ensemble (Figs. 4 and 5).

This field adds more structure, since it will need to specify which instrument is playing which pitch of the chord. As a result it has over fifteen hundred nodes—much more than fourteen times thirty-one. This already gives three ways of looking at what the "states" of the musical structure are. The harmony field is the most central in the definition of *Ensemble*'s structure, but the other fields are relevant to its operation, too. A full analysis of the possibilities of performance could of course include much more information about the piece's states; including dynamics, colouring, articulation and so on. All of these would lead to vaster fields.

Additionally, there are time fields that sketch the conditions for ending a movement. These indicate possible sequences of events towards fulfilling these conditions. Thus, all four pitches have to be heard in unison—meaning we move from a starting situation (the first pitch played) to a situation in which two pitches have been heard in unison; to one, where three pitches have been heard, and then to the final unison, at which point the piece may stop. These fields can be given in the form of directed graphs.

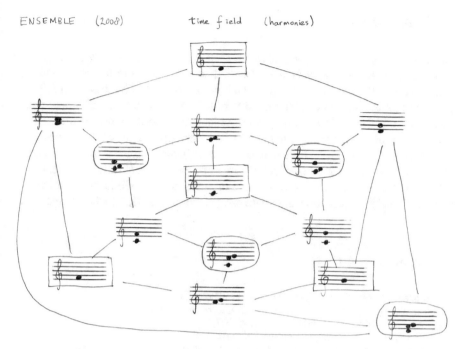

Fig. 4 *Ensemble*'s time field (harmonies) representation

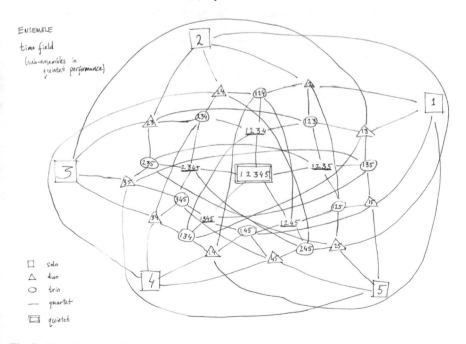

Fig. 5 *Ensemble*'s time field sub-ensembles

3 Conclusions

An adequate analysis of the form of works such as *For One, Two or Three People* and *Ensemble*, needs the tools that can deal with this multiplicity of time fields, and their relationship to the actions themselves. The pieces are shaped by these two different perspectives: the bottom-up, individual subjectivity given by the individual action grammars, and the bird's eye-view, group subjectivity described by the time fields; as well as the real-time process of negotiating the interdependence of these two levels.

The performer's task in these pieces is to coordinate individual actions, to arrive at a specific collective action, articulated in—or even, as—real time. The creation of more rigorous mathematical tools to develop these notions will help to outline new spaces of musical exploration, and to investigate how human interaction shapes time.

References

1. Beuger, A.: ba da duos. Edition Wandelweiser, Haan (2006)
2. Jedrzejewski, F., Johnson, T.: Looking at Numbers. Springer, Basel (2014)
3. Johnson, T.: Networks. Editions 75, Paris (2007)
4. Johnson, T.: Other Harmony. Editions 75, Paris (2014)
5. Mazzola, G., Andreatta, M.: Diagrams, gestures and formulae in music. J. Math. Music **1**(1), 23–46 (2007)
6. Pareyon, G.: On Musical Self-Similarity. The International Semiotics Institute, Imatra-Helsinki (2011)
7. Vriezen, S.: Ensemble (2008). https://sqv.home.xs4all.nl
8. Vriezen, S.: Action Time. The Ear Reader (2013). http://earreader.nl
9. Winter, M.: On Minimal Change Musical Morphologies (2016). In the present volume
10. Wolff, C.: For One, Two or Three People. C. F. Peters, New York (1964)
11. Xenakis, I.: Formalized Music. Pendragon Press, New York (1992)

On Minimal Change Musical Morphologies

Michael Winter

Abstract In this paper, we examine a number of minimal change musical morphologies. Each morphology has an analogous representation in mathematics. Our mathematical objects of study are Gray codes, de Bruijn sequences, aperiodic necklaces, disjoint subset pairs, and multiset permutations with musically motivated constraints that result in several open problems.

1 Introduction and Preliminaries

Several different minimalist trends exist in art and music. In this paper, we focus on minimal change musical morphologies where the word "minimal" primarily denotes "minimal change" between adjacent elements in a given morphology. Each morphology has an analogous representation in mathematics. Our mathematical objects of study are Gray codes, de Bruijn sequences, aperiodic necklaces, disjoint subset pairs, and multiset permutations with musically motivated constraints that result in several open problems.

First, we discuss a taxonomy of "morphological constraints" used to contextualize the definition of each morphology. Next, we review previous research in order to show the genesis of our current formalization. Section 2 focusses on examples of minimal change musical morphologies and the open mathematical problems that result from musically motivated constraints on the analogous mathematical representations. We conclude with an overview of the open problems and suggestions for further research.

1.1 Scope and Complexity of Morphological Constraints

In this paper, the morphologies and their analogous mathematical representations are defined by a subset of four types of morphological constraints: Combinatorial

M. Winter (✉)
Independent Researcher, Los Angeles, CA, USA
e-mail: mwinter@unboundedpress.org

© Springer International Publishing AG 2017
G. Pareyon et al. (eds.), *The Musical-Mathematical Mind*,
Computational Music Science, DOI 10.1007/978-3-319-47337-6_31

Constraints (CC), Local Morphological Constraints (LMC), Global Morphological Constraints (GMC), and Optimal Global Morphological Constraints (OGMC). This taxonomy shows a hierarchy of scope (from elements of a morphology to its large-scale form) and computational complexity (from easy to difficult to compute). The complexity at each hierarchical level is based on the satisfaction of that constraint in conjunction with all lower-level constraints.

1.1.1 Combinatorial Constraints (CC)

A CC is a constraint that defines all the elements of a morphology. As no further constraints are imposed (between adjacent/pairwise elements or among sets and sequences of elements), computing a set of elements defined by a CC is generally easy/efficient. For example, in Sect. 2.1.1, we discuss a morphology where the CC is that each element must be a subset of a set of n sounds. The set of all subsets (the powerset) can be represented mathematically by all binary words of size n where each bit position corresponds to one of the sounds.

1.1.2 Local Morphological Constraints (LMC)

A LMC is a constraint at the next-higher hierarchical level; i.e., between adjacent elements in a morphology. A morphology that satisfies a LMC with no higher-level constraints is generally easy to compute. However, the computation is likely to take more time and resources than generating a set of elements using a CC alone unless the known fastest algorithm that satisfies the CC also satisfies the LMC. Continuing with the example that will be discussed in Sect. 2.1.1, the LMC is that from subset to subset, only one sound can be added or removed; or framed mathematically, only one bit can flip from word to word.

1.1.3 Global Morphological Constraints (GMC)

A GMC constrains a statistical property of the morphology. For most of the morphologies detailed below, the GMC is that each element defined by the CC occurs only once; e.g., any given subset of sounds or binary word is never repeated. Unless the known fastest algorithm that satisfies all lower-level constraints also satisfies the GMC, finding a morphology that satisfies a GMC is harder than just a LMC and/or CC. Often, the difficulty increases exponentially with respect to the number of elements defined by the CC. For example, some of our morphologies can be generated by finding Hamiltonian paths (where each vertex is traversed only once) in representative graphs. Finding Hamiltonian paths is known to be NP-complete for arbitrary graphs as brute-force search times typically explode exponentially with the size of the graph. Section 1.1.5 further explicates this taxonomy's relation to graph theory.

1.1.4 Optimal Global Morphological Constraints (OGMC)

An OGMC constrains sets or sequences of three or more elements in the morphology (as opposed to just adjacent elements as with the LMC) thus defining a subset of morphologies satisfying all lower-level constraints. In most of the examples in Sect. 2, the OGMCs are satisfied such that the order of elements minimizes or maximizes some feature/characteristic of the morphology (hence the use of the word "optimal"); e.g., codes that have maximally uniform, long run-lengths and sequences where the running sum is minimized. Depending on the OGMC, finding a satisfactory morphology can be extremely hard with complexity on the order of solving difficult games and puzzles.

1.1.5 Relation to Graph Theory and Constraint Programming (CP)

Two methods used to generate some of the morphologies detailed in this paper have cogent relations and near-analogs to the above taxonomy: finding paths in representative graphs and searches using Constraint Programming (CP; see [20]).

A graph with vertices defined by CCs and edges induced by LMCs is essentially a structural representation of the morphology. The graph can be used to generate the morphology by finding a path that satisfies any defined GMCs and OGMCs. Generating morphologies using this technique illustrates an important, if not fundamental, link between morphology (or shape) and structure.

In CP, a solver searches for a solution that satisfies a programmed set of "binary" and/or "global" constraints applied over a "domain" by optimizing a set of "objectives" (minimizing or maximizing a set of functions). A domain in CP is equivalent to a set of elements defined by a CC. Binary constraints are similar to LMCs as they both involve only two variables. Global constraints and GMCs relate because they both involve more than two variables and are constraints at a hierarchical level higher than binary constraints and LMCs, respectfully. However, global constraints also relate to OGMCs as both constrain sequences or sets of elements. Thus, a global constraint could be considered as something between a GMC and an OGMC. Notwithstanding the connection between global constraints and OGMCs, CP objectives clearly relate to OGMCs because of the optimization process.

To summarize, Table 1 shows the morphological constraint taxonomy in relation to hierarchical scope, difficulty, and analogs to graph theory and CP.

1.2 Precedence of Musical Thinking with Respect to Morphological Constraints

The two pieces described in this section exemplify a compositional process where musical morphologies are defined by morphological constraints. Both have

Table 1 Summary of morphological constraint taxonomy

Constraint	Hierarchical Scope	Difficulty	Graph Theory	CP
CC	elemental	easiest	vertices	domain
LMC	pairwise	easy	edges	binary constraint
GMC	global	hard	path	global constraint
OGMC	global optima	hardest	optimal path	objective

well-defined CCs, LMCs, and GMCs, but do not have OGMCs. While several of the pieces described later have OGMCs, the concept was theoretically formalized only recently for the purposes of this paper.

The mathematical implications of the following two examples are investigated more thoroughly in "Chordal and timbral morphologies using Hamiltonian cycles" [1], where the authors show the conditions that admit a Hamiltonian path or cycle[1] in representative graphs derived and generalized from the pieces. Section 2 is a focussed extension of the ideas in the aforementioned article: focussed in that we look exclusively at minimal change morphologies and extended in that we also look at examples with OGMCs. "Chordal and timbral morphologies using Hamiltonian cycles" also provides a historical context connecting this work to the work of James Tenney and Larry Polansky among others (specifically, Tenney's definition of form as shape and structure in *Meta+Hodos* [26] and Polansky's definitions of "morphological metrics" [16]). These writings along with the author's dissertation "Structural Metrics: an epistemology" [29] further illustrate the genesis of compositional thinking detailed throughout this paper.

1.2.1 Maximally Smooth Chordal Cycles

In a "maximally smooth cycle", as defined by Richard Cohn [6], one part moves by a semitone or whole step while the other parts remain on the same pitch. Tom Johnson's piece *Trio* (2005)[2] is a variant of this idea that exemplifies well-defined morphological constraints. In *Trio*, each pitch in a four-octave chromatic set is represented by a number 0 to 48 where middle C equals 24. The musical morphology enumerates through all three-note chords satisfying the CC that the numbers representing the pitches within each chord are distinct integer partitions without repetitions of 72. The LMC is that from chord to chord, one pitch must remain the same while the

---

[1] A Hamiltonian cycle is basically the same as a Hamiltonian path except it returns to the start vertex.
[2] Score to this piece available at http://www.editions75.com (accessed January 2015).

Trio

288 three-note chords with sums of 72 (middle C = 24)

Tom Johnson

Fig. 1 Excerpt from score of *Trio*

other pitches move by a semitone in contrary motion. The GMC is that each chord occurs only once. The morphology is analogous to a Hamiltonian path (which by definition satisfies the GMC) on a graph where the vertices represent the chords defined by the CC and edges are induced by the LMC. Figure 1 shows the first system of the score to *Trio*.

1.2.2   Maximal Change Timbral Morphologies

In the author's piece *maximum change* (2010),[3] the elements are all timbral possibilities of a chord with 4 pitches using 4 instruments assuming that each instrument can play up to all of the pitches at once (the CC), which framed mathematically are all 4-tuples where the position in the tuple represents the pitch to which an instrument, represented by a number at that position, is assigned. The LMC is that from chord to chord, each pitch is played by a different instrument; or mathematically, from tuple to tuple, each position is assigned a different number. That is, the same chord is repeated but the mapping of instruments to pitches changes as maximally as possible. The GMC is that each timbral possibility occurs only once. The morphology is analogous to a Hamiltonian path in a graph where the vertices represent the timbral possibilities (or mappings of instruments to pitches) defined by the CC and edges are induced by the LMC.[4] Fig. 2 shows the first 10 measures of the score to *maximum change*.

---

[3]Score to all the author's pieces available at http://www.unboundedpress.org (accessed January 2015). Unless otherwise specified, all works discussed are that of the author.

[4]This is essentially the opposite of the types of morphologies discussed in this paper. However, in [1], the problem and corresponding graph are generalized such that the number of elements that stay the same from tuple to tuple is specified. *maximum change* is a specific instance where the number of elements that stay the same from tuple to tuple is 0.

Fig. 2 Excerpt from score of *maximum change*

2 Minimal Change Musical Morphologies: Applications and Resulting Mathematical Problems

2.1 Gray Codes

A Gray code (after Frank Gray [10]) is an enumeration of all binary words of a given length n (the CC) such that only one bit changes from word to word (the LMC) and each word occurs only once (the GMC). We reserve a discussion of OGMCs for the following subsection. For a comprehensive overview of Gray codes, see Carla Savage's "A survey of combinatorial Gray codes" [21].

2.1.1 Maximally Balanced, Maximally Uniform Long-Run Gray Codes

The musical composition *gray codes* (2009) is an exploration of all subsets of a set of sounds such that the overall sound changes as minimally/gradually as possible over time. Each instrument (or sound) follows one bit position in a Gray code. An instrument is sounding when 'on' (or 1) and not when 'off' (or 0). The score gives the following description of an OGMC desired to generate a version of the piece.

> "Ideally, a particular type of Gray code is desired to achieve this effect [of minimal change]. That is, a Gray code where the standard deviation of all run-lengths plus the standard deviation of bit flips across the positions is as close to 0 as possible."

The score then includes a version for orchestra with an 8-bit Gray code. Every subset of instruments from 8 groups—flutes, oboes, clarinets, bassoons, horns, vibraphones, strings I (violins and violas), and strings II (cellos and basses)—sounds together once at some point in the piece. A realization is played exclusively on one pitch with a gradually changing overall timbre.

While Gray codes are defined by the LMC that only one bit changes from word to word, standard Gray codes are not at all balanced (in that each bit position in the enumeration has similarly many bit flips as all others). The musical consequence of using an unbalanced code is that some sounds will come in and out more frequently than others. Also, run-lengths may vary tremendously in standard Gray codes. For examples of both these factors, see Fig. 3a as compared to Fig. 3b and/or Fig. 3d from Donald Knuth's *The Art of Computer Programming* [12]. The OGMC given in the score aims to ensure that each part changes as infrequently as all others by assuming that balanced Gray codes generally have uniformly long run-lengths.

Knuth's long-run Gray code (Fig. 3d) is the one used for the orchestral version of the piece *gray codes*. The canonical transition sequence for Knuth's code is given in Fig. 4. Each number represents the bit position where the bit flip occurs in each successive binary word. For the orchestral realization of *gray codes*, 0, 1, 2, 3, 4, 5, 6, 7 represent clarinets, strings I, flutes, vibraphones, oboes, strings II, bassoons, and horns, respectively.

Another possible Gray code that could be used to generate the piece is called a Beckett-Gray code (after the playwright Samuel Beckett). Beckett defined this particular type of Gray code for his work *Quad* (1981), where he wanted all combinations of performers to be on stage at some point throughout the work such that the one who has been on stage the longest will always be the next to exit. Mathematically speaking, the OGMC of a Beckett-Gray code is that the position with the current longest 'on' bit run will always be the next to flip 'off'. By definition, a Beckett-Gray code should be quite balanced and have reasonably uniform, long run-lengths.

It turns out that a 4-bit Beckett-Gray code does not exist, which is why Beckett was unable to implement his original idea and altered it in order to finish the piece. Recently, an 8-bit Beckett-Gray code was found by Brett Stevens, et al. [24]. Shortly after, a fast algorithm to generate Beckett-Gray codes was defined by Joe Sawada, et al. [22]. The canonical transition sequence for the 8-bit code presented in [24] is given in Fig. 5.

Several open questions arise from the need of a Gray code that is highly balanced and has uniformly long run-lengths such as how to define and encode the OGMCs. It is unclear if the OGMC given in the score of the piece *gray codes* is adequate as it relies on the assumptions that balance will result in uniformly long run-lengths and that minimizing the standard deviation of the number of bit flips across the positions will balance the code. The difficulty of the optimization problem is compounded by the fact that there are three potential optima: maximal uniformity of run-lengths, maximality of run-lengths, and balance.[5] How these optima might relate or conflict both perceptually in the resulting sound and with respect to modelling the problem warrants further investigation.

---

[5]In CP, this is called a multi-objective problem (see [11, 18, 19]). For example, one could weight and sum the objectives to prioritize conflicting optima.

Fig. 3 Examples of 8-bit
Gray codes from Donald
Knuth's *The Art of Computer
Programming*. **a** standard;
b balanced; **c**
complimentary; **d** long-run;
e nonlocal;
f monotonic; **g** trend-free

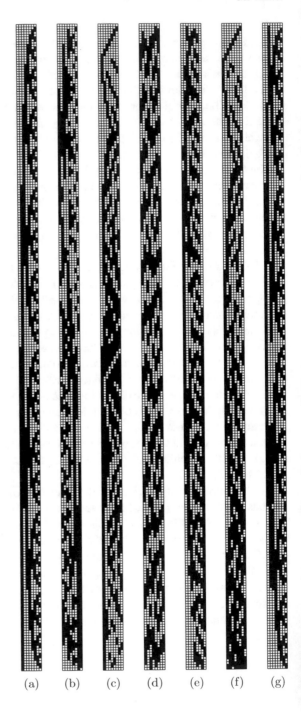

(a)   (b)   (c)   (d)   (e)   (f)   (g)

10623517425016352071452631502735146205173250164523710526315427 05
13620517425316052371452601532705146235170253164520713526015427 35
10623517425016352071452631502735146205173250164523710526315427 05
13620517425316052371452601532705146235170253164520713526015427 35

Fig. 4 The canonical transition sequence for Knuth's long-run Gray code

01234560701213243565760710213534626701537412362567017314262065 70
13421465605731024645375710204353761407363046427370356402713275 05
41210275641502403654250136025416156043125760325720431576243217 60
45204175163547670356475706254372421326241615234175143671431643 14

Fig. 5 The canonical transition sequence for the 8-bit Beckett-Gray code found by Stevens, et al.

2.2 De Bruijn Sequences

Next, we examine de Bruijn sequences (after Nicolaas Govert de Bruijn [7]). Formally defined, a de Bruijn sequence $B(k, n)$ is a cyclic sequence of a given alphabet A of size k in which every word of length n in A appears uninterrupted only once. Essentially, using a de Bruijn sequence is the fastest way to brute force hack a combination lock[6] with combination size n because the last $n - 1$ symbols of a word in the sequence will always overlap with the first $n - 1$ symbols of the next word.

The morphological constraints of a de Bruijn sequence are nicely illustrated by a particular type of directed graph referred to as a de Bruijn graph. In a de Bruijn graph, the vertices are all words of length n from a given alphabet A of size k (the CC) and two vertices are connected by a directed edge if the last $n - 1$ symbols of the out-vertex overlap with the first $n - 1$ symbols of the in-vertex (the LMC; for an example, see Fig. 6). The sequence itself can be constructed by finding a Hamiltonian cycle (the GMC) on such a graph.

Fig. 6 $B(3, 2)$ with Hamiltonian cycle (indicated by *dashed lines*)

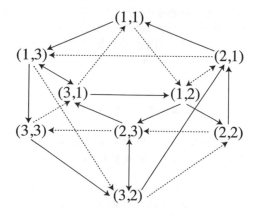

---

[6]This is assuming that one does not need to reset anything after entering each combination.

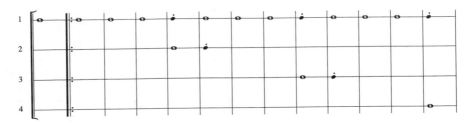

Fig. 7 Excerpt from score of *room and seams*

There are several known algorithms to generate de Bruijn sequences. Notably, Harold Fredricksen and James Maiorana have defined an algorithm that generates a lexicographic least de Bruijn sequence by concatenating the lexicographic sequence of Lyndon words of length divisible by n [9]. A Lyndon word is a string that is smaller in lexicographic order than all its rotations. Not only are Lyndon words useful in efficient generation of de Bruijn Sequences, they are also representatives of aperiodic necklaces; the topic of Sect. 2.3.

2.2.1 Spatial de Bruijn Sequences

In the piece *room and seams* (2008), 4 groups of performers are located in a room as far as possible from all other groups. The piece enumerates all spatial sequences of size 4 in the shortest morphology possible. The representative de Bruijn sequence has both an alphabet size and word length of 4. Each symbol in the alphabet represents a location in space articulated by the sounding of a tone from the group at that location. As no OGMCs were defined, the de Bruijn sequence used to generate the piece was computed with a program written by Hakan Kjellerstrand that implements the algorithm by Fredricksen and Maiorana mentioned above.[7] An excerpt from the score is provided in Fig. 7. Despite lacking an OGMC, this example still demonstrates one of many ways even a standard de Bruijn sequence can be of interest musically. Our next example extends the standard de Bruijn sequence with an OGMC necessitated by a musical practicality.

2.2.2 Space-Limited Contour de Bruijn Sequences

The piece *dissection and field* (2008) enumerates all melodic contours of size 6 in the shortest morphology possible. The representative de Bruijn sequence has an

---

[7]The program for this de Bruijn sequence generator written in Java is available at http://www.hakank.org/comb/deBruijn.java (accessed January, 2015). It is a port of Frank Ruskey's C and Pascal versions, which are available upon request from the Combinatorial Object Server at http://theory.cs.uvic.ca/inf/neck/NecklaceInfo.html (accessed January, 2015).

alphabet of three numbers $\{-1, 0, 1\}$ that indicate direction in a pitch morphology: down, same, up, respectively.

In order for pitches not to get extremely high or low, the range of the sequence's running sum is constrained. We denote a de Bruijn sequence constructed from an alphabet of integers that sum to 0 with an OGMC that constrains the range of the running sum as "space-limited". Due to this added constraint, the algorithm used for *room and seams* was not suitable. The final contour morphology for *dissection and field* was found by Kjellerstrand. After unsuccessfully trying to find a solution by brute-force searches for Hamiltonian cycles in a de Bruijn graph, Kjellerstrand turned to CP in order to limit the solution space for greater efficiency by defining an objective that minimized the difference between the extremal values in the running sum while satisfying the fundamental de Bruijn sequence constraints.[8] The final pitch morphologies[9] for *dissection and field* were created from several composed melodic fragments such that, when pieced together, ultimately conformed to the contour sequence as a whole. That is, the contour sequence was reconstructed from the melodic fragments. Both the sequence and an excerpt from the score are provided in Figs. 8 and 9, respectively.

While Kjellerstrand found a solution satisfactory for the creation of *dissection and field*, it remains an open question whether or not it is optimal. That is, whether it is the de Bruijn sequence $B(3, 6)$ with alphabet $A = \{-1, 0, 1\}$ where the difference of the extremal values of the running sum (which is 8 for the sequence used in *dissection and field*) is smaller than all other such de Bruijn sequences. As an extension, the general case of lower bounds on the range of extremal values in the running sum of optimal solutions for space-limited de Bruijn sequences also remains open.

---

[8] The program that generated the final solution for *dissection and field* is available at http://www.hakank.org/minizinc/debruijn_space_limited.mzn (accessed January, 2015). It was written in a CP language called MiniZinc available at http://www.minizinc.org/ (accessed January, 2015). Kjellerstrand has several other implementations using various CP languages to find traditional de Bruijn sequences with the running sum constraint relaxed at http://hakank.org/common_cp_models/#debruijn (accessed January, 2015). Also, while MiniZinc is considered a general high- or medium-level CP language, there is also a music specific CP language written by Torsten Anders called Strasheela available at http://strasheela.sourceforge.net (accessed January, 2015). Anders and others have produced interesting results modelling music theories using CP (see [2, 3, 27]).

[9] The piece integrates the space-limited de Bruijn sequence with other formal concerns. Two morphologies were constructed from the contour sequence; one for each of two groups. That is, the groups play the same contour but on different notes (with one group always higher than the other). In the score (see Fig. 9), the notes and rests (the latter of which were arbitrarily inserted using the caret symbol) have numbers and tick marks that indicate general durations (which were also arbitrarily/intuitively assigned). Each performer plays independently of the others, which blurs the sequence to some extent. Also, one of the performers from the first group departs from the sequence for a significant portion of the piece to sustain a high-pitched tone.

```
+  +  x  -  -  +  +  -  -  x  -  -  +  x  x  -  x  +  +  +  x  +  -  -  -  +  -
-  -  +  x  x  +  +  x  x  x  +  -  x  x  -  x  x  -  -  -  +  +  -  x  -  +  +
+  -  x  +  -  +  -  +  +  -  +  -  x  -  x  x  x  x  x  x  x  +  -  -  +  x  x  x
x  -  x  x  +  x  -  +  +  +  x  +  +  x  +  +  -  -  x  x  +  -  -  -  x  -  -
x  x  +  +  -  +  x  +  x  -  +  x  -  +  -  -  x  x  -  x  +  x  x  x  -  x  -
x  -  x  x  +  +  x  +  -  x  x  x  x  +  +  -  x  -  x  -  +  -  -  +  +  +  +
-  -  x  +  +  +  +  +  +  x  x  -  x  x  x  +  -  +  -  -  x  +  x  x  -  +  -
+  +  x  -  -  -  x  x  +  x  x  +  x  +  x  x  -  -  x  x  -  -  +  -  +  -  -
+  x  -  +  x  +  +  +  x  -  x  +  -  x  +  -  x  -  +  -  +  -  +  -  x  +  +
x  +  x  x  x  x  x  -  +  -  x  x  x  +  x  x  -  x  -  -  x  +  -  x  x  +  x
+  -  -  x  +  -  -  +  +  x  +  -  +  +  -  x  x  +  -  +  x  -  -  -  +  -  +
+  +  x  -  +  -  +  x  -  x  x  -  +  x  +  -  -  +  -  +  x  +  -  +  -  x  x
-  -  x  +  x  +  x  -  -  x  x  x  -  -  -  -  x  -  +  -  x  +  x  x  +  -  x
-  x  +  -  +  x  x  +  -  +  +  +  +  x  +  x  +  +  x  -  -  x  -  x  -  -  -
+  x  +  x  +  x  +  -  +  x  +  +  -  x  +  x  -  +  -  x  -  -  +  +  +  -  +
x  x  x  -  -  +  +  x  -  +  x  x  -  -  +  x  -  x  +  +  -  +  -  +  x  x  -
+  x  -  -  +  -  x  +  -  -  x  x  x  +  +  x  -  +  +  x  -  x  x  x  -  +  +
x  x  -  -  -  x  -  x  +  +  x  -  x  -  -  +  -  -  +  -  -  x  -  +  +  x  +
+  +  +  -  +  +  -  -  -  x  x  -  +  -  -  -  x  +  -  +  +  x  x  +  +  +  x
x  x  x  +  x  -  -  -  -  +  x  -  -  x  +  +  x  x  +  -  -  x  -  x  x  -  x
-  +  +  -  x  +  +  +  -  x  x  -  +  +  -  -  +  x  +  -  x  -  -  -  x  +  +
-  -  +  -  x  x  +  +  +  -  -  +  +  -  +  x  -  +  +  -  +  +  +  -  -  -  -
-  -  x  +  x  -  x  x  +  -  x  +  x  +  +  -  +  +  x  +  x  -  x  +  x  +  -
x  +  +  -  x  x  x  -  x  +  -  -  -  -  +  -  x  -  +  x  +  x  x  +  +  -  -
-  +  +  x  x  x  -  +  x  x  +  x  -  x  -  +  x  -  x  -  x  -  x  +  x  -  -  +  x
+  +  x  x  -  +  +  +  +  +  -  x  -  -  x  -  -  -  -  -  +  +  +  x  x  +  x
x  x  +  x  +  +  +  -  +  -  -  -  -  x  x  x  x  -  -  x  -  +  x  x  x  +  +
+  +  x  -  -  -
```

Fig. 8 de Bruijn sequence used for *dissection and field* where -1, 0, and 1 map to $-$, $\times$, and $+$, respectively

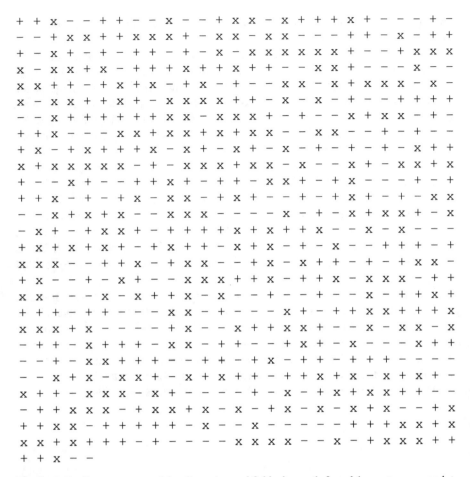

Fig. 9 Excerpt from score of *dissection and field*

2.3 Aperiodic Necklaces

The piece *necklaces* (2014) is a minimal change ordering of unique picking patterns using a set of plucked strings where the resultant pitch of each string is the same as all others. Each pattern in the piece is analogous to a representative of an aperiodic necklace, which is an equivalency class on aperiodic strings under rotation and permutation of the symbols. For example, $(0, 1, 0, 2, 3, 3)$ is equivalent to $(1, 0, 2, 3, 3, 0)$ under rotation and $(2, 3, 2, 1, 0, 0)$ under symbol permutation. Therefore, they are all representatives of the same aperiodic necklace. In the last section, we discussed how de Bruijn Sequences can be generated efficiently by concatenating Lyndon words ordered lexicographically. The morphology for *necklaces* is related because each aperiodic necklace contains one Lyndon word which means Lyndon words form representatives of aperiodic necklaces.

To demonstrate how equivalencies of aperiodic necklaces relate to the uniqueness of picking patterns, lets map the example above to a traditionally tuned soprano ukulele: IV $\rightarrow$ G_4, III $\rightarrow$ C_4, II $\rightarrow$ E_4, I $\rightarrow$ A_4. $(0, 1, 0, 2, 3, 3)$ could represent $(I_0, II_5, I_0, III_9, IV_2, IV_2)$ which all sound A_4.[10] Under permutation of symbols (the strings of the ukulele in this case), $(III_9, IV_2, III_9, II_5, I_0, I_0)$ results in the same pattern because all the strings still sound A_4 and the rhythm remains unchanged. In the piece, each pattern may be repeated several times successively. This is why equivalency under rotation is also necessary. When repeated successively, the rhythmic character of the repeated pattern remains the same regardless of which rotational representative is used.

necklaces enumerates through all unique picking patterns of length 6 or less using 4 strings (the CC) such that from pattern to pattern one element is added, removed, or changed (the LMC). Not considering the immediate repetitions, each pattern occurs only once (the GMC) except for the patterns of length 1 and 2 (explained below). The OGMC is that the morphology submit to an "arc" form where the lengths of the patterns generally increase then decrease. An excerpt of the score is provided in Fig. 10.

A solution that satisfies the morphological constraints outlined above can be generated by finding a Hamiltonian path on a graph where the vertices are representatives of the aperiodic necklaces and edges are induced by the LMC. Note that the graph can change substantially based on which representatives are chosen. By definition, adjacent necklaces cannot differ in length by more than one because of the LMC. Clearly the graph cannot submit a Hamiltonian cycle since the trivial necklace, (0) and its equivalencies, can only connect to one other necklace: that of length 2, $(0, 1)$ and its equivalencies. However, the graph does submit a Hamiltonian cycle if the

---

[10]In this example, the letters indicate note names with subscripts that indicate octaves whereas the Roman numerals indicate string numbers of the ukulele with subscripts that indicate frets; 0 being the open string. It is also understood that this fingering, even on the ukulele, is very difficult. We use it just for demonstrative purposes. All the performances to date (as of January 2015), have been with a ukulele where the open strings are tuned to the same pitch. Minor variations in tuning, string tension, and string gauge contribute to the overall sound of the piece even though conceptually, the strings are assumed to be equivalent.

Fig. 10 Excerpt from score of *necklaces*. Each cell indicates a picking pattern given by a tablature where each line represents one of 4 plucked strings and the horizontal axis shows order. The ring around a cell is the necklace representation where the numbers, starting from the number centered above the tablature and moving clockwise, correspond to the strings of the picking pattern

trivial case is excluded. Intuitively, it can be seen that a Hamiltonian cycle instead of just a path is more likely to satisfy the OGMC because the only way to return to the start vertex would mean that you would have to generally increase then decrease the length of the necklaces. Otherwise, all vertices adjacent to the start vertex would already be traversed.

The final morphology of the piece was found by brute force (implemented in the programming language Mathematica) as follows. Generate a graph where a vertex represents all representatives of a given aperiodic necklace (excluding the trivial case) and two vertices are connected if any of their representatives satisfies the LMC (note that this graph is highly connected). Starting at the vertex representing the necklace of length 2, randomly choose one of its representatives. Then from all adjacent vertices remove any representatives that no longer satisfy the LMC (which might eliminate some of the edges altogether). Then randomly choose one of the remaining adjacent representatives. Repeat this process until either a Hamiltonian cycle is found or until the path cannot extend any further, in which case, start over.

As a Hamiltonian cycle was found, the necklace of length 2 is repeated since it was the start and end vertex. Then, the trivial case was added at the beginning and end.

It is unknown to the author whether or not enumerations through aperiodic necklaces similar to the one above can be generated more efficiently. Also, it is unknown under what conditions there exists a graph that submits a Hamiltonian cycle when the minimum and maximum lengths of the necklaces are changed.

2.4 Disjoint Subset Pairs

Our final example is the piece *partition and gate* (2014) for sustaining instruments and computer. While the morphology is not constrained by an OGMC, it has a LMC of minimal change and uses an algorithm to iterate through the elements defined by the CC that suggests an interesting non-deterministic method to search for Hamiltonian paths in a graph.

The piece works as follows. Two microphones are placed equidistant from a single speaker such that performers, who repeatedly play long sustained tones, can move freely in the space among the microphones and the speaker. At any given time, the microphones map to disjoint subsets (including the empty set) of four sources: a high frequency sine tone, a low frequency sine tone, and two recordings. The subset of sources mapped from the microphone with the louder signal (as tracked by an amplitude follower) is output to the speaker while the other subset is muted. Every 15–30 seconds, the mapping from one of the microphone changes such that a source is added or removed (minimal change) while the other microphone maps to the same subset of sources (no change); always favoring mappings that have occurred less. In this case, the musical motivations are both situational and perceptual. By changing the mappings over time, the players' expectations of how they are effecting the system are continually shifting while different combinations of the sources are promoted.

Mathematically, the CC defines all disjoint pairs of subsets of the superset {1, 2, 3, 4} (where the numbers indicate the sources). The LMC is that between any two pairs, one subset must either add or remove a number while the other stays the same. The GMC is that pairs that have occurred less are favored.

In a realization of the piece, a computer generates the morphology in real-time by a quasi-random walk with statistical feedback on a graph where the vertices are the subset pairs and edges are induced by the LMC (see Fig. 11). The algorithm is derived from James Tenney's dissonant counterpoint algorithm (see [17]), which he used as a defacto quasi-random element chooser for many of his computer generated pieces after 1985. Tenney's algorithm works as follows. A set of elements are initialized to some arbitrary set of probabilities. After each trial, the probability of the chosen element is set very low or to 0 and the probabilities of all other elements are incremented. Simply put, the longer an element has not been chosen, the more

Fig. 11 Graph of *partition and gate*

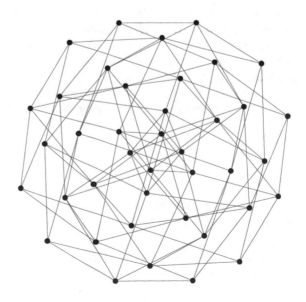

Table 2 Example walk of
partition and gate

mic 1	mic 2
$\{1, 2\}$	$\{3, 4\}$
$\{1\}$	$\{3, 4\}$
$\{1\}$	$\{2, 3, 4\}$
$\{1\}$	$\{2, 3\}$
$\{1, 4\}$	$\{2, 3\}$
$\{1, 4\}$	$\{3\}$
$\{1, 2, 4\}$	$\{3\}$
$\{1, 2, 4\}$	$\{\}$
$\{1, 2, 3, 4\}$	$\{\}$
$\{1, 3, 4\}$	$\{\}$
$\{1, 3, 4\}$	$\{2\}$
$\{3, 4\}$	$\{2\}$
$\{3, 4\}$	$\{1, 2\}$

likely it will be chosen. Based on the increment function, the algorithm can generate evenly distributed quasi-random choices over a limited number of trials.

Similarly in *partition and gate*, when a vertex in the graph is chosen, its probability is set to 0 and the probabilities of all other vertices in the graph are incremented. Therefore, the walk is generally directed towards vertices depauperate in the morphology up to that point. An example sequence of choices is given in Table 2.[11] Note that the *partition and gate* algorithm would be exactly the same as Tenney's if the graph were completely connected.

We leave as an open question whether or not this algorithm (or something similar) might be of use in trying to find Hamiltonian paths in arbitrary graphs.

3 Conclusion

We have examined several types of minimal change musical morphologies. These investigations, particularly with respect to the additional, musically motivated con-

---

[11] The vertices of the graph (shown in Fig. 11) represent the subset pairs such that order of subsets within the pair does not matter. However, the computer program that generates the random walk tracks which microphone is mapped to which subset. This does not prohibit any particular subset being mapped from either of the microphones. For example, the first and last mapping in Table 2 are represented by the same vertex in the graph.

straints on the analogous mathematical representations, have provided several open questions.

1. How exactly would OGMCs be defined mathematically and encoded computationally for a maximally balanced, maximally uniform long-run Gray Code? How do the three optima of maximal run-length uniformity, run-length maximality, and balance relate to or conflict with each other?
2. What are the lower bounds on the range of extremal values in the running sum of optimal solutions for space-limited de Bruijn sequences?
3. Given a graph with vertices that are representatives of aperiodic necklaces of length n to m (one representative per necklaces) with edges induced between two representatives if one element is added, removed, or changed, under what conditions is the graph Hamiltonian? Further, if the graph does submit a Hamiltonian cycle, does there exist an efficient algorithm to generate the enumeration?
4. Can Tenney's dissonant counterpoint algorithm be used and/or altered to non-deterministically find Hamiltonian paths in arbitrary graphs?

To add, the author is currently working on a piece derived from multiset permutations where only one transposition/swap occurs between adjacent permutations in the morphology (the LMC) and specifically where only one element in the multiset repeats (the CC). As with most of the morphologies we have discussed, the GMC is that each element occurs only once. Similarly to the piece *gray codes*, a highly balanced morphology with maximally uniform, long run-lengths is desired (the OGMC). And like standard Gray codes, there exists several algorithms for generating minimal change mulitset permutations (e.g., see [14, 15, 23, 25, 28]), but the resulting morphologies have highly varying run-lengths and are unbalanced. While the example was not included in detail because the composition is yet unfinished, the idea has already led to an interesting discussion about the character of multiset permutations and all the questions posed for Gray codes apply.

Finally, it might be of interest to investigate these ideas and objects more deeply with respect to algorithmic complexity [4, 5, 13], graph metrics [8, 30], Polansky's morphological metrics [16], other structural metrics [29], and music perceptual measures (such as perceived rate of change as discussed by Tenney in [26]). Does the number of morphologies that satisfy a given set of morphological constraints relate to complexity? How do morphologies that satisfy the same set of morphological constraints compare under various metrics? Can the taxonomy of morphological constraints presented in this paper as a generative tool also prove useful as an analytical tool? Addressing such questions might give us a better understanding of these types of morphologies and suggest further research.

Acknowledgements I would like to thank Hakan Kjellerstand. His generous help has led to a number of new pieces and he was essential in the formulation of the ideas presented in this paper, especially with respect to CP paradigms and implementations. I would also like to thank Larry Polansky for helping me clarify many of the ideas. Finally, I would like to thank Juan Sebastián Lach Lau and Gabriel Pareyón for encouraging me to submit these ideas for presentation at the 2014 International Congress on Music and Mathematics.

References

1. Akhmedov, A., Winter, M.: Chordal and timbral morphologies using Hamiltonian cycles. J. Math. Music **8**(1), 1–24 (2014)
2. Anders, T., Miranda, E.R.: A Survey of constraint programming systems for modelling music theories and composition. ACM Comput. Surv. **43**(4) (2011)
3. Anders, T., Miranda, E.R.: Constraint application with higher-order programming for modeling music theories. Comput. Music J. **34**(2), 25–38 (2010)
4. Chaitin, G.: Meta Math!: The Quest for Omega. Vintage (November) (2004)
5. Chaitin, G.: A Theory of program size formally identical to information theory. J. ACM **22**(3), 329–340 (1975)
6. Cohn, R.: Maximally smooth cycles, hexatonic systems, and the analysis of late-romantic triadic progressions. Music Anal. **15**(1), 9–40 (1996)
7. de Bruijn, N.G.: A combinatorial problem. Koninklijke Nederlandse Academie van Wetenschappen **49**, 758–764 (1946)
8. Fernández, M.L., Valiente, G.: A graph distance metric combining maximum common subgraph and minimum common supergraph. Pattern Recognit. Lett. **22**(6–7), 753–758 (2001)
9. Fredricksen, H., Maiorana, J.: Necklaces of beads in k colors and k-ary de Bruijn sequences. Discret. Math. **23**(3), 207–210 (1978)
10. Gray, F.: Pulse code communication. U.S. Patent 2,632,058 Nov 1947
11. Klamroth, K., Jørgen, T.: Constrained optimization using multiple objective programming. J. Glob. Optim. **37**(3), 325–355 (2007)
12. Knuth, D.: The Art of Computer Programming. Addison Wesley, New York (1997)
13. Kolmogorov, A.: Three approaches to the quantitative definition of information. Int. J. Comput. Math. **2**(1), 157–168 (1968)
14. Korsh, J., LaFollette, P.: Loopless array generation of multiset permutations. Comput. J. **47**(5), 612–621
15. Ko, Ch.W., Ruskey, F.: Generating permutations of a bag by interchanges. Inf. Process. Lett. **43**(5), 263–269 (1992)
16. Polansky, L.: Morphological metrics. J. New Music Res. **25**(4), 289–368 (1996)
17. Polansky, L., Barnett, A., Winter, M.: A few more words about James Tenney: dissonant counterpoint and statistical feedback. J. Math. Music **5**(2), 63–82 (2011)
18. Rollon, E., Larrosa, J.: Constraint optimization techniques for exact multi-objective optimization. In: Barichard, V., Ehrgott, M., Gandibleux, X., T'Kindt, V. (eds.) Multiobjective Programming and Goal Programming, Lecture Notes in Economics and Mathematical Systems, vol. 618, pp. 89–98. Springer, Berlin (2009)
19. Rollon, E., Larrosa, J.: Multi-objective propagation in constraint programming. In: Proceedings of the 17th European Conference on Artificial Intelligence, pp. 128–132 (2006)
20. Rossi, F., van Van Beek, P., Walsh, T.: Handbook of Constraint Programming (Foundations of Artificial Intelligence). Elsevier Science Inc., Amsterdam (2006)
21. Savage, C.: A survey of combinatorial gray codes. SIAM Rev. **39**(4), 605–629 (1997)
22. Sawada, J., Chi-Him Wong, D.: A fast algorithm to generate Beckett-Gray codes (Extended Abstract). Electron. Notes Discret. Math. **29**, 571–577 (2007)

23. Shen, X.S., Williams, A.: A hot potato gray code for permutations. Electron. Notes Discret. Math. **44**, 89–94 (2013)
24. Stevens, B., Cooke, M.: North, Chap. Beckett-gray codes, Discrete Mathematics (submitted, 2007)
25. Takaoka, T.: Time Algorithm for Generating Multiset Permutations. Lecture Notes in Computer Science, vol. 1741, pp. 237–246. Springer, Berlin (1999)
26. Tenney, J.: META + HODOS and META Meta + Hodos, Frog Peak Music, 2nd edn. (1986)
27. Truchet, C., Assayag, G.: Constraint Programming in Music. Wiley, New York (2011)
28. Vajnovszki, V.: A loopless algorithm for generating the permutations of a multiset. Theor. Comput. Sci. **307**(2), 415–431 (2003)
29. Winter, M.: Structural Metrics: An Epistemology. University of California, Santa Barbara, PhD Diss. (2010)
30. Zager, L.A., Verghese, G.C.: Graph similarity scoring and matching. Appl. Math. Lett. **21**(1), 86–94 (2008)

Restoring the Structural Status of Keys Through DFT Phase Space

Jason Yust

Abstract One of the reasons for the widely felt influence of Schenker's theory is his idea of long-range voice-leading structure. However, an implicit premise, that voice leading is necessarily a relationship between chords, leads Schenker to a reductive method that undermines the structural status of keys. This leads to analytical mistakes as demonstrated by Schenker's analysis of Brahms's Second Cello Sonata. Using a spatial concept of harmony based on DFT phase space, this paper shows that Schenker's implicit premise is in fact incorrect: it is possible to model long-range voice-leading relationships between objects other than chords. The concept of voice leading derived from DFT phases is explained by means of *triadic orbits*. Triadic orbits are then applied in an analysis of Beethoven's *Heiliger Dankgesang*, giving a way to understand the ostensibly "Lydian" tonality and the tonal relationship between the chorale sections and "Neue Kraft" sections.

1 Long-Range Voice-Leading Structure Without Reduction

1.1 Schenker's Implicit Premise

As a voice-leading based approach that can address large-scale tonal structure, Schenkerian theory is widely regarded to be amongst the most sophisticated extant theories of tonality. However, when Schenker claimed that his theory of levels would supplant traditional notions of form and key, he overplayed his hand, creating conceptual tensions that persist in Schenkerian theory today. Schachter's [8] insightful deconstruction of the Schenkerian perspective on keys stops short of denying their reality even as he claims that Schenkerian structures override them.

J. Yust (✉)
School of Music, Boston University, 855 Commonwealth Ave., Boston, MA 02215, USA
e-mail: jason.yust@gmail.com

© Springer International Publishing AG 2017
G. Pareyon et al. (eds.), *The Musical-Mathematical Mind*,
Computational Music Science, DOI 10.1007/978-3-319-47337-6_32

329

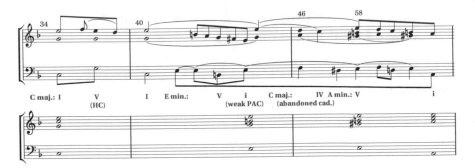

Fig. 1 A reduction of the subordinate theme in Brahms's Op. 99 Cello Sonata

The conflict of Schenker's theory with the traditional idea of keys as the objects of long-range structure comes from an implicit premise: that voice leading can only be a relationship between individual harmonies. This implies that a theory of long-range tonal structure based on voice leadings must posit that certain "structural" chords can be isolated from their contexts to relate directly at a deeper level. Specific tonic chords then must stand in for keys, undermining the important distinction between chord and key. While such an analytical approach often seems serviceable despite the underlying conceptual flaws, in certain circumstances it prevents an accurate analysis.

One such instance is the exposition of Brahms's F Major Cello Sonata, Op. 99, which Schenker analyzes in *Der Freie Satz* [9, Fig. 110d2]. The reduction in Fig. 1 illustrates the problem posed by the piece. The subordinate theme begins in the standard subordinate key of C major. After a momentary deflection to E minor (confirmed by a weak cadence) the music enters a cadential phase where it toys with the possibility of ending in A minor or C major, tipping just at the last minute into A minor. Only an analysis that can show how C major is in play up to the last few measures can accurately reflect Brahms's tonal rhetoric here. A reductive analytical method, however, must procede by first selecting out the most structural harmonies of the passage. As shown on the lower staff, the presence of cadences in E minor and A minor all but demand that these chords be selected as the most structural ones, which is exactly how Schenker analyzes the passage.

A spatial concept of tonality can serve us better in this situation, as shown in Fig. 2. The space used here is the *DFT phase space* described in [1, 13], and based on the DFT on pcsets discussed in [2, 6, 7], and elsewhere. The vertical axis of the space is the phase of the fifth Fourier coefficient of a pitch-class set, and the horizontal axis is the phase of the third coefficient. Dashed lines show the tonal regions derived in Sect. 8 of [13]. Any pitch-class set or multiset has a position in the space, including chords, scales, and single pitch classes. As Fig. 2 shows, a trajectory may be drawn by plotting chords used in the passage, which are somewhat spread out in the space. However, averaging over multiple chords restricts the range of activity, and each pair of chords averages to a location within the appropriate tonal region. The summary of the progression—i.e., a long-range picture of tonal motion through

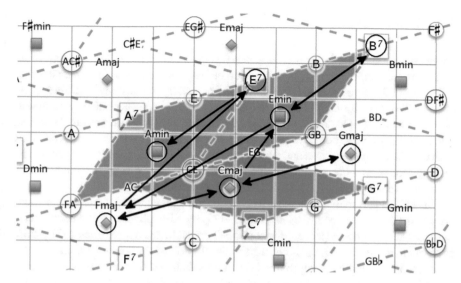

Fig. 2 Significant chords in the subordinate theme of Brahms's Op. 99 Cello Sonata

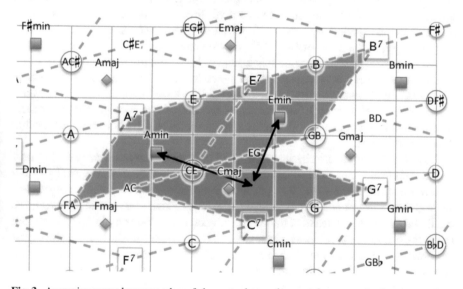

Fig. 3 Averaging over a larger number of elements shows a large-scale progression between regions

the passage—shows the key of C major acting as an intermediary between excursions to the bordering regions of E minor and A minor, as Fig. 3 shows. C major is central to the harmonic content of the entire passage, making this the principal key of the theme.

1.2   Triadic Orbits

One of the compelling features of Schenkerian theory is its grounding of tonal structure in voice leading. Positions in Fourier phase space can also be understood to reflect voice-leading relationships. Since one can analyze harmony at different levels using the space by summarizing the harmonic content of larger or smaller amounts of music, as illustrated above, this means that Schenker's implicit premise is wrong. It is in fact possible to conceive of voice-leading relationships between objects other than chords, including scales and keys. Therefore, we can theorize large-scale voice-leading processes without making reductive assumptions—that is, without asserting direct relationships between remote harmonic objects.

The DFT reparameterizes a pcset by modeling it with sinusoidal pitch-class distributions that divide the octave evenly into 1–11 parts. The phase of these components indicates which perfectly even distribution the pcset best approximates. The third component approximates the pcset with a distribution of three evenly spaced peaks, while the fifth component approximates it with a distribution of five or seven evenly spaced peaks. Motion between relatively even three-note chords (triads) in the horizontal dimension of the phase space in Figs. 2 and 3 reflects the direction of the most efficient voice leading. Motion in the vertical dimension on relatively even seven-note collections (scales) reflects the direction of scalar voice leading, or change of key signature, in the sense of Hook [4] or Tymoczko [11]. However, since pcsets of any cardinality appear in the same space, near to harmonically related pcsets of other cardinalities, we can also speak of scalar voice leadings between chords, or triadic voice leadings between scales. Roughly speaking, a scalar voice leading between chords is the average change between scales that contain each chord. Similarly a triadic voice leading between scales is the average voice leading between the possible tonics of that scale.

It is useful to interpret phases of the third component as *triadic orbits*. Figure 4 shows a sinusoid for the third component of a C diatonic scale. The peaks of the sinusoid are aligned as closely as possible to notes of the scale while the troughs avoid them. The troughs of the sinusoid divide the pitch class circle up into three triadic orbits, with the peaks at the center of each orbit. We can interpret notes that fall in the center of the orbits as triadically stable, and notes towards the periphery as unstable, drawn to the center of their respective orbits by a force of triadic resolution. A voice leading within orbits shifts them in the direction of the voice leading, but a voice leading that crosses orbit boundaries shifts them in the opposite direction.

In analytical application of Fourier phase space we may relate pcsets in two ways: a path from A to B may indicate "A in the context of B" or a motion from A to B. Mathematically these are equivalent: if a motion from A to B has a descending voice leading, then A has an upper-neighbor quality in the context of B (its notes tend to be high in the triadic orbits of B). Conceptually, however, these two kinds of relation are quite different and tend to apply to different kinds of objects. For example, if A is a single pitch class and B is a scale, we are more likely to talk about A in the context

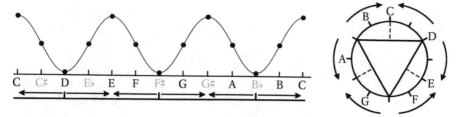

Fig. 4 The third Fourier component for a C diatonic scale, and its triadic orbits

of *B*. Nonetheless, it is theoretically possible to speak of a triadic voice leading from a single pitch class to a scale (e.g., to explain a common-tone modulation).

The idea of orbits can also be applied to the fifth Fourier coefficient (the vertical dimension of the phase spaces in Figs. 2 and 3). The tendency tones of a key (such as leading tones) and chromatic notes would be unstable, with their orbits indicated by the letter name of their spelling.

2 Beethoven's *Heiliger Dankgesang*

2.1 Tonal Contexts and Triadic Orbits

Beethoven's *Heiliger Dankgesang*, the third movement of his late A minor String Quartet, op. 132, remains inspiring yet enigmatic nearly two centuries after its composition. Its biographic resonances, play of musical topics, and misprision of antiquated contrapuntal styles have all been well explored (e.g., [3, 10]). But current theories of harmony are not well tooled to address one of its most puzzling features, the status of tonality in this nominally "Lydian mode" work. The piece begins *in* F major, but ends only *on* F major, because its tonal context has shifted to C major. The meaning and purpose of this unusual tonal design is inaccessible to a theory that reduces the tonal contexts out of the middleground representation. We can overcome the problem without throwing the proverbial baby—the idea of deep structural voice leading—out with the bathwater with the spatial concept of tonality and harmony outlined above and the use of triadic orbits to construe this space in voice-leading terms.

The C-D interval is a prominent motivic element of the movement, manifest at many levels, and brought the forefront especially in the final chorale section. The status of this interval constitutes one of the most significant differences between the triadic orbits of the F major tonality suggested by the initial intonation of the chorale sections, and the C major tonality established in the later phrases of the chorale. This shift is already indicated by the first chorale phrase, as shown in Fig. 5. In the initial contexts of {FGAC} in the intonation and {CDEFGA} in the first part of the chorale,

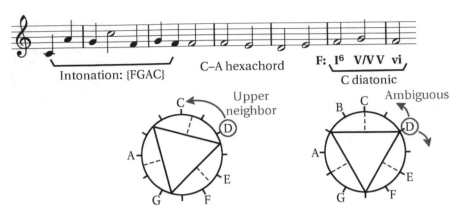

Fig. 5 Triadic orbits and the status of D in the first intonation and chorale phrase

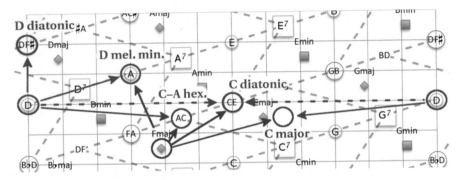

Fig. 6 The position of D and the F major triad in different tonal contexts. An arrow pointing to the *right* indicates an object in the *upper* part of the triadic orbits of the context, and to the *left* in the *lower* part

D is an upper neighbor in the triadic orbit of C. The first cadence introduces the full C diatonic context, in which D is ambiguously positioned between two orbits. In later chorale phrases, the greater centrality of C major and its dominant shift the triadic orbits further in this direction so that D crosses over into an orbit where it strives upward, away from C. This is also reflected in the melodic shape of the first two chorale phrases, where D resists the descent to C that would complete the F–C tetrachordal space.[1]

The fourth phrase of the chorale ends on an F major triad after having decisively shifted into a C major tonal context. As one can see from the phase space relationship of the F major triad to these contexts (Fig. 6), the shift to C diatonic (and further to a more central C major context) puts F major in the upper periphery of the triadic orbits, giving this cadence its feeling of suspension, of coming to a standstill in a precariously

---

[1] Korsyn's [5] motivic analysis highlights motives of the chorale tune involving D, including C–C–D and D–E–F, the latter representative of D's resistance to downward resolution.

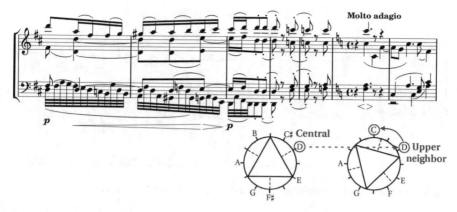

Fig. 7 The transition from the *Neue Kraft* interlude back to the chorale

unstable place. (The effect is highlighted especially in Vitercik's commentary [12], which focuses on the reorientation of stability in the E-F melodic interval.) The transition into the contrasting *Neue Kraft* sections moves through a D melodic minor context to D diatonic, where the note D changes from peripheral position in its orbit to a stable central position.

One place where the motivic C-D comes strongly to forefront is the retransition from the first *Neue Kraft* section (Fig. 7) to the second chorale section. Here the stable position of D at the center of its triadic orbit evaporates like a daydream as the melodic voice descends to C and back to the tonal universe of the chorale.

2.2 Strength and Weakness

A central metaphor to the *Heiliger Dankgesang* movement is the concept of strength versus weakness. Beethoven finds a musical analogue to this dichotomy ready at hand: strong and weak harmonic progressions, particularly cadential progressions. The feeling of strength or lack thereof in a tonal harmonic progression can be understood through triadic orbits: if the voice leading of the progression crosses triadic orbits, it is strong. The model of harmonic strength, the PAC, does just this: the melodic descent from $\hat{2}$ crosses a triadic orbit. If voices instead remain within their orbits, acquiescing to their gravitational forces, the progression will tend to feel weak, as in the typical neighboring I–V–I or I–IV–I progressions so often used to tread water at the beginning of a Classical theme.

The third and last chorale section of the piece, an extended contrapuntal interweaving of the first intonation and first chorale phrase, ends with a C major cadence that is set up by the most intense dynamic and registral crescendo that the ensemble can muster. This willful cadence in C major, despite the clear formal requirements of a beginning in F, are an unprecedented tonal representation of human agency.

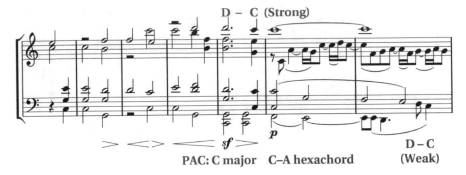

Fig. 8 Strong and weak C–D motions at the end of the third chorale section

Korsyn [5] notes how this section synthesizes registral, dynamic, and rhythmic traits of the *Neue Kraft* music with the material of the chorale sections. The D–C motion that has operated motivically at numerous levels in the piece is set in relief at the cadence by its register, the dramatic triple suspension, and the use of double stops in all instruments other than the first violin (Fig. 8). The chorale tune returns in the cello after this cadence, and here, for the first time, the melody completes the F–C tetrachord with a D–C descent. The context, however, has changed: following the cadence, the note B is completely absent for the remaining ten measures of the movement. The D–C motion in the cello is weak in this tonal context, the same one shown for the beginning of the movement in Fig. 5. This juxtaposition of two D–C motions, one strong, the other weak, exhibits the distance that has been traveled by the tonal process of the chorale, and summarizes the main idea of the movement, gratitude for the strength granted in life, expressed at its fullest in the final moments before the surrender to death.

References

1. Amiot, E.: The Torii of Phases. In: Yust, J., Wild. J., Burgoyne, J.A. (eds.) Mathematics and Computation in Music: 4th International Conference, MCM 2013, pp. 1–18. Springer, Berlin (2013)
2. Amiot, E., Sethares, W.: An algebra for periodic rhythms and scales. J. Math. Music **5**(3), 149–69
3. Brandenburg, S.: The historical background to the 'Heiliger Dankgesang' in Beethoven's a minor quartet Op. 132. In: Tyson, A. (ed.) Beethoven Studies, pp. 161–91. Cambridge University Press, Cambridge (1982)
4. Hook, J.: Signature transformations. In: Hyde, M., Smith, C. (eds.) Mathematics and Music: Chords, Collections, and Transformations, pp. 137–160. University of Rochester Press, Rochester (2008)
5. Korsyn, K., Sullivan J.W.N., Heiliger Dankgesang.: Questions of meaning in late beethoven. In: Lockwood, L., Webster, J. (eds.) Beethoven Forum 2, pp. 133–74. Lincoln, Neb.: University of Nebraska Press (1993)

6. Lewin, D.: Special cases of the interval function between pitch-class sets X and Y. J. Music Theory **45**(1) (2001)
7. Quinn, I.: General equal tempered harmony. Perspectives of New Music 44/2–45/1, 114–159, 4–63 (2006–2007)
8. Schachter, C.: Analysis by key: another look at modulation. Music Anal. **6**(3), 289–318 (1987)
9. Schenker, H., Free Composition: Volume III of New Music Theories and Phantasies, translated and edited by E. Oster. Longman (1979)
10. Theurer, M.: Playing with time: the Heiliger Dankgesang and the evolution of narrative liberation in Op. 132. J. Musicol. Res. **32**(2–3), 248–265 (2013)
11. Tymoczko, D.: Scale networks and debussy. J. Music Theory **52**(2), 251–272 (2004)
12. Vitercik, G.: Structure and expression in Beethoven's Op. 132. J. Musicol. Res. **13**, 233–253 (1993)
13. Yust, J.: Schubert's harmonic language and fourier phase space. J. Music Theory. **59**(1): Forthcoming

Mazzola, Galois, Peirce, Riemann, and Merleau-Ponty: A Triadic, Spatial Framework for Gesture Theory

Fernando Zalamea

Abstract This contribution connects Mathematical Music Theory (MaMuTh) with Peircean semiotics, identifying general grounds for Gesture Theory (in Mazzola's sense). In order to make clear this connection, some of the contributions included in this volume are refered, unveiling a common framework for semiotics in MaMuTh.

1 Introduction

Guerino Mazzola has proposed a perspicuous dialectics, formed by Galois connections, or adjunctions, between formulas and gestures [1, 3–5]. The dialectics extends his earlier, profound contributions to music theory presented in *The Topos of Music* [2], and opens up a new range of analysis, where musical interpretation *dynamizes* the complex spectrum of musical *life*. The full triadic range of sounds, partitions-formulas, and gestures becomes then suitable for complex, multilayered conceptualizations. We can profit from earlier semiotic, philosophical and mathematical constructions to enrich Mazzola's approach. Three main lines of thought seem interesting:

1. Peirce's triadic sign (object-representamen-interpretant) helps to multiply, or extend continuously in space [9], dyadic polarities (such as Galois connections or adjunctions [10]). An adequate use of Peirce's triadic semeiotics should help then to expand Mazzola's multilayered conception of music. Moreover, Saint-Victor's definition of "gesture" (movement and figuration with an aim, fostered by Mazzola) is fully pragmatic(ist) in Peirce's sense.
2. A long tradition in French philosophy of mathematics has acknowledged the importance of gestures in knowledge. Mazzola has reckoned [3] the importance of Merleau-Ponty, Cavaillès, Deleuze, Châtelet, Alunni, but those brief mentions may be expanded to a wider underlying philosophical corpus for gesture theory. On another hand, Mazzola's compression/unfolding functors between formulas

F. Zalamea (✉)
Universidad Nacional de Colombia, Cra. 45, Bogotá, Colombia
e-mail: fzalameat@unal.edu.co

© Springer International Publishing AG 2017
G. Pareyon et al. (eds.), *The Musical-Mathematical Mind*,
Computational Music Science, DOI 10.1007/978-3-319-47337-6_33

and gestures recall the dual processes uniformization/ramification in Riemann surfaces [11], dear to many French philosophers of mathematics.

3. Merleau-Ponty's "entrelacs" and "chiasme" [6, 7] postulate a gluing of subject/object, being/world, mind/body where the chiasma (crossing of optic nerves on the brain) helps to explain passages between visibility and non-visibility. A similar chiasmatic experience is in act in music, along the Galois connection formulas-gestures.

In what follows we will address these issues. The first section sketches Mazzola's main problem and his triangular set-up (sounds, scores, gestures) for Gesture Theory. The second section explains how the (bilateral) Galois' connections become pairs of natural adjunctions for "horosis". The third section shows that Mazzola's (degenerate) triangle can be extended to a true Peircean Triad. The fourth section suggests how the (triadic) *horos* may become a ramification point in a multilayered, Spatial Riemann surface. The fifth, and final, section presents Merleau-Ponty's (visual) "entrelacs" as a ground for chiasmatic musical experience.

2 Mazzola: The Problem and the Triangular Set-Up (Sounds, Scores, Gestures)

In 2002, after his gigantic *The Topos of Music* [2], Guerino Mazzola could well have rested on his laurels. A all New Continent of musical theory had being unraveled. Nevertheless, as happens with many great thinkers, Mazzola continued to explore even further, studying exactly the opposite paths to the ones he had already offered. *La vérité du beau dans la musique* [3] courageously opens a new problem (in a courageously anti-postmodern monograph, where the supposedly dead "Truth of Beauty" guides our knowledge). Mazzola has reckoned there his "crise du 18 mai 2002" [3, 145] —reminiscent of Valéry's Genoa crisis (1892)— where, preparing himself for a jazz improvisation at IRCAM, he suddenly discovered that the musical theory of *The Topos of Music* (addressed to extremely sophisticated analysis of scores and interpretations), was of little use when it came to the understanding of the forces unleashed in musical improvisation. The *gestures* seemed to have a life of their own, that had to be studied as seriously as *The Topos of Music* had laid out a mathematical framework for scores, formulas and interpretations. The result has been a very ambitious program on *gesture theory* [3–5], which will produce in the next few years a pendulum-sequel (*The Topos of Music II*) to Mazzola's sweeping ideas on the architecture of music. In *La vérité du beau dans la musique*, Mazzola presents a diagram [3, 146] where he shows the problem of correlations between sounds ("événements sonores"), scores ("partitions") and gestures ("gestes"). The initial, founding diagram (2002) is of course tentative, but one can point out to two characteristics that have not yet been completely elucidated in the following publications. First, some dialectics between pairs are indicated, but they are not fully explored. For example, between the polarity gestures/scores, Mazzola situates a folding/unfolding dialectics ("geler/dégeler") [3, 146], but the other correlations (between gestures/sounds, or between scores/sounds) are only half-way reckoned ("℘", "instrumentaliser") [3, 146]. Second, the dialectics are bipolar (between two nodes), but no really *triadic* relation between sounds, scores and gestures is mentioned or imagined.

3 Galois: How the (Bilateral) Dialectic Pairs Become Natural Adjunctions for Horotics

Going a little further in the completion of Mazzola's diagram, one can render fully explicit the several dialectics at work, which can be considered as *Galois connections* or *Galois adjunctions* (Mazzola is well aware of such general settings, see [1] or [3, 5]). Between scores and sounds we have compression/decompression functors, between scores and gestures folding/unfolding functors occur, and between sounds and gestures the instrumentalisation/improvisation functors guide musical imagination (see below, Fig. 1). Now, when a Galois mathematical framework is elucidated (something which has yet to be done formally), the truly important concepts become the *invariants* of the back-and-forth Galois connections. They form a boundary, a border, a mediation, a middle ground —what we may call a horos, following the Greek etymology— where *musical* action develops.

As interesting, complementary questions where such conceptualizations (polarities, multiplications, invariances) are intertwined with music, we may consider parts of Octavio Agustín-Aquino and Samuel Vriezen presentations at ICMM 2014.[1]

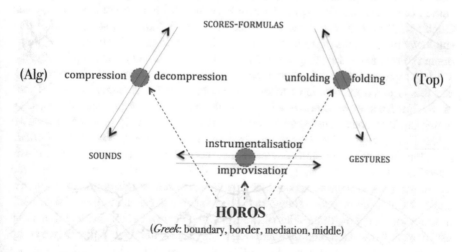

Fig. 1 The triangle of Galois connections: sounds – scores – gestures

---

[1] Both presentations included in the present volume. Other following contributions to ICMM–Puerto Vallarta, mentioned by the author, are also included in this book.

Agustín-Aquino mentioned the use of computations and combinatorics in Counterpoint Theory, and wondered about some more general settings for the mathematics involved. In fact, some of the tools can be extended to (Janowitz) residuated lattices, an abstract algebraic framework for Galois connections, and even to (Scott) continuous lattices, where the central problem between discrete interpretation and continuous musical sense can be better stated. On another hand, Samuel Vriezen asked if it was at all possible to understand time as some sort of consequence of action (inverting our usual, classical understanding). The answer is positive: when one extends both the idea of a singular action to many actions and the setting of group actions to *monoid* actions, one enters the realm of an elementary topos of monoid actions (Lawvere), where one can prove (it is a theorem) that the intrinsic logic is classical if and only if the underlying monoid is a group. Thus, if one considers monoid, *non-group*, actions, an intrinsic intuitionistic logic appears, and, as is well known, intuitionistic logic forces the development of time, through its natural Kripke models.

4 Peirce: How the (Degenerated) Triangle Becomes a True Triad

Peirce's triadic sign [9, 59–61] is defined as a *full triadic relation* between an object, its representamen (a representation of the object) and its interpretant (an interpretation of the representamen). Imagine a sunflower (object), the figure of a sunflower in van Gogh's mind (representamen), and the actual sunflower painted in van Gogh's canvas (interpretant). Or, beginning a new semiosis, imagine yourself seeing van Gogh's sunflower in an Amsterdam Museum (object), fixing the suggestion of that sunflower in your mind (representamen), and finally writing on that experience (interpretant). From a musical perspective, the true triad is obtained when sounds (objects) are duly represented in scores (representamens), and then interpreted (interpretants). But this is just a *first circle* of musical experience (fully accounted in The Topos of Music). Another, *second circle* begins with gestures and improvisation (objects), which produce unexpected sounds (representamens), to finally produce innovative scores (interpretants). Here we enter the realm of the ongoing *The Topos of Music II*. A bunch of "Dante's circles" would then come into the picture if the semiosis were carried sufficiently enough. The sign is *degenerated* if only two of the three components are taken into account (for example, in Saussure's approach). The triangle of Galois connections shown in Fig. 1 is such a degenerated triad, "flattened" in some sort along the three sides of the triangle. A *true triad* for musical dynamics has to expand into a tetrahedron where the three "faces" of musical experience come together (see Fig. 2). There, *spatial 3-dimensional* relations (R,O,I), which may be *iterated* in time and *projected* onto the planar Galois connections, configure musical complexity. From a mathematical point of view, the Galois connection between sounds (objects) and scores (representamens) (*algebraically* studied in The Topos of Music) and the Galois connection between scores (representamens) and gestures

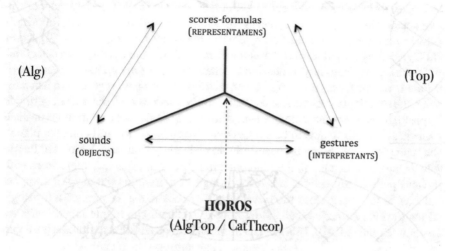

Fig. 2 Peirce's true triad of musical complexity

(interpretants) (*topologically* studied in *The Topos of Music II*) blend together in a superior dimension, through techniques in *algebraic topology* and category theory [4, 5].

Two good examples of the eventual use of Peirce's triad came along Juan Sebastián Lach-Lau, and Silvia Pina-Romero and Gabriel Pareyon talks in Puerto Vallarta. Addressing the compositional techniques and harmonic facets of Julian Carrillo's *Leyes de Metamorfosis Musicales*, Lach studied the problem of pitch *saturation* in multidimensional lattices which tended to approach the pitch continuum. The setting is perfectly adapted to Peirce's ideas, where the multidimensionality of the *iterated triadic* sign approaches Peirce's continuum [9]. On another hand, Pina-Romero and Pareyon studies of phase synchronization of the *teponaztlis* (Mexican native drums), through representations in Hilbert space, may be seen as a good, natural example of a *non-standard* emergence of a Peirce triad. In fact, the hits of the strips in the *teponaztlis* (sounds, objects) are *not* associated to scores, and, thus, the Hilbert transforms of pulses are proposed instead (formulas, representamens), to finally obtain a percussion classification (gestures, interpretants). Against simple dualities, multidimensionality and continuity are fostered by the three young Mexican scholars.

5 Riemann: How the (Triadic) *Horos* Becomes Ramified in a Multilayered Surface

A hundred and fifty years before Elaine Chew's self-proclaimed "discovery" of "tipping points" at Puerto Vallarta, Bernhard Riemann had invented the fantastic idea of a *ramification point* in a multilayered surface in the theory of functions of a complex variable (now duly called a Riemann surface [11]). Well beyond the "tipping points" (just degenerated, planar projections of ramification points), Riemann's ideas involve imaginary numbers, where the *horos* between (1) sounds, (2) scores and (3) gestures truly folds/unfolds along Peirce's three cenopythagorean categories [9, 56–58]. This imaginary dimension is related to *one of the greatest ideas offered at Puerto Vallarta*, Mazzola's suggestion to think about *imaginary time* to patch together the seemingly inescapable contradiction between actual, discrete fingering at the piano (to be represented along an *imaginary axis*), and modal, continuous musical pianistic accomplishment (to be *conformally* projected on the full complex plane). The "complex circle marvels" beautifully explained by Emmanuel Amiot at Puerto Vallarta have still to unleash their full potential. Along unexpected connections with previous parts of this article, it is astonishing to observe that the logic of complex variables is also very close to *Peirce's logic of the existential graphs*, a topological logic where *both* classical and intuitionistic calculi can be presented (work by Oostra, as explained in [9, 126–129]). The musical improvisation gestures at the piano are not far away from the wandering logical gestures on the phemic sheet of the existential graphs. In fact, both actions *trigger creativity* in unexpected, non-standard ways. As Mazzola suggests, we may be in presence of new fields of imagination that are waiting to be duly represented by the geometry of the imaginary numbers. Here, a sophisticated *meta-triadic* sign may also be at work, where piano fingerings (objects) are to be represented by complex, imaginary numbers (representamens), and to be projected, afterwards, into the realm of true, continuous pieces of music (interpretants).

6 Merleau-Ponty: How the (Visual) Entrelacs Becomes a Chiasmatic Musical Experience

A basic study of the phenomenological spectrum (the "phaneron") lead Peirce to the construction of an original *phaneroscopy*, weaved around his three essential cenopythagorean categories (1: immediacy, 2: action-reaction, 3: mediation). Beyond dualities and (classical) arguments developed in two dimensions (yes/no), 3-dimensionality, and a multiplicity of logical values, seem to be important keys to approach general knowledge, not just geometry or logic. In a similar vein, Teresa Campos presented at Puerto Vallarta some ideas to extend linear Braille musical notation, to planar Braille, which should even be extended to 3-dimensional Braille, in order to fully capture the richness of musical gestures. Stockhausen, Berio, and many other composers of the 20th Century had already proposed to extend the classical score notations to multidimensional figures (see [8]). From a phenomenological, philosophically abstract perspective, the tetrahedron of musical semiosis (preceding Sects. 3 and 4) can enter Merleau-Ponty's framework for understanding the union of opposites [6, 7]. What we have denoted *horos* in Figs. 1 and 2, is what Merleau-Ponty calls the "entrelacs" or "chiasme", the place where both neural optics and phenomenological transits break the distances between mind and body, between intelligibility and sensibility. The musical tetrahedron *evolves then in three dimensions* (with its three main triangles around the horos/entrelacs/chiasme) producing the increasing spatial spiral (with many leaves in a Riemann surface) of musical architecture. A multilayered *transit* between the paths, faces and leaves of a complex geometrical structure (close also to Tarkovsky's levitation in *Stalker*, or to Valéry's abstraction in the *Cahiers*) may help then to explain the ambiguities, richness and variety of musical experience. Sounds compress in formulas, which unfold in gestures, which produce sounds, which generate new musical signs and actions, continuing along Riemann's ramifications and Peirce's infinite semiosis. Merleau-Ponty's *entrelacs* weaves many forms of reflexivity between objects, representamens and interpretants.

References

1. Mazzola, G.: Towards a *Galois* Theory of Concepts, preprint, post-scriptum to The Topos of Music (2002)
2. Mazzola, G.: The Topos of Music. Birkhäuser-Verlag, Berlin (2002)
3. Mazzola, G.La.: vérité du beau dans la musique. Delatour-IRCAM, Paris (2007)
4. Mazzola, G., Andreatta, M.: Diagrams, gestures and formulae in music. J. Math. Music 1(3), 199–200 (2007)
5. Mazzola, G.: Categorical gestures, the diamond conjecture, Lewin's question, and the Hammerklavier Sonata. J. Math. Music 3(1), 31–58 (2009)
6. Merleau-Ponty, M.: L'oeil et l'esprit. Gallimard, Paris (1964)
7. Merleau-Ponty, M.Le.: visible et l'invisible. Gallimard, Paris (1964)
8. Villa-Rojo, Jesús.: Notación y grafía musical en el Siglo XX. Iberautor, Madrid (2003)
9. Zalamea, F.: Peirce's Logic of Continuity. Docent Press, Boston (2012)
10. Zalamea, F.: La emergencia abismal de la matemática moderna. (I) Galois (1811–1832), Mathesis (to appear), Mexico City, p. 20 (2015)
11. Zalamea, F.: La emergencia abismal de la matemática moderna. (II) Riemann (1826–1866), Mathesis (to appear), Mexico City, p. 24 (2015)

Printed in the United States
By Bookmasters